U0826573

双语教材

基础地球科学

FOUNDATIONS OF EARTH SCIENCE

焦述强　孔　华　编著

中国地质大学出版社

内容简介

本书是用于地质、地理、工程地质，以及岩土工程、勘察、矿产、土建和国土资源等相关专业双语教学的基础地球科学（普通地质学）教材，是为了适应和满足国内地学类专业双语教学要求而编写。全书用英文编写。为了适应学科的发展，对传统的普通地质学内容有所取舍；为了适应学科融合的趋势，并扩大学生的知识面，加入了一些传统意义上属于别的学科的内容。本书在版面形式上吸取国内外优秀教材的优点，在每一章的前面都列出关键词，在每一章后面有回答问题和多项选择练习，对学生掌握所学内容有很大帮助。

图书在版编目(CIP)数据

基础地球科学＝Foundations of Earth Science/焦述强，孔华编著. —武汉：中国地质大学出版社，2006.12　(2019.7 重印)
ISBN 978－7－5625－2158－7

Ⅰ.基…
Ⅱ.①焦…②孔…
Ⅲ.地球科学—双语教学—高等学校—教材—英、汉
Ⅳ.P

中国版本图书馆 CIP 数据核字(2006)第 157346 号

基础地球科学(FOUNDATIONS OF EARTH SCIENCE)	焦述强　孔　华　编著
责任编辑：谌福兴	责任校对：张咏梅
出版发行：中国地质大学出版社(武汉市洪山区鲁磨路 388 号)	邮编：430074
电　话：(027)87482760　传真：87481537	**E-mail**：cbb@cug.edu.cn
经　销：全国新华书店	**http**://www.cugp.cn
开本：787mm×1092mm 1/16	字数：500 千字　印张：19.875　彩图：4
版次：2006 年 12 月第 1 版	印次：2019 年 7 月第 3 次印刷
印刷：荆州鸿盛印务有限公司	印数：2701—3700 册
ISBN 978－7－5625－2158－7	定价：35.00 元

如有印装质量问题请与印刷厂联系调换

前　言

本书是用于地质、地理、工程地质及岩土工程、勘察、矿产、土建和国土资源等相关专业双语教学的基础地球科学(普通地质学)教材。

由于教育部的倡导和目前全球化的形势,双语教学已经是大势所趋。我们在教学的过程中深感目前双语教材的不足。首先国外教材由于版权和费用等原因很难推广;其次国外教材,尤其是基础学科教材,一般都不系统,或者说其系统不符合我国的教学习惯和模式,而目前国内也没有合适的地球科学(普通地质学)双语教材。这是编此本教材的初衷。

本书主要有如下特点:

(1)为了适应学科的发展,对传统普通地质学的内容有所取舍。因为普通地质学主要是地质学和其他相关学科的入门课程,但在某些内容上面过于艰深。比如对地球演化和岩浆分异的过多叙述,实际上连专业人士都感觉到困难。我们取舍的原则是对某些在后续课程中可以深入学习的内容,我们只作最简单和通俗的描述,避免使用过多的专业术语。

(2)为了适应学科融合的趋势,并扩大学生的知识面,我们也在本书中加入了一些传统意义上属于别的学科的内容。比如随着社会的进步和发展,地表作用在人类社会中越来越重要,所以我们加入了大气循环等一般不属于普通地质学的内容。另外也加强了环境问题的介绍,比如臭氧洞、厄尔尼多现象、水土流失和土地利用及地质灾害等。同时考虑到现在社会中能源的重要性,我们也加入了一般普通地质学教材中不予介绍的地球的资源一章,使得学生能够对资源的一般分类和地质特征有个基本的认识。

(3)本教材在版面形式上吸取国内外优秀教材的优点。在每一章前面都列出了相关的关键词,在章后添加了相关练习,练习包括问答和选择题。此类练习对学生掌握所学内容有很大的帮助。

(4)因为是英文教材,为了方便教学,我们不但给出了关键词的汉语翻译,同时对几乎所有专业术语和中文地名都给了相应的汉语翻译和注释。

(5)地球科学是描述性很强的科学。我们在编写教材的时候所用例子尽量取自国内。

本书的编写前后约用了五年时间。初稿作为校内教材也已经使用了五届。期间,根据学生的反映和教学的需要,不断地进行修改,并数易其稿。编写工作可以说是字斟句酌,呕心沥血。但由于本人无论专业水平还是英语水平都很有限,错误和不妥之处在所难免,敬请有关专家和同行批评指正。

本书的编写和出版过程中,谌福兴编辑、周辉博士给予了很大的帮助和支持。

谌福兴编辑、俞姗博士、简文彬教授提供了部分照片。唐德平教授、杨坤光教授提出了宝贵的意见。勘察专业2004级学生廖宗湖多次对本文进行了全面的文字校对和版面修正。此外本书采用了大量"NOAA"和"NASA"等机构的属于"Public Domain"(免版权,自由版权)的照片。NOAA的全称是National Oceanic & Atmospheric Administration (US)。NASA的全称是National Aeromautics and Space Administration (US)。少量来自于其他机构的"Public Domain"照片,在书中也一一注明。在此对上述人员和机构一并致谢。最后需要说明的是,本书采用的照片和插图除明确注明外,其版权均为本书作者所有。

最后需要特别提出的是,本书在内容安排和编写方式上都参照了南京大学夏邦栋所编汉语版《普通地质学》(1995,第二版)。这里对夏教授表示感谢。为了教学的方便,在章节安排上力求与上述汉语版《普通地质学》基本对应。因此建议大家在使用本教材的时候与上述汉语版《普通地质学》对照使用,从而方便学生的学习和教学的实施。

焦述强　孔　华

2006年8月3日

Contents

Chapter 1 Introduction .... (1)

1 Earth Sciences .... (1)

2 Why Study the Earth? .... (2)

3 How to Study the Earth? .... (3)

4 Earth Physical Characteristics .... (5)

Chapter 2 Minerals .... (11)

1 Elements .... (11)

2 Minerals .... (12)

3 Physical Property of Minerals .... (14)

4 Common Minerals .... (19)

Chapter 3 Igneous Rocks .... (25)

1 Magma .... (25)

2 Intrusion and Intrusive Body .... (26)

3 Volcano and Volcanism .... (29)

4 Igneous Rocks .... (33)

Chapter 4 External Geological Process and Sedimentary Rocks .... (41)

1 Atmosphere .... (41)

2 Hydrologic Sphere .... (44)

3 Biosphere .... (45)

4 Formation of Sedimentary Rocks .... (45)

5 Minerals of Sedimentary Rock .... (47)

6 Texture of Sedimentary Rock .... (47)

7 Structure of Sedimentary Rock .... (49)

8 Classification of Sedimentary Rocks .... (52)

Chapter 5 Metamorphic Rocks .... (57)

1 Intruduction .... (57)

2 Factors Controlling Metamorphism .... (59)

3 Types of Metamorphism .... (61)

4 Classification of Metamorphic Rocks .... (63)

5 Rock Cycle .... (66)

Chapter 6 Geologic Time Scale ······ (70)

1 Stratigraphic Principles ······ (71)
2 Principle of Cross-Cutting Relationship ······ (73)
3 Principle of Biologic Succession ······ (73)
4 Unconformity ······ (75)
5 Radiometric Dating ······ (77)
6 Geologic Time Scale ······ (78)

Chapter 7 Earthquakes ······ (85)

1 What is Earthquake? ······ (85)
2 Types of Earthquake ······ (86)
3 Earthquake Wave ······ (88)
4 Seismograph ······ (89)
5 Magnitude and Intensity of Earthquake ······ (93)
6 World Wide Distributions of Earthquake ······ (95)
7 Effects of Earthquakes ······ (97)
8 Prediction and Prevention of Earthquake ······ (99)

Chapter 8 Geological Structures ······ (103)

1 Orientations ······ (104)
2 Fault ······ (105)
3 Joints ······ (109)
4 Fold ······ (110)

Chapter 9 Plate Tectonics ······ (118)

1 The Continental Drift Theory ······ (118)
2 Sea-floor Spreading Theory ······ (119)
3 Plate Tectonics ······ (120)
4 Types of Plate Boundaries ······ (121)
5 Earth's Major Plates ······ (125)
6 Continental Margins ······ (125)
7 Wilson Cycle ······ (127)
8 Cause of Plate Tectonics ······ (129)

Chapter 10 Weathering and Soil ······ (132)

1 Physical Weathering ······ (132)
2 Chemical Weathering ······ (134)
3 Biological Weathering ······ (135)
4 Factors that Influence Weathering ······ (136)

5 Products of Weathering ······ (138)
6 Regolith and Soil ······ (140)

Chapter 11 Streams ······ (144)

1 Introduction ······ (145)
2 Stream Erosion ······ (149)
3 River Transportation ······ (156)
4 Stream Deposition ······ (157)
5 Degradation ······ (159)
6 River System Pattern（水系型式）······ (161)
7 Landscape Evolution（地貌演化）······ (163)

Chapter 12 Groundwater ······ (169)

1 Instruction ······ (169)
2 Recharge and Discharge of Groundwater ······ (172)
3 Storage Condition of Groundwater ······ (173)
4 Geothermal Water ······ (178)
5 Karst Topography ······ (179)
6 Groundwater and Engineering ······ (182)
7 Groundwater Contamination ······ (184)

Chapter 13 Ocean and Coastal Processes ······ (187)

1 Ocean Movements ······ (188)
2 Coast ······ (192)
3 Continental Margin and Ocean Basin ······ (201)

Chapter 14 Glacier ······ (207)

1 The Formation of Glacier ······ (207)
2 The Budget of a Glacier ······ (209)
3 Movement of Glacier ······ (209)
4 Glacial Classification ······ (210)
5 Glacial Erosion ······ (211)
6 Glacial Transportation ······ (215)
7 Glacial Deposition ······ (215)
8 Glaciation and Its Effects ······ (219)
9 Causes of Ice Ages ······ (220)

Chapter 15 Wind Action and Desert ······ (225)

1 Wind Erosion ······ (225)
2 Wind Transportation ······ (227)

3 Wind Deposits ······ (228)
4 Deserts ······ (231)
5 Desertification and Dust storm ······ (233)
6 Loess ······ (233)

Chapter 16 Mass Wasting ······ (237)

1 Factors controlling Mass Wasting Hazards ······ (238)
2 Types of Mass Wasting: Slope Failures ······ (240)
3 Types of Mass Wasting: Sediment Flows ······ (243)

Chapter 17 Lake and Wetland ······ (249)

1 A Genetic Classification of Lakes ······ (249)
2 Hydrology of Lake ······ (252)
3 Geological Process of Lake ······ (254)
4 Wetland ······ (256)

Chapter 18 The Solar System ······ (260)

1 Introduction ······ (260)
2 Terrestrial Planets ······ (264)
3 Jovian Planets ······ (271)
4 Origin and Evolution of the Solar System ······ (273)

Chapter 19 Resource and Sustainable Development ······ (277)

1 Oil and Gas ······ (278)
2 Coal ······ (280)
3 Metals and Ores ······ (283)
4 Nonmetallic Resources ······ (288)
5 Resources and Sustainable Development ······ (289)

Chapter 20 Atmosphere and Environment ······ (293)

1 El Nino and La Nina ······ (293)
2 Air pollution and Ozone Zone ······ (295)
3 Tropical Cyclone ······ (297)
4 Global warming ······ (301)

References ······ (308)

图 版

Chapter 1 Introduction

Key words

Geological process	Historical geology	Internal Process	Mantle	Core	Moho
地质作用	地史学	内动力地质作用	地幔	地核	莫霍面
Geoid	Physical geology	External Process	Uniformitarianism		Crust
（大地水准面、地球体）	普通地质学	外动力地质作用	均变论、将今论古说		地壳
Magnetic declination	Hypothesis	Earth's Magnetic Field	Scientific method		Theory
磁偏角	假说	地磁场	科学方法、原理		理论

1 Earth Sciences

Earth sciences study everything about Earth. It encompasses everything from outer space to the Earth's fiery core, from the most ancient fossils to the landslides and earthquakes of today. Topics in the Earth sciences include Earth's structure and composition, climate changing, mineral resources and so on.

The scientific study of the Earth has never been so important a concern for all the people as it has become today. The four essential determinants of human well-being — energy, food, health and water are all topics directly or indirectly discussed in the course of Earth sciences. The Earth provides us with all the essential elements for human endeavor, but the activity of human being has exerted such a huge impact on the Earth that we are now talking about "future of human being". Earth is responsible for man, but to what extent is man responsible to the Earth? The global warming (全球变暖), the energy crises indicated by soaring oil price, flood, ozone hole (臭氧洞), all the dismal facts show that we stand at a critical moment in Earth's history, a time when humanity must choose its future. And therefore, a knowledge and understanding of our planet is critical to our social well-being and indeed, vital to our survival.

The core of Earth sciences is geology. In addition, Earth sciences also deal with astronomy (the study of the universe), meteorology (the study of the atmosphere and the processes that produce weather and climate) and environment sciences to a certain extent.

Geology is derived from two Greek roots, Gaia meaning mother Earth and logos meaning the love of wisdom of science, and consequently geology is the study of the Earth — how it was born, how it evolved, how it works, and how we can help it. Traditionally, geology is divided into two broad areas:

- **Historical Geology**

Historical geology (地史学) examines origin of the Earth, origin of life, and changes in Earth and life through time. Historical history has been recorded in the layered rocks.

• **Physical Geology**

Physical geology(普通地质学) examines materials the Earth is made of and processes operating on and beneath the surface of the Earth. Earth materials include oil, gas, rocks, minerals, salt, coal, sand, clay, soil, atmosphere, hydrosphere and organisms. What we call a geological process is the natural process by which the Earth's materials and various features are modified and reformed due to application of natural forces resulting from gravity, temperature changes, earthquakes and volcanic activities. All of dynamic actions or events that occur at the Earth's surface or with the Earth's interior are geological processes. And all of the geological features, such as mountains, plains and ocean basins are the results of geological process. Due to the action of geological process, the Earth's surface and crust are constantly in changing. Earthquake, erosion and volcanic activity are some examples of geological processes.

According to the energy related to geological processes, geological processes can be divided into internal (operating in Earth's interior) and external (operating on Earth's surface). External processes are driven by heat energy from the sun and gravity, including weathering, erosion and deposition. Internal processes are driven by heat energy from the interior of the Earth, most from radioactive decay — nuclear energy, including earthquakes, volcanic activity and mountain building(造山运动).

2 Why Study the Earth?

Understanding Earth materials and processes is of great significance to the modern sciences. For example, the problems of origins of life and Earth, which belong to basic problems in modern natural sciences, are impossible to be solved without the help of geological knowledge. Physical geology also plays an important role in our daily life.

2.1 Resource and Energy

Geologic knowledge is very helpful to prospect the natural resources. Nearly all of the natural resources, such as petroleum, coal and other metal ores are associated with various kinds of rock in the Earth, and are the products of various geologic processes. Groundwater(地下水) is another kind of important natural resource and is key to agricultural and industrial development. The geologic knowledge is the basis for the explorations of all of the above-mentioned resources.

2.2 Geologic Hazards

Geologic hazards are geological disasters occurring when natural geological processes impact on our activities, either through loss of life or injury, or through economic loss. Earthquakes, volcanic eruptions and landslides could affect us at any time. Knowledge of these natural phenomena and extent to which they can be predicted are necessary to prepare for these eventualities.

2.3 Engineering Geology

Engineering geology (工程地质) is the application of geological principles to civil engineering. Engineering Geologists are found worldwide working on problems as varied as foundation and slope design to the assessment of seismic and landslide risk. During the construction of dams, tunnels and highways, the strength of rock and soil must be taken into consideration.

3 How to Study the Earth?

3.1 Field Work

Geological phenomena are the products of geological processes. Careful observation in the field is of great importance to geological work. So, the heart and soul of geology is fieldwork and geology is the only one science where your entire life need not be spent in a lab.

Much of the knowledge about the Earth comes from the direct observation on the Earth's surface. But a comprehensive understanding about " how the Earth works" needs more information about the interior of the Earth. However, there is no way to peer hundreds or thousands kilometers into Earth's center to observe the process in action. Currently, the earthquake wave is used to detect the mineral composition and the structure of Earth's deep interior.

3.2 Lab

There are lots of experimental ways to study the geology. Microscope is the most commonly used apparatus to observe and identify the Earth materials. The TEM (Transmission electron microscopes, 透射电镜) and SEM (Scanning electron microscopes, 扫描电镜) can magnify a sample up to 50 million times its size, with a spatial resolution (分辨率) of 0.1 nanometer (nm, 纳米). Other physical and chemical experiments can also be used to analyze the related properties of minerals and rocks.

3.3 Theoretical Study

Any kind of scientific research should be done according to scientific methods. The scientific method is the process by which scientists, collectively and over time, endeavor to construct an accurate (that is, reliable, consistent and non-arbitrary) representation of the world. The scientific method is the best way yet discovered for winnowing the truth from lies and delusion. It is difficult to comprehend any aspect of science without first understanding the scientific method. This is a very important problem solving technique which has been extensively used by scientists in their quest to build their knowledge base and gradually gain insight into the workings of nature. The scientific method typically has the following steps (Figure 1.1).

• **Observation**

The first step of scientific method is observation. Observation is the description of a natural phenomenon or group of phenomena. Gathering as much evidence as possible about the phenomenon is essential for further research. Observation should be done as objectively as possible. In geology, observation can be done either in the field or in the lab.

• **Hypothesis**

The second step is to formulate a hypothesis to explain the phenomena. It is a rational explanation or tentative explanation or idea of the observed phenomenon and can be used to predict the existence of other phenomena. Hypothesis is a provisional idea which can be either confirmed or disproved, but it can guide you in further work

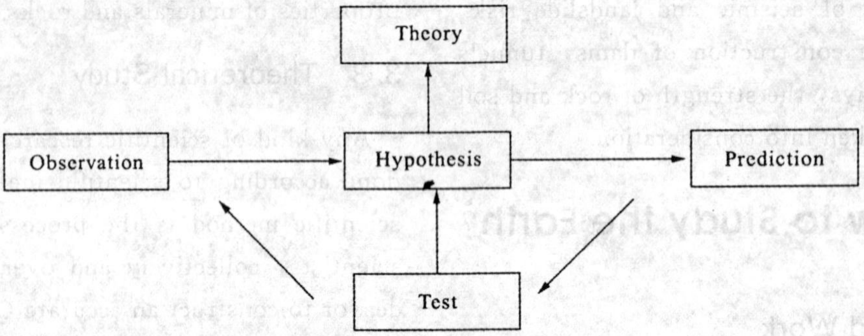

Figure 1. 1 Flow chart for scientific method

to get a better answer.

• **Test**

The third step is to design an experiment or series of experiments to test the hypothesis. If the experimental result supports the hypothesis, the hypothesis can be called a theory. At this moment, you can publish your paper on the scientific journals. Otherwise, the hypothesis is just discarded. For example, "The Earth is a flat" is a hypothesis. It should be proved or disproved. How to test this hypothesis? Perhaps the best way is to construct an aircraft to go to the out-space and view the Earth directly.

• **Theory**

The word"theory" in normal usage means a guess or a vague and fuzzy sort of fact. But in science, a "theory" is a belief that has been verified by actual experimentation and/or observation. It is a conceptual framework that explains all related existing phenomena with considerable facts to support it and predicts new ones. A theory is a unifying hypothesis that has survived all the tests. It cannot be proved, but only be disproved. One scientist cannot create a theory; he can only create a hypothesis. If a single piece of available information does not support a theory, then the theory is disproved. For example, the plate tectonics is taken by nearly all geologists as theory because it can explain nearly all the geological observations by a very simple model.

• **Law**

The well-established theory is called a law. Scientific law is a statement of fact. It is used to give a concise explanation of an action or set of actions. Laws generally can be expressed as mathematical equations. A law actually is considered by most scientists as established fact. For example, most biologists believe that evolution is more than a theory; it is an established fact. Newton's laws of motion were taken as facts by essentially all the scientists because those laws are supported by so much evidence and they have always been observed to be true.

One of widely used and accepted law in geology is the principle of Uniformitarianism. It was proposed by James Hutton (1726 – 1797), now known as the father of modern scientific geology. This principle states that everything must move slowly in a repetitive, continuous cycle and the same geological processes we observe today have been operating throughout Earth's history. As present processes are thought to explain all past events, the slogan of Uniformitarianism is"the present is the key to the past". According to the principle of Uniformitarianism, if we observe those processes occurring today, we can infer that the same processes operated in the past and will take place in the future. For example, at present, the river can transport the sand and mud to the sea and finally form sediments with typical features, so we can infer that the rivers in the past would transport and deposit the sediments in the same ways. Most scientists place Uniformitarianism at the principal or"law" level rather than theory, meaning that this principle is an undisputed fact and does not need to be proved.

4 Earth Physical Characteristics

4.1 Shape and Size

One of the oldest Earth mysteries was what shape it is? Ancient Chinese thought the heaven was a dome-shaped tent covering on the square-shaped land. Eratosthenes, a Greek geographer (about 276 to 194 B.C.) was the first to prove the sphericity of the Earth and estimate accurately the diameter of the Earth. Aristotle proposed that the inevitable circularity of the shadow on the moon is because the Earth itself is a sphere. However, most Europeans believed that the Earth was flat until the explorations of the Renaissance.

Now we know the Earth approximates an ellipsoid. A reference ellipsoids(参考椭球) of revolution defined by semi-major (equatorial radius) and flattening (the relationship between equatorial and polar radii) is usually used to represent the shape of the Earth. But actual Earth's surface is far from a geometrically perfect ellipsoid. It is highly irregular and constantly in change. For example, the Mount Qomolangma(Everest) is as high as 8 844.43 meters, and the Mariana trench has a maximum depth of 10 911 meters below sea level. Even the climate-induced large-scale movement of ocean water can change the shape of the Earth.

In order to describe the shape of Earth more accurately, the concept"Geoid"(大地水准面,地球体) is introduced(Figure 1.2).

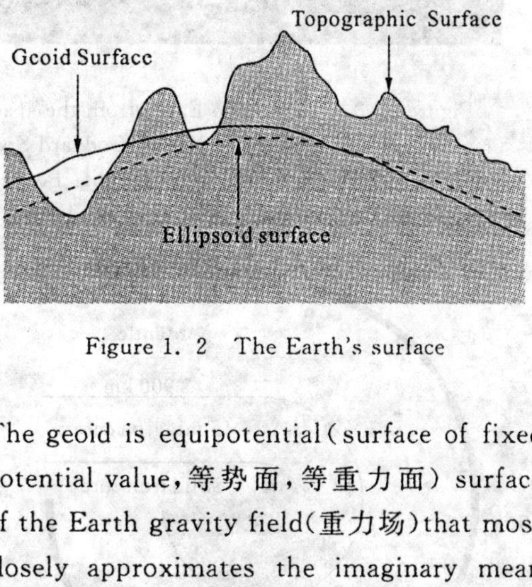

Figure 1. 2 The Earth's surface

The geoid is equipotential(surface of fixed potential value,等势面,等重力面) surface of the Earth gravity field(重力场)that most closely approximates the imaginary mean sea surface. In other words, it is the"mathematical figure of the Earth". This topography of an equipotential surface is about ±100m with the reference ellipsoid of revolution, and it is a clear representation of the lateral variations in the Earth density! At every point the geoid surface is perpendicular to the local plumb line (the direction of gravity). The geoids coincides with mean sea level in the oceans. However, it is important to note that a significant difference exists between the geoid and the real object.

Because the Earth rotates, the actual sh-

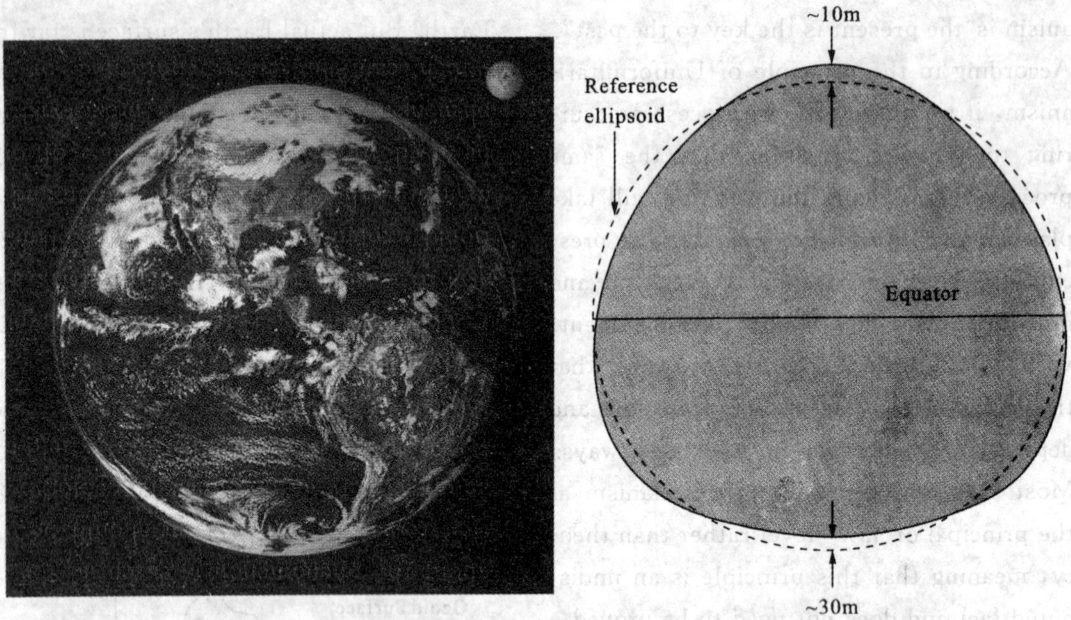

Figure 1. 3 Left: The Earth from the space (satellite image). The moon is at the upper - right corner. Image credit: NASA Goddard Space Flight Center (NASA - GSFC), Public domain
Right: Pear - shaped Earth (geoid)

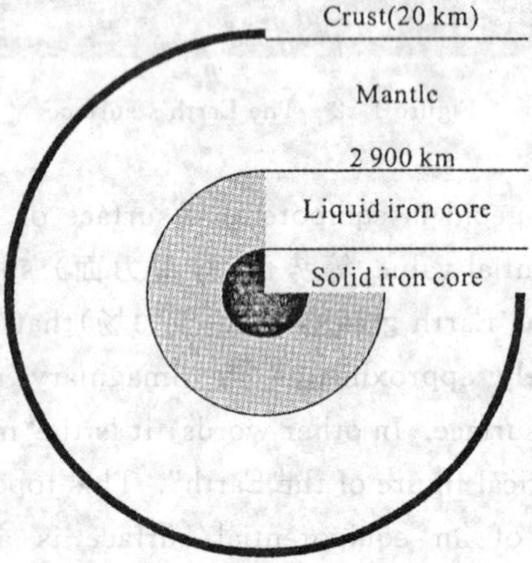

Figure 1. 4 The structure of Earth

ape of the Earth is a pear-shaped oblate with a slightly higher radius 6 378 km at Equator and 6 357 km at the poles(Figure 1. 3). The average radius of planet Earth is about 6 371 km. It is slightly elevated at the Antarctic Ocean and slightly depressed at the Arctic Ocean. And it is also slightly squashed at the Poles and bulging at the Equator.

4. 2 Internal Structure of the Earth

The Earth consists of several layers. The three main layers are the core, the mantle and the crust(Figure 1. 4).

- **Crust**

The crust is the Earth's hard outer shell, the surface on which we are living. The crust is very thin (average 20 km) and is 1 percent of Earth's volume. The crust is made up of solid material but these material is not everywhere the same, which is thinner under the oceans, thicker under the continents. In continental areas, the crust's average thickness is 35 km. The continental crust in the Himalayas is some 75 kilometers deep. In oceanic areas, the crust's average thickness is 5 km. The density of Earth's crust is 2. 7 g /cm^3.

The base of the crust is defined as the

Mohorovicic discontinuity(莫霍不连续面), or Moho, named after Andrija Mohorovicic, a Croatian(克罗地亚) geologist.

• **Mantle**

The mantle is the layer sandwiched between the core and the crust. It is about 2 900 km thick and makes up nearly 80 percent of the Earth's total volume.

The upper part of the mantle is solid but at greater depth the heat causes the rock to behave more like a liquid. The high pressure, however, stops the rock from melting.

• **Core**

The Gutenberg discontinuity(古登堡面) is the boundary between the core and the mantle.

The core begins just before halfway to Earth's center, that is, it extends from 2 900 to 6 367 km. The outer core is the layer beneath the mantle. It is made of liquid iron and nickel. Complex convection in liquid metal outer core gives rise to the Earth's magnetic field. The inner core is made of solid iron and nickel. Temperatures in the core are thought to be in the region of 5 000 ～6 000 ℃ and it's solid due to the massive pressure.

4.3 Earth's Magnetic Field

The Earth is surrounded by its magnetic field. This magnetic field can be sensed by migratory birds as well as many animals. The magnetic field around the Earth's core is like the field of a bar magnet and magnetic field lines radiate between Earth's north and south magnetic poles. The magnetic field is strongest at magnetic pole because it is the place where the magnetic field lines are the closest together. The magnetic field stretches way out into space — at least 60 000 km — to form a protective bubble known as the magnetosphere. The presence of the magnetic field shields the Earth from the charged particles from the Sun (solar wind,太阳风). Auroras(极光)(Figure 1.5) are glowing streaks in the sky seen at night in the far north and south. Auroras are cau-

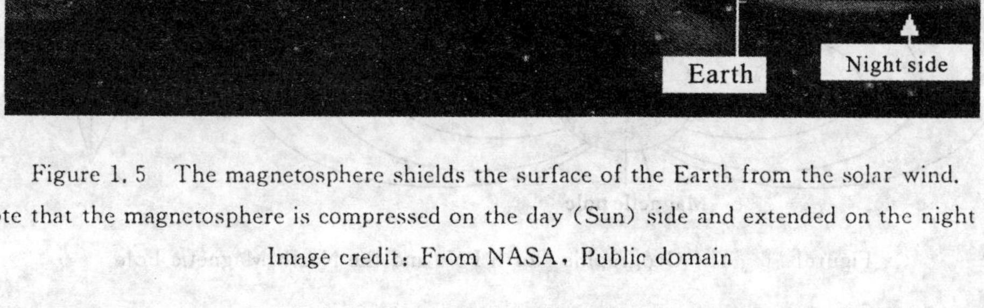

Figure 1.5 The magnetosphere shields the surface of the Earth from the solar wind. Note that the magnetosphere is compressed on the day (Sun) side and extended on the night side

Image credit: From NASA, Public domain

sed by high-energy particles from the solar wind that are trapped in the Earth's magnetic field.

• Origin of the Earth's Magnetic Field

The origin of the Earth's magnetic field is not well understood. According to the dynamo theory(发电机理论), the magnetic field is supposed to be caused by the convection of molten iron and nickel, within the outer core, along with the Earth's rotation. For example, the magnetic field of a bar magnet results from the motion of negatively charged electrons in the magnet. The electrical charges in liquid metal core are supposed to circulate unevenly, producing an electric current, thus generating the magnetic field.

• Magnetic Declination

Long before GPS (Global Positioning System) satellites and other high-tech navigational aids, the compass gave humans an easy and inexpensive way to orient themselves. But the compass needle points to the North Magnetic Pole rather than the geographic North Pole. The true north, or the geographical north, is the surface point located on the Earth's axis of rotation. Geographic North Pole and the North Magnetic Pole are not in exactly the same position because Earth's magnetic field is slightly tilted with respect to the planet's spin axis. The angular difference between these two directions is called magnetic declination (磁偏角). There is currently a difference of about 11.5° between the two. Compass directions must be adjusted by fixed amounts at given points on the surface of the Earth to yield true directions.

Different places have different magnetic declinations. And the location of the magnetic north pole also changes slowly over time. Current study shows that there is a gradual northward migration of the north magnetic pole in the past century and a half.

• Magnetic Field Reversals(地磁极倒转)

The Earth's magnetic field reverses (the north and south magnetic poles switch, 反转) at intervals, ranging from tens of thousands to many millions of years, with an average interval of approximately 250 000 years. That magnetic field reversals have occurred in the past is confirmed in the geological record. The exact mechanism responsible for magnetic field reversals is not well understood. At present, the magnetic field is decaying at 5% per century and may disappear thousands of years later. The weakening of Earth's magnetism is one of the factors believed to be predictive of a pole reversal.

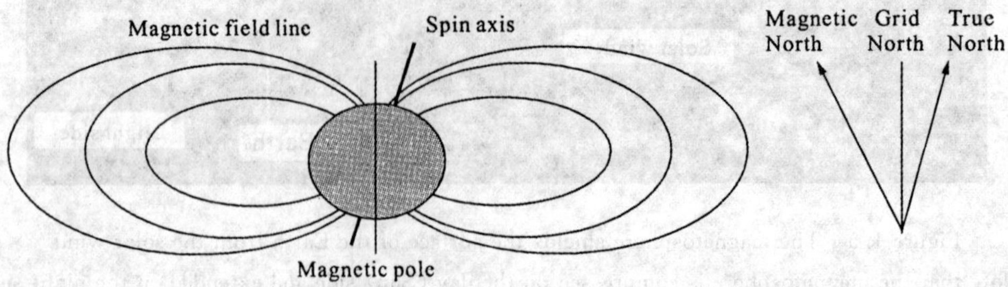

Figure1. 6 The Geographic North Pole and the North Magnetic Pole

1. Answer the Following Questions

(1) What is geology?
(2) What is Uniformitarianism?
(3) List and describe in sketch the different layers of Earth.
(4) Define the geological process.
(5) Define the geoid.
(6) Differentiate the "internal geological process" and "external geological process".
(7) What is magnetic declination?
(8) Explain the scientific method.
(9) Differentiate the hypothesis and theory.
(10) What is dynamo theory?

2. Multiple Choices

(1) The density of Earth's crust is ______.
A: 5.7 /cm^3 B: 1.0 g /cm^3
C: 1.0 g /cm^3 D: 2.7 g /cm^3

(2) Hypotheses generally make predictions that will allow them to be tested.
A: True B: False

(3) Which of these phrases is not true for magnetic field?
A: The axis of the magnetic North and South Poles does not coincide with the geographical axis of the Earth's rotation.
B: Earth's magnetic field is decaying.
C: Magnetic fields are produced by the motion of electrical charges in the liquid metal core.
D: Geographical north is not the true north.

(4) Which layer with the Earth is liquid?
A: Crust B: Mantle
C: Outer core D: Inner core

(5) The boundary between the core and mantle is called ______.
A: The Moho
B: The Gutenberg discontinuity
C: Both of them
D: None of them

(6) The geoid coincides with reference ellipsoid of rotation everywhere on the Earth's surface.
A: True B: False

(7) If the experiments bear out the hypothesis, it may come to be regarded as a theory.
A: True B: False

(8) Which one belongs to the external geological process?
A: Earthquake
B: Mountain building
C: Weathering
D: Volcano eruption

(9) Geoid is a perfect sphere.
A: True B: False

(10) Which statement about the scientific method is not true?
A: A scientific law is an accepted scientific principle taken to be correct and universally applicable.
B: A hypothesis must be a testable one by experimentation.
C: In general, both a scientific theory and a scientific law are accepted to be true by the scientific community as a whole.
D: A hypothesis implies that something has been proven and is generally accepted as being true.

(11) A theory can be proved or disproved.
A: True B: False

(12) The crust's average thickness is:
A: 30 km B: 35 km
C: 75 km D: 5 km

(13) Which one belongs to the internal geological process?
A: Deposition
B: Erosion
C: Weathering
D: Volcano eruption

(14) Which sentence about the geoid is not true?
A: The geoid is a representation of the surface of the Earth that it would assume if the sea covered the Earth.
B: The geoid is defined as a surface on which the Earth's attractive (i. e. gravitational) forces are everywhere equal.
C: Geoid on continents lies below continents

and corresponds to level of nearly massless fluid if narrow channels were cut through continents.

D: Geoid is a regular and smooth ellipsoid.

(15) Current magnetic declination is about ______.

A: 0°　　B: 11.5°

C: 30°　　D: 18°

(16) Auroras are caused by interaction between solar wind and the Earth's magnetic field.

A: True　　B: False

(17) In oceanic areas, the crust's average thickness is ______.

A: 30 km　　B: 35 km

C: 75 km　　D: 5 km

(18) A law actually is considered by most scientists as established fact.

A: True　　B: False

(19) Mount Qomolangma is as high as ______.

A: 8 844.43 m　　B: 8 848.43 m

C: 8 844.13 m　　D: 8 848.13 m

(20) Which sentence about Uniformitarianism is not true?

A: Uniformitarianism holds that changes in the Earth's surface that occurred in past geologic time are referable to the same causes as changes now being produced upon the Earth's surface.

B: Uniformitarianism states that present processes can be used to interpreter past patterns.

C: Uniformitarianism suggested that the landscape developed over long periods of time through a variety of slow geologic processes.

D: The massive extinction of dinosaur is an excellent example of Uniformitarianism.

Chapter 2 Minerals

Key wrods

<table>
<tr><td>Element</td><td>Atomic number</td><td>Abundance</td><td>Clark value</td><td colspan="2">Crystal</td><td>Nucleus</td></tr>
<tr><td>元素</td><td>原子数</td><td>丰度</td><td>克拉克值</td><td colspan="2">晶体</td><td>原子核</td></tr>
<tr><td>Isotope</td><td>Radioisotope</td><td colspan="2">Crystal habit</td><td colspan="2">Rock forming minerals</td><td>Luster</td></tr>
<tr><td>同位素</td><td>放射性同位素</td><td colspan="2">晶体习性</td><td colspan="2">造岩矿物</td><td>光泽</td></tr>
<tr><td>Streak</td><td>Cleavage</td><td colspan="2">Mohs scale of hardness</td><td>Fracture</td><td>Fluorescence</td><td>Mineral</td></tr>
<tr><td>条痕</td><td>解理</td><td colspan="2">摩氏硬度计</td><td>断口</td><td>荧光</td><td>矿物</td></tr>
</table>

Earth materials are materials from which the Earth is made. It is a general term including the rocks and minerals. Mineral is composed of elements and rock is composed of minerals. To identify a particular rock, the minerals that form the rock must first be identified. There are about over 4 000 minerals identified by mineralogists.

1 Elements

1.1 Elements and Isotopes

Chemical elements are the fundamental materials of which all matter is composed. An element consists entirely of atoms of the same sort (with the same atomic number). It cannot be separated into anything simpler by chemical means. There are 116 known elements, only 91 of which are naturally occurring. An atom is made up of positive-charged protons, negative-charged electrons, and neutrons, which are electrically neutral. Atomic nuclei consist of protons and neutrons. The atomic number of an element is the number of protons in one atom of the element. Each element has a different atomic number.

Isotopes are atoms of a chemical element whose nuclei have the same atomic number, Z, but different atomic weights, A. The word isotope, meaning at the same place, comes from the fact that isotopes are located at the same place on the periodic table.

Most isotopes are stable, but some of them are unstable. Radioisotope is an isotope of chemical element having an unstable nucleus that emits alpha, beta, or gamma rays until stability is reached. The process that radioisotope gives off energy and particles spontaneously are called radioisotope decay. In alpha decay an atomic nucleus emits an alpha particle (helium nucleus). Since an atom loses two protons during alpha decay, it changes from one element to another. For example, after undergoing alpha decay, an atom of uranium (with 92 protons) becomes an atom of thorium (with 90 protons). Beta (β) particles are high-energy electrons. Gamma rays are an energetic form of electromagnetic radiation.

1.2 Abundances of the Elements in the Earth's Crust

The abundance of a chemical element is how common the element is, or how much of the element there is. The abundance of the elements in the Earth's crust is called Clark value of crust. From the table following (Table 2.1), you can find that the elements in the Earth's crust are not distributed uniformly.

2 Minerals

2.1 Definition of Minerals

A mineral is a naturally-occurring, homogeneous solid, inorganically formed substance, with a definite chemical composition.

• "Naturally-occurring" means that it forms in nature by geological process. Those synthetic compounds cannot be regarded as minerals. They are often being called artificial minerals or synthetic minerals.

• "Homogeneous" means that it must be chemically and physically homogeneous. Granite is not a mineral because it contains more than one compounds and therefore it is chemically and physically inhomogeneous.

• A liquid or a gas does not belong to minerals. Ice is a mineral, but liquid water and water vapor are not.

• "Inorganic" substances do not contain carbon-hydrogen bonds. Organic compounds are mostly produced by organisms, however organisms can produce minerals such as calcite. Calcite is made as part of the shells of clams or corals. Actually, calcite can also be made by a chemical way in the nature.

• "Definite chemical composition" means that their chemical compositions can be expressed by generally fixed chemical formula. For example, coal is not a mineral because it has not a fixed chemical formula.

• "Ordered atomic arrangement" means it must be a crystal (晶体). Crystals are those materials which have a typical crystal structure composed of a set of atoms arranged in a particular way, which is periodically repeated in three dimensions. For example, halite has a definite crystal structure, thus halite is a mineral. Glass is not a mineral because the atoms are randomly distributed.

2.2 Crystal

A crystal is a solid in which the constitu-

Table 2.1 Abundances of the elements in the Earth's crust

(approximate percent, by weight, After Xia Bangdong, 1995)

Oxygen	O	46.30	Sodium	Na	2.36
Silicon	Si	28.15	Potassium	K	2.09
Aluminum	Al	8.23	Magnesium	Mg	2.33
Iron	Fe	5.63	Hydrogen	H	0.15
Calcium	Ca	4.15	All others		0.15

ent atoms, molecules, or ions are packed in a regularly ordered, repeating pattern extending in all three spatial dimensions. All minerals are crystals. Chemical bond is electrical force responsible for binding together the atoms, ions, and ionic groups that constitute crystalline solids. Chemical bonds include ionic bond(离子键), covalent bond(共价键), metallic bond(金属键) as well as hydrogen bond (氢键).

The regular repeating pattern of atoms in a crystal is called the crystal lattice. Crystal lattice (晶格) is three-dimensional configuration of points connected by lines used to describe the orderly arrangement of atoms in a crystal. Unit cell(晶胞) is the smallest part of a crystal lattice structure which has all the elements of the structure of that lattice. The unit cell can be thought of as a single "brick" which when stacked together in three-dimensions produces the crystal lattice. Crystals are formed by the repetition of unit cells in 3-D space.

Under favorable conditions, minerals may grow as well-formed crystals, characterized by their smooth plane surfaces and regular geometric forms (Figure 2. 1). The outward appearance of crystal form is a reflection of its internal lattice structure. For example, rock salt usually assume a cubic form and calcite generally take a form of rhombohedron(菱面体).

However, most solid objects are made of many tiny interlocking crystals. An aggregate of crystalline intergrowths, such as granite, that does not show well-defined crystal forms. It is only occasionally that the individual crystals are large enough to catch our attention.

Some chemical substances form more than one kind of crystal. Minerals which have the same chemistry but different crystal structures are called polymorphs (同质多象). Diamond (cubic) and graphite (hexagonal) are examples of polymorphs of carbon. They are chemically identical — both are composed of carbon (C), but physically, they are very different.

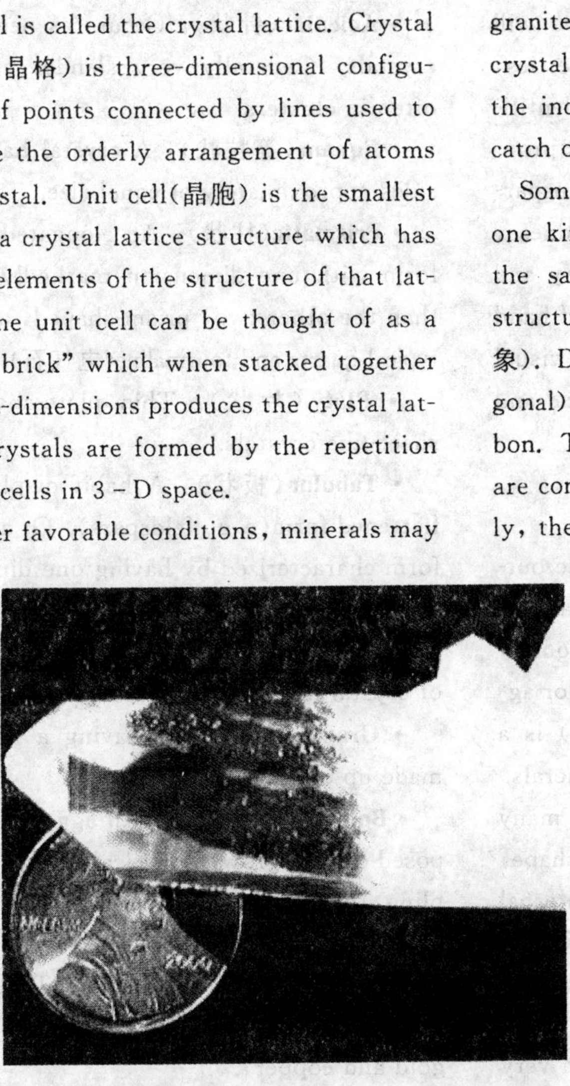

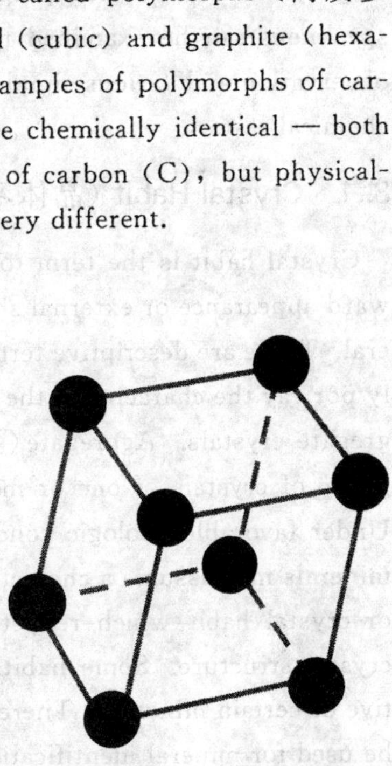

Figure 2.1 Left: Quartz crystal. Credit: U. S. Geological Surrey, Department of the Interior/USGS. Public Domain. Right: Simple unit cell

A noncrystalline material, which has no long-range order, is called an amorphous(非晶体), or glassy material(玻璃质). Glass and flint (燧石) are examples of amorphous solid.

3 Physical Property of Minerals

Due to the differences in chemical composition and internal structure, different mineral has different property. Here we will only introduce some physical properties, which are most commonly used for identification. There are ten major diagnostic properties of minerals: crystal habit, transparency, luster, color, streak, hardness, cleavage, fracture and specific gravity and magnetism. The physical properties of hand specimens can be examined through visual observation and various simple physical and chemical tests.

3.1 Crystal Habit (晶体习性)

Crystal habit is the term to indicate outward appearance or external shape of a mineral. There are descriptive terms that quickly portray the character of the crystal or aggregate crystals. Aggregate(集合体) is a group of crystals of one or more minerals. Under favorable geologic conditions, many minerals may assume a characteristic shape, or crystal habit which reflects its internal crystal structure. Some habits are distinctive of certain minerals. Therefore they can be used for mineral identification.

For example, quartz often grows in very characteristic columnar hexagonal prisms capped by a hexagonal pyramid. Galena and halite occur as cubes and garnets develop 12- or 24-sided equidimensional(等轴状) forms. On the other hand, the habit can be described as a very descriptive way. For example, the habit of olivine, pyroxene, and garnet is granular. In rock, quartz crystals are typically irregular in outline whereas feldspar is tabular shaped, amphibole is needle shaped, and mica is platy (Figure 2.2).

Terms used to describe the crystal habit of single crystals are listed below:

- **Acicular**(针状): Occurs as needle-like crystals. Generally very slender crystals, often in clusters.
- **Equant**(等粒状): A crystal having all sides approximately of equal size.
- **Prismatic**(柱状): An elongated crystal form with one dimension markedly longer than the other two, prism-shaped, as in the typical quartz and tourmaline(电气石).
- **Platy** (片状): Thin plate-shaped or sheet-like crystals.
- **Tabular** (板状): A habit displaying a flattened form(e.g. feldspars). Or a crystal form characterized by having one dimension markedly smaller than the other two.

Terms used to describe the crystal habit of crystal aggregates are listed below:

- **Granular** (粒状): Having a structure made up of small grains.
- **Botryoidal**(葡萄状): An aggregate composed of intergrown spherical shapes resembling the outside surface of a bunch of grapes.
- **Arborescent** (树枝状): "Tree like" growths of branched systems (e.g. silver, gold and copper).
- **Massive**(块状): A homogeneous texture characterized by an absence of any directional structure or preferred orientation

Figure 2.2 Crystal habit.

Image from: U. S. Geological Surrey, Department of the Interior/USGS

of constituent minerals. They are massive formless lumps, having no discernable regular crystal form.

• **Radiating**(放射状): Groups of elongated crystals that extend from a common central point.

• **Nodular**(结核状): A habit displaying small, irregular lumps.

• **Stalactitic**(钟乳石状): Shaped like pendant columns as stalactites or stalagmites (e.g. calcite).

• **Fibrous**(纤维状): Crystals made up of fibers.

• **Disseminated**(浸染状): Occurs in small, distinct particles dispersed in matrix.

• **Earthy**(土状): Dull, clay-like texture with no visible crystals.

3.2 Luster

Luster refers to the way in which light is reflected from the surface of a mineral. So, it is a description of appearance of a mineral in reflected light. Luster is described by quality and intensity of reflected light. The terms used to describe luster are self-explanatory. For example, some minerals look metallic because they are.

There are three main catagories of luster. These are metallic, submetallic and nonmetallic luster.

Metallic luster(金属光泽): The mineral looks shiny like a metal or has the appearance of metal when light reflects off it. The opaque minerals have appearance similar to a bright chrome bumper on an automobile.

Submetallic luster(半金属光泽): Mineral having a luster between metallic a nonmetallic is said to have submetallic luster. For example, chromite (铬铁矿) has a submetallic luster. The opaque minerals have either metallic luster or submetallic luster.

Nonmetallic luster(非金属光泽): This is a general term used to described a variety of different lusters. It is important to note that nearly all the common rock forming minerals (造岩矿物)have a nonmetallic luster. That is, they do not look like metals or submetals. Nonmetallic luster is a typical feature of transparent menerals. The most common nonmetallic lusters are listed here:

• **Pearly luster**(珍珠光泽): Pearl-like luster. The exterior of all the oysters has the beautiful pearl-like luster. Example: Selenite(透石膏), a colorless and transparent variety of gypsum, can show pearl-like luster.

• **Resinous luster**(树脂光泽): Looks resinous-examples: amber (琥珀), sphalerite and sulfur.

• **Silky luster**(丝绢光泽): Silk-like luster. Muscovite in tiny flakes looks silky. Asbestos(石棉) show a typical silky luster.

• **Vitreous luster**(玻璃光泽): Vitreous luster describe a mineral like quartz that has the appearance of glass.

• **Waxy lusters** (蜡状光泽): Fairly dull luster like wax.

• **Greasy**(油脂光泽): Mineral appears to be covered with a thin layer of oil. Quartz also shows greasy luster sometimes.

• **Earthy luster** (土状光泽): Luster is best described from fresh and unweathered surfaces. Weathered minerals are typically described as earthy in luster. For example, clay has a typical earthy luster.

3.3 Transparency

Transparency (透明度) is the degree of transmitting of visible light through minerals. It describes how well light passes through a mineral sample. Minerals can be transparent(透明), translucent(半透明)or opaque(不透明). A transparent mineral transmits essentially all of the visible light spectrum.

Broadly speaking, all the metal minerals are opaque and all the nonmetal minerals are translucent or transparent.

3.4 Color

Color is one of the most obvious and striking physical properties of a mineral. But unfortunately color is not a good property to be used in the identification of minerals.

Color in minerals is caused by selective absorption or reflection of certain wavelengths of light by the mineral during light transmission or reflection. For example, a white mineral looks white because it reflects essentially the entire visible spectrum and a black mineral absorbs all wavelengths of light. A colored mineral will selectively absorbs certain wavelengths of light and transmits or reflects the remaining light that our eyes see. Minerals with metallic luster usually have characteristic colors.

For example, the yellow color of pyrite and white gray color of galena can be used as a diagnostic tool.

The selective absorption depends on the presence of certain atoms, especially transition metals (过渡金属), such as iron or chromium that strongly absorb portions of the light spectrum. Most minerals have different colors due to impurities caused by chemical substitution. For example, Corundum(刚玉)can be sapphire (blue)(蓝宝石) or ruby (red)(红宝石).

3.5 Streak

Streak is the color of a mineral that has been crushed to a powder. It is a more reliable indication of a mineral than its outward colour, as it is more constant.

Unlike the color of a mineral, the streak is generally very consistent from specimen to specimen for a given mineral. For example, a translucent mineral, such as fluorite (萤石) that has variable outward colors will almost certainly have a white streak.

The proper way to test for streak is to rub a mineral across a white unglazed porcelain and to examine the color left behind.

Streak is a very improtant diagnostic feature for some of the opaque and metallic minerals. For example, the hematite and galena can be easily confused because they often have a gray color. However, the blood-red streak of hematite is totally differnet from the lead gray streak of galena. Unforturanately, translucent minerals usually have a rather undiagnostic white streak.

3.6 Hardness

The hardness of a mineral describes its resistance to scratching. It depends on the strength of chemical bonding. Minerals with relatively strong chemical bonds have greater resistance to scratching and are thus harder than minerals with relatively weak chemical bonds.

We use the Mohs scale of hardness (Table 2.2) to measure the hardness of other minerals. A Mohs scale of hardness is made up of ten common minerals ranked in order of increasing hardness. Mohs scale was established by Friederich Mohs in the early nineteenth century. It is a relative hardness scale. This means that if one mineral is able to scratch another, it is said to be harder.

On this scale, you can find that talc is one of the softest; nearly every other mineral can scratch it. Talc feels smoothing as a powder becaue it is softer than your skin and cannot scratch skin. Diamond is the hardest mineral, because it can scratch all others. To determine the hardness of an unknown mineral, we compare it to the harnesses of a standardized set of minerals in Mohs scale of hardness. For example, if an unknown mineral scratches calcite but does not scratch a fluorite, its hardness must be greater than 3.0 and less than 4. We can use some of the common objects listed in the Table 2.2 to determine the relative hardness of an unknown mineral. The finger nails (H = 2 or higher), copper pennies (H = 3), plate glass (H = 5.5), a steel knife or hammer (H = 6.5) are especially useful to decide the hardness of an unknown mineral in the field.

It is important to note here that the Mohs hardness scale is a relative scale other than an absolute hardness scale. This means that we cannot determine the absolute hardness of a mineral by Mohs hardness scale . For

Table 2.2 Mohs scale of hardness

Mohs' Hardness	Mineral	Some common tools useful for measuring hardness
Talc(滑石)	1	Your skin is 1.5
Gypsum(石膏)	2	Finger nail (2.2)
Calcite(方解石)	3	Copper Penny (3+)
Fluorite(萤石)	4	
Apatite(磷灰石)	5	Pocket knife is about 5.1
Orthoclase(正长石)	6	Glass plate or window glass (5.5) is solfter than quartz
Quartz(石英)	7	Steel file = ~6.5 Quartz (harder than window glass)
Topaz(黄玉)	8	Unglazed porcelain streak plate ($H=7.5$) is harder than quartz
Corundum(刚玉)	9	
Diamond(金刚石)	10	

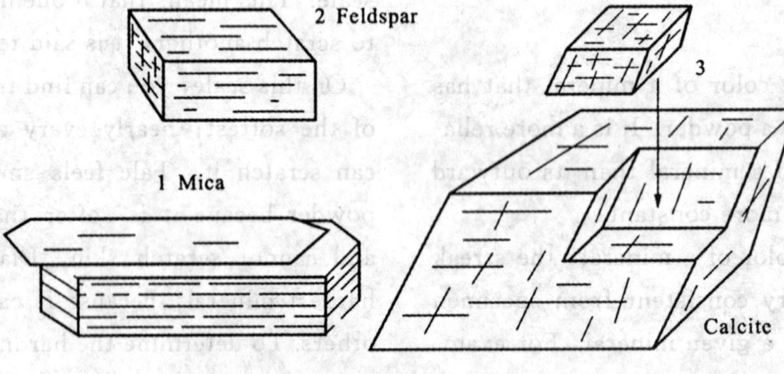

Figure 2.3 Cleavage directions of mineral

example, quartz is not 7 times harder than talc as would be the case with an absolute hardness scale. Acturally, the absolute hardness of quartz is 130 times harder than talc.

3.7 Cleavage

Cleavage is the tendency of many minerals to split along definite smooth planar surfaces. Cleavage is determined by mineral's internal structure. Cleavage planes are parallel to the planes with weaker chemical bonds in the minerals internal structure and are always parallel to a possible crystal face. The weaker chemical bond means a wider spacing between atoms. The force between atoms is not great enough to attract adjacent planes of ions closely together, and therefore, the mineral is easily to split along those weak planes.

Cleavage is one of the most diagnostic properties of mineral because it is so strongly dependent on mineral structure. One of the ways to describe cleavage is the number of directions. Multiple cleavage planes those are parallel to each other count as one cleavage direction. Mica has perfect cleavage in one direction. Feldspar has two cleavage directions and the calcite crystal has three directions of cleavage (Figure 2.3).

The quality of cleavage may be described in terns of how easy the cleavage is produced. From easiest to hardest to produce the terms are: perfect, good, distinct and poor.

3.8 Fracture

Some minerals, such as quartz, olivine and garnet, do not exhibit cleavage. That is because the strengths of the chemical bonds in their internal structures are approximately the same in all or some directions. When these minerals are struck by external forces, they will break along random directions. Failure along such directions will produce an irregular broken surface called a fracture. Be sure that the names of these fractures are descriptive and need no other explanation.

• **Conchoidal** (贝壳状): The most common fracture type is conchoidal. Mineral breaks along smooth curved surfaces like the interior surface of the conch shell. Quartz and glass exhibit distinct conchoidal fractures.

• **Hackly**(锯齿状): Jagged fractures with sharp edges. Usually metals and metal alloys have this fracture.

• **Uneven or irregular** (参差状): Irregular surfaces.

3.9 Specific Gravity (相对密度)

Specific gravity is the ratio of the weight of a given volume of the mineral to the weight of an equal volume of water, measured at 4℃. Actually specific gravity is a measure of the density of a mineral. It depends on the chemical composition and crystal structure of the mineral.

By "hefting" hand samples, minerals can be put into groupings of low, average and heavy specific gravities.

• Light: with a specific gravity less than 2.5.

• Average: with specific gravity from 2.5～4.

• Heavy: with a specific gravity greater than 4.

3.10 Other properties

Other properties that may be diagnostic include magnetism, fluorescence, taste and so on.

• **Magnetism.** Magnetism is an important diagnostic tool for a few minerals. Magnetite(磁铁矿) is strongly attracted to a magnetic field.

• **Fluorescence** (荧光). Some minerals will emit visible light as a result of exposure to ultraviolet and X-rays. This is known as fluorescence. One of the most common fluorescent minerals is fluorite. Fluorite will emit violent glow under ultraviolet light. Manganese-bearing calcite looks pink under UV(紫外线). The colour can also vary according to whether the ultraviolet radiation is long-wave or short-wave.

• **Phosphorescence**(磷光) is the ability of a mineral to glow after the initial activating ultraviolet light is removed. Some calcite hand specimens are featured by phosphorescence.

• **Taste**: Halite (salt) tastes salty most notably and sylvite(钾盐) tastes bitter.

4 Common Minerals

There are about 4 000 minerals in the Earth's crust. Only 20 to 30 of these minerals are common.

4.1 Rock-forming Minerals

Common rock-forming minerals are

those, which are relatively abundant and are the main components of rocks. The followings are the 7 commonest rock forming minerals, of which, the olivine, pyroxene, amphibole and biotite are dark-colored (深色) minerals, and quartz, feldspar, muscovite and calcite are light-colored (浅色) minerals.

• **Olivine**(橄榄石)

Olivine has an apple green or yellowish green color. Crystals are transparent to translucent. Commonly, the olivine grains in rocks show a granular crystal habit. They have a poor cleavage in two directions at 90°(Figure 2.4). Specific gravity varies from 3.2 to 4.3 according to the variation of chemical composition. Olivine is one of major components of mantle. On the Earth's surface, it is commonly found as xenoliths (包体,包裹体) in volcanic rock.

• **Pyroxene**(辉石)

The pyroxene is a general name for a group of minerals having similar chemical compositions and structures.

In general, minerals of pyroxene group have a crystal habit of short prismatic(短柱状). In aggregate, their crystal habit is usually descried as granular. The color of pyroxene varies in between dark green, brown and black. Hardness is 5.5～6. They usually have two directions of cleavage with an angle of 87°.

• **Hornblende**(普通角闪石)

The color of hornblende is almost always black to dark green. It usually has a prismatic or acicular crystal habit. Crystals can have a hexagonal cross-section. Hardness is 5～6. Specific gravity is approximately 3.02 ～3.45. Cleavage is good in two directions at 56° and 124°.

• **Orthoclase**(正长石)

Pink color is typical, 2 directions of cleavage at 90°, no striations on its crystal surfaces.

• **Plagioclase**(斜长石)

Green, gray and white. Have 2 cleavages at approximatly 90°. May have striations, but not always. Similar properties to orthoclase.

• **Feldspar** (长石)

Approximately 60%(长石, Figure 2.5) of the Earth's crust is composed of feldspar. Feldspar is generally dull to opaque with a porcelain-like appearance. Feldspar is also hard but can be scratched by quartz. Feldspar is the most common mineral in rocks.

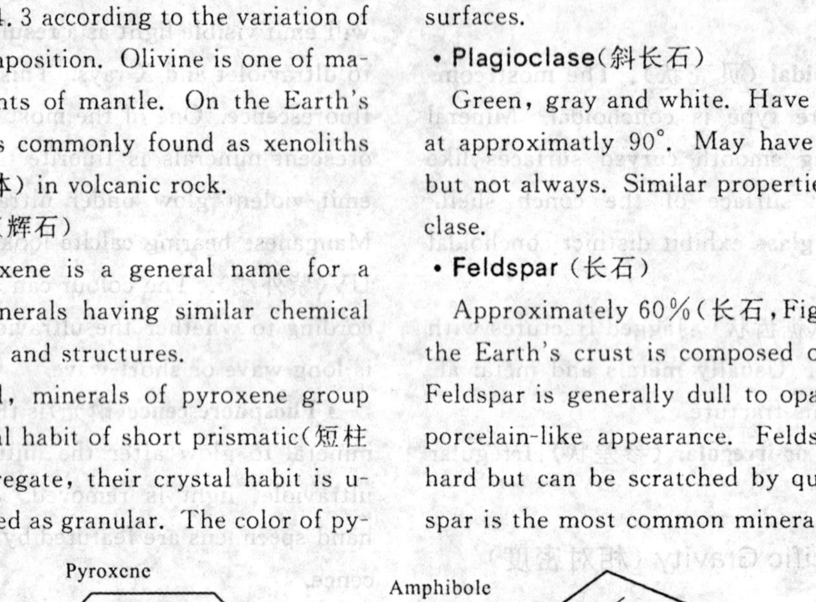

Figure 2.4 The square cross sections of pyroxene and amphibole and their two directions of cleavage

Figure 2.5 Some rock-forming minerals. Credit: U. S. Geological Surrey, Department of the Interior/USGS

The two commonest varieties are orthoclase and plagioclase.

• **Mica**(云母)

Mica is easily distinguished by its characteristic of breaking into many thin flat smooth sheets or flakes. Mica may be white (muscovite) (白云母) or dark (biotite) (黑云母).

• **Quartz**(石英)

Quartz is one of the most common minerals on Earth. Quartz is a glassy looking, transparent or translucent mineral. Durable, and no cleavage. Color is typically white or clear but can be pink, red, and purple, black. Quartz is an extremely hard mineral. It is so hard that it can actually scratch steel!

• **Calcite** (方解石)

Commonly white to gray in color. Transparent. Calcite can be scratched easily by a steel knife blade. Rhomb-like in shape (Due to its 3 directions of cleavage). Calcite's major distinguishing characteristic is its vigorous reaction with dilute hydrochloric acid (HCl, 稀盐酸).

4.2 Other Important Minerals

Here are some minerals that are either economically important or accessory minerals (副矿物) in rocks.

• **Graphite** (C)(石墨)

It is black and has a black streak. Cleavage is perfect in one direction. Graphite usually occurs as scaly (鳞片状) granules in rocks. These flexible flakes can easily slip over one another. This is the cause of graphite distinctive greasy feel. Its hardness is 1～2. The specific gravity is 2.2.

• **Pyrite (FeS_2, Iron Sulfide)(黄铁矿)**

There are shiny brassy yellow minerals. They often occur in massive aggregates. The well-formed crystal has a perfect cubic form. Striations can be found on cubic faces. Streak is greenish black and hardness is 6～6.5. The specific gravity is 5.

The brassy-yellow metallic colour of pyrite has in many cases lead to people mistaking it for gold, hence the common nickname"Fools Gold".

• **Chalcopyrite ($CuFeS_2$, Copper Iron Sulfide)(黄铜矿)**

Chalcopyrite often occurs as massive or granular aggregates. It has a bright golden color and greenish black streak. By its low hardness of 3～4, chalcopyrite can be easily distinguished from pyrite. Striations(条纹) are common on most crystal faces. Tarnished surface often show beautiful to iridescent blues, greens, yellows and purples. The specific gravity is 4.1～4.3.

• **Galena (PbS, Lead Sulfide)(方铅矿)**

Galena has a typical cubic form. But it may have octahedron and combinations of the two. Its color is lead to silver gray and its streak is lead gray. The hardness is 2～3. The specific gravity is 7.4～7.6. Galena has three cleavage directions.

Galena is the most common ore of lead.

• **Sphalerite (ZnS, Zinc Sulfide) (闪锌矿)**

Its color is typically black but can be brown and yellow and its streak is yellow to light brown. Massive forms are common. The hardness is 3.5～4, slightly harder than galena. The specific gravity is 3.5～4.

Sphalerite is a mineral that is the chief ore of zinc.

• **Hematite (Fe_2O_3, Iron Oxide)(赤铁矿)**

Its color is steel or silver gray and its streak is blood red. They often occur as massive (块状)or earthy form. Hardness is 5～6 and specific gravity is 4.0～5.3.

Hematite is mined as the main ore of iron.

• **Magnetite (Fe_3O_4, Iron Oxide)(磁铁矿)**

It has a black color and black streak. They are commonly found as massive or granular aggregates. Hardness is 5.5～6.5 and specific gravity is 5.1.

Magnetite is a natural magnet.

• **Fluorite (CaF_2, Calcium Fluoride) (莹石)**

The most striking feature of fluorite is its color. The range of common colors varies from the purple, blue, green, yellow, colorless, brown, pink, black to reddish orange. Fluorite often takes a form of cubic or octahedron. It has three perfect cleavage directions. Hardness is 4 and specific gravity is 3.18.

Fluorite is frequently fluorescent and its fluorescent colors are extremely variable. Typically it fluoresces blue.

• **Gypsum ($CaSO_4 \cdot 2(H_2O)$, Hydrated Calcium Sulfate)(石膏)**

Color is usually white, colorless or gray. Tabular and blocky crystals are the most commonly found shapes. Hardness is 2 and specific gravity is 2.30～2.37.

1. Self-assessment Questions

(1) What is a unit cell? What is a crystal?

(2) What is crystal lattice?

(3) What is a mineral?

(4) How to distinguish pyroxene and hornblende?

(5) How to determine the relative hardness of an unkown mineral?

(6) What is the difference between color and streak of a mineral?

(7) List the main particles of an atom and explain how they differ from one another.

(8) What is the difference between cleavage and fracture?

(9) What is the specific gravity of a mineral?

2. Multiple Choices

(1) Which of the following is not a mineral?

A: Quartz B: Glass

C: Diamond D: Gold

(2) Which of the following is a mineral?

A: Liquid water B: Water vapor

C: Protein D: Silver

(3) The streak of the hematite is ______.

A: White B: Green

C: Blood red D: Gray

(4) Which of the following is a fluorescent mineral?

A: Quartz B: Fluorite

C: Diamond D: Graphite

(5) Pyroxene usually has a crystal habit of ______.

A: Botryoidal B: Platy

C: Acicular D: Short prismatic

(6) Which of the following is a not dark-colored mineral?

A: Quartz. B: hornblende

C: Olivine D: Pyroxene

(7) Quartz and glass exhibit distinct ______ fractures.

A: Hackly. B: Conchoidal

C: Uneven D: Irregular

(8) Mica and calcite have ______ and ______ direction(s) of cleavage respectively.

A: 1, 2 B: 2,3

C: 1,3 D: 2,1

(9) The hardest mineral is ______.

A: Quartz B: Glass

C: Olivine D: Diamond

(10) The two most common elements in Earth's crust are ______.

A: Oxygen and Aluminum

B: Aluminum and Silicon

C: Silicon and Oxygen

D: None of Above

(11) Diamond is 10 times harder than talc.

A: True B: False

(12) Which of the following is a mineral?

A: Amber B: coal

C: Graphite D: Glass

(13) Which of the following is not part of an atom?

A: Electron B: Neutron

C: Molecule D: Proton

(14) The most abundant element in the Earth's crust is ______.

A: Water B: Gas

C: Oxygen D: Silicon

(15) Which mineral can react vigorously with dilute hydrochloric acid?

A: Calcite B: Limestone

C: Quartz D: Silicon

(16) The angle between two cleavage directions of hornblende is ______.

A: 56° B: 87°

C: 90° D: 60°

(17) Crystal habit of mineral is decided by ______.

A: Chemical composition

B: Internal structure

C: Outward appearance

D: Crystal form

(18) Chalcopyrite is harder than pyrite.

A: True B: False

(19) Which mineral is a natural magnet?

A: Fluorite B : Gypsum

C: Sphalerite D: Magnetite

(20) Which mineral has one perfect cleavage direction?

A: Fluorite B : Gypsum

C: Graphite D: Quartz

(21) Which of the following is 7 on the Moh's scale of hardness?

A: Fluorite B: Gypsum

C: Apatite D: Quartz

(22) Crystalline materials differ from amorphous materials by ______.

A: containing different chemical elements.

B: having periodic spacing of atoms.

C: being natural rather than man-made.

D: all of the above.

(23) Which of the following minerals fizzes when it comes in contact with hydrochloric acid (HCl)?

A: Fluorite B: Gypsum

C: Calcite D: Quartz

(24) Conchoidal fracture is common in minerals that ______.

A: contain 2 cleavages

B: contain 1 cleavage

C: do not have a cleavage

D: None of the above

Chapter 3 Igneous Rocks

Key words

Magma	Igneous rocks	Glassy texture	Aphanitic texture	Phaneritic texture	Texture
岩浆	火成岩	玻璃质结构	隐晶质结构	显晶质结构	结构
Lava	Intrusive rocks	Extrusive rocks	Magmatic Differentiation		Dike
熔岩	侵入岩	喷出岩	岩浆分异		岩脉
Sill	Batholith	Cinder cone	Tephra	Bomb	Tuff
岩席	岩基	火山渣锥	火山喷出物	火山弹	凝灰岩
Blocks	Plutonic rock	Pillow structure	Xenolith	Lapilli	Structure
火山块	深成岩	枕状构造	包体	火山砾	构造

Rock is a naturally occurring, solid substance composed of one mineral or a mixture of minerals or other solid substances such as glass. Based on their mode of origin or how they formed, rocks can be divided into three groups. The groups are igneous, sedimentary and metamorphic. The word "igneous" is derived from the Latin ignis, meaning "fire". Igneous rocks are formed when molten rock cools and solidifies, with or without crystallization, either below the surface as intrusive rocks or on the surface as extrusive rocks. All igneous rock starts deep in the Earth as hot, molten magma. If the magma cools and hardens inside the Earth it is called "intrusive" rock. When the magma comes out of the Earth's crust through a volcano, it is called "extrusive". Igneous rocks make up approximately ninety five percent of the upper part of the Earth's crust, but their great abundance is hidden on the Earth's surface by a relatively thin but widespread layer of sedimentary and metamorphic rocks.

1 Magma

Magma is naturally occurring molten rock located inside a magma chamber(岩浆房) beneath the surface of the Earth. Magma usually originates at a depth of 50 to 200 km. Magma is a complex high-temperature solution. It consists primarily of the elements found in silicate(硅酸盐) minerals (oxygen, silicon, and smaller amounts of aluminum, iron, magnesium, sodium, and others), along with gasses, mostly water vapor. Temperature of magmas ranges from 650 to 1 200 ℃.

One of its important features is viscosity. Viscosity(粘度) is the resistance to flow. Viscosity depends primarily on the composition of the magma and temperature. Viscosity increases with increasing SiO_2 concentration in the magma and decreases with increasing temperature of the magma.

Igneous rocks are the results of cooling and solidification of molten rocks. If the magma cools and hardens inside the Earth it is called "intrusive" igneous rock. Because magma is less dense than the surrounding solid rocks, it will work its way toward the surface. When magma appears above the Earth, it is called lava (Figure 3.1). The resulting rock is called extrusive rock or volcanic rock.

2 Intrusion and Intrusive Body

Intrusion is the process of emplacement (侵位) of magma in pre-existing rock before it reach to Earth's surface. The resulting rock is intrusive rock, or simply intrusive. The three dimensional entity of intrusive rock is referred to as intrusive body (侵入体). The pre-existing rock surrounding the intrusive body is called country rock(围岩).

When the magma forces its way into the country rock, it will react with surrounding rocks in many different ways.

2.1 Assimilation and Magma Mixing

Assimilation (同化作用) is the process of incorporation of surrounding rock bodies into a magma. The intensity and extent of assimilation depend on the magma's temperature, composition, as well as the type of country rock. Xenolith(包体) is the fragment of a pre-existing rock that has been incorporated into magma without melting. Xenoliths [Greek, *xeno* (foreign) and *lith* (rock)] form when rising magma forces its way through channels and cracks, tearing off fragments of their walls and incorporating them into rising magma. Xenoliths can originate hundreds of kilometers underground, thus conveying valuable data from the depths.

Magma mixing (混染) is the mixing of two or more magmas or magma batches to form a hybrid magma. Due to the assimilation and magma mixing, a variety of rock types can be produced from only one type of

Figure 3.1 Lava in Walnut Canyon, Arizona
Photo by Dr. Yu Shan(俞姗)

magma.

2.2 Magmatic Differentiation(岩浆分异)

Sometimes the magma will stop on its way to surface. In this case, the magma may begin to cool and solidify in place. As it does so, the originally freely moved ions in magma will slow down and begin to arrange themselves into orderly patterns. This process is called crystallization, where ions are confined to an orderly crystalline arrangement by the forces of chemical bonding.

During the course of cooling, some minerals will crystallize first and some minerals later. Early-formed crystals will often react with the magma around them to later-forming crystals, producing the changing of chemical composition of remaining magma. If early formed crystals sink (crystal settling) to the bottom and the magma crystallizes or is then injected upward into overlying rocks, the resulting rock will have a different compostion from the original magma. This process is called magmatic differentiation. That is, magmatic differentiation is the process whereby one parent magma composition can produce a number of igneous rock compositions through mineral crystallization.

Actually the minerals in magma crystallize in orders according to Bowen's reaction series. Bowen's reaction series(鲍文反应序列)(Figure 3.2) is an experimental demonstration of magmatic differentiation.

According to Bowen's reaction series, minerals with higher melting point temperatures crystallize before the lower melting point ones. The early-formed mineral will react with the melt to produce the next mineral in the series. For example, olivine, the first mineral to form, will react with remaining magma to produce pyroxene. In Bowen's reaction series, the quartz is the last mineral to crystallize.

Bowen's reaction series can be thought of as an idealized order of crystallization of a cooling magma. However, not all of these minerals will be crystallized together in the same rock.

2.3 Occurrence of Intrusive Body

Igneous rock occurrences (产状) can be

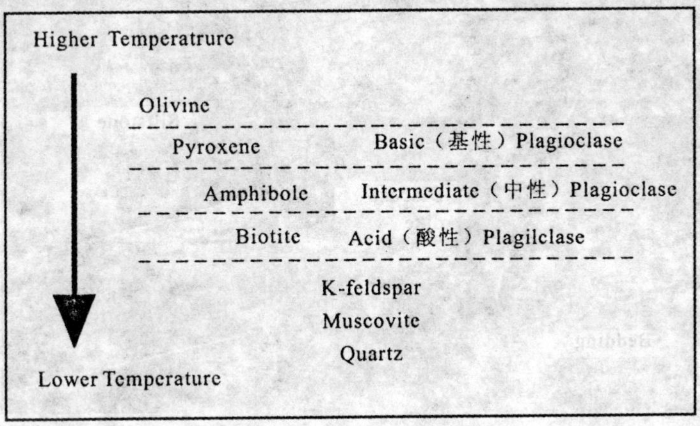

Figure 3.2　Bowen's reaction series

described in terms of shape and size of the intrusive body, contact relationship with country rocks, and original geological setting in which it forms. Typical occurrences of intrusive rock bodies are as follows:

• **Dike** (岩脉)

A dike (岩脉) in geology refers to a small tabular intrusive igneous body. The thickness, usually much smaller than the other two dimensions, can vary from a few centimeters to several tens of meters. The lateral extension of dike ranges from a few meters to tens of kilometers. They usually show a crosscutting relationship to the country rocks (Figure 3. 3). This means that a dike normally cuts across or through other pre-existing layers or bodies of rock. Because it is usually near vertical in orientation, a dike(岩墙)may form a wall-like feature on the surface when the surrounding rocks are eroded.

Dikes often form as either radial or concentric swarms around other geological bodies, such as large intrusive body and volcano vent.

• **Sill** (岩床)

Sill (Figure 3. 4) is a small tabular intrusive body that has been intruded laterally between layers of country rocks. Therefore, it is parallel to pre-existing features. The thickness of sill varies from a few meters to hundreds of meters.

• **Laccolith**(岩盖)

When the magma is injected between nearly horizontal sedimentary layers, the force of magma results in uplifting and folding of the pre-existing rocks above the intrusion, thus giving dome or mushroom-shaped intrusive body called laccolith (Figure 3.5).

• **Lopolith**(岩盆)

Lopolith is a bowl-shaped large intrusive body with contact surface being parallel to the bedding of the country rock. Both the roof and floor of lopolith are concave upward.

• **Stock**(岩株)

Stock refers to a medium sized igneous intrusion with outcrop area less than about 100 km^2.

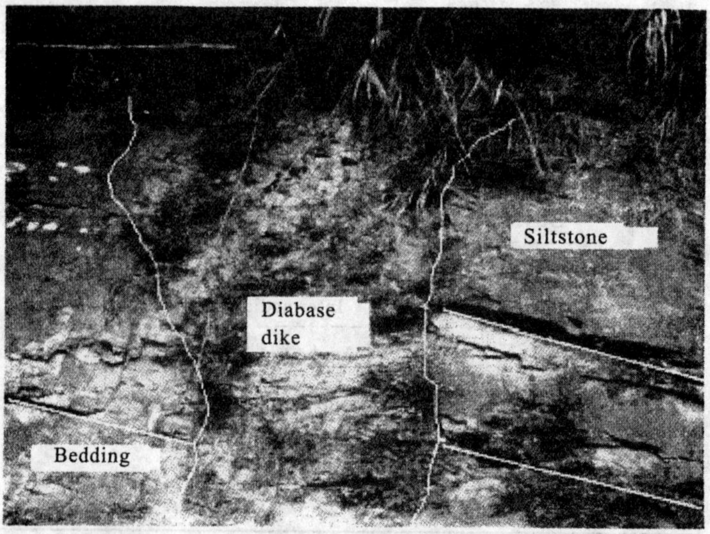

Figure 3. 3 A diabase(辉绿岩) dike intruded into siltstone

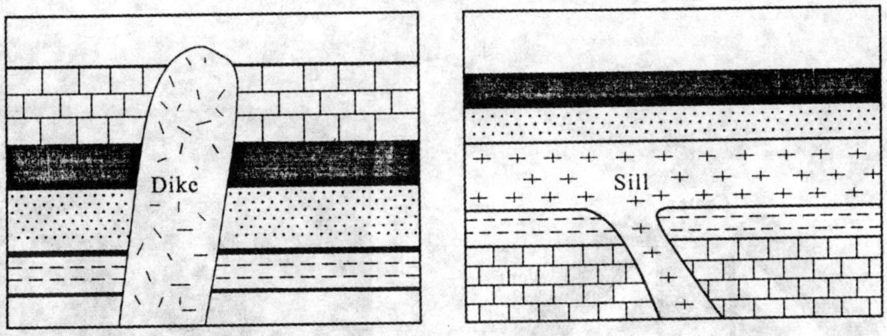

Figure 3.4 Dike and sill

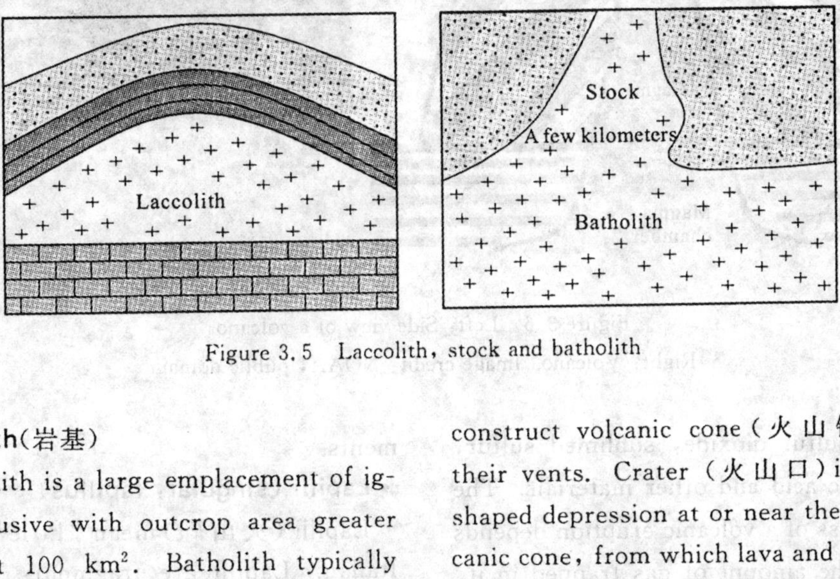

Figure 3.5 Laccolith, stock and batholith

- **Batholith(岩基)**

A batholith is a large emplacement of igneous intrusive with outcrop area greater than about 100 km^2. Batholith typically forms mountains when it is exposed on the surface.

3 Volcano and Volcanism

The most obvious effect of igneous activity is a volcano. A volcano is a geological landform where the solid and molten materials and hot gases erupt through the surface of the planet. It includes the surrounding cone of built-up material. Volcanic vent (火山通道)(Figure 3.6) is an opening on a volcano through which lava is released and solid and molten materials are ejected. With the material they emit, some volcanoes will construct volcanic cone(火山锥) around their vents. Crater(火山口)is a funnel-shaped depression at or near the top of volcanic cone, from which lava and other materials are ejected. Crater is connected to a magma chamber via vent.

Volcanism(火山作用)is the process of volcanic activity. It is part of the process of bringing material from the deep interior of a planet.

3.1 Volcanic Products

The materials discharged by a volcano include liquid lava, solid fragments of rock and gas.

When the magma reaches to Earth's surface, the gases dissolved in magma will be given off as a result of pressure decreasing. The gaseous released by volcano include super-heated water steam, carbon dioxide,

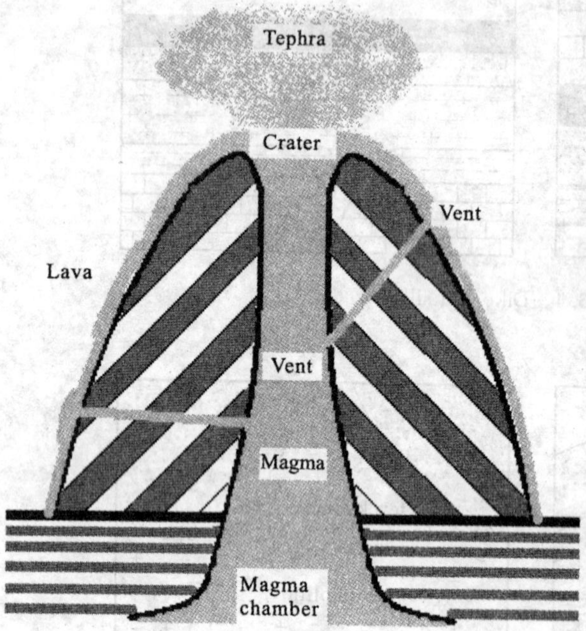

Figure 3.6 Left: Side view of a volcano
Right: Volcano. Image credit: NOAA, public domain

nitrogen, sulfur dioxide, sublimed sulfur, hydrochloric acid and other materials. The explosiveness of a volcanic eruption depends partly on the amount of gas trapped in it. Porous volcanic rock is the result of gas expanding as lava solidified.

Lava is the liquid ejected by a volcano.

Tephra (Figure 3.7) is the general term for airborne volcanic ejecta of any size. They consist of clots of lava and fragments of solid that are erupted from a crater or volcanic vent and deposited from the air. The tephra can be described by its size.

• **Ash** (火山灰)

Volcanic ash is the term for very fine fragments less than 2 mm in diameter that are ejected from a volcanic vent. Ash is generally dominated by broken glass shards, broken crystals and rock fragments.

• **Lapilli** (**singular: lapillus**)

Lapilli (火山砾) mean "little stones" in Italian. Lapilli are fragments between 2 and 64 mm in average diameter. They may be ejected in either a liquid or solid state. Lapilli can be spheroid, teardrop, dumbbell, or button shaped. Droplet-shaped fragments are ejected in a very fluid condition and solidified in the air.

• **Blocks** (火山块)

Blocks are ejected as solid fragments (>64 mm) with angular to sub-angular shapes. The angular shape of block is caused by the fracturing of solid material during the eruption. It is important to note that blocks are ejected during the eruption in a solid state.

• **Bombs** (火山弹)

Volcanic bombs are ejected as lava frag-

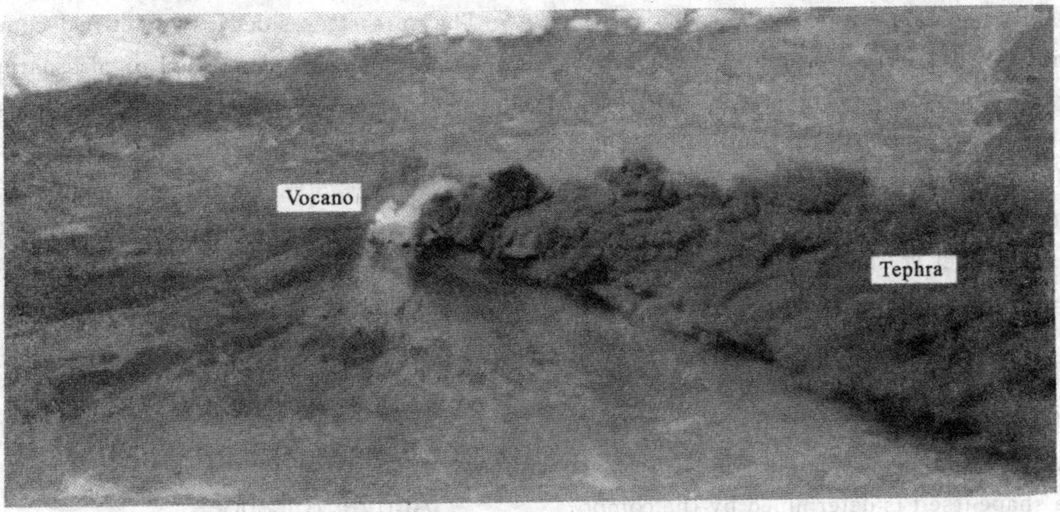

Figure 3.7 Eruption of Mount Etna, Sicily (satellite image)
Image credit: NASA-JSC-ES&IA, public domain

ments (>64 mm) in semi-solid or plastic condition. Bombs are thrown from volcano in a semi-molten condition and solidify during flight or shortly after they land, thus inheriting streamlined, aerodynamic shapes. The common shapes of bomb include spindle, ribbon, spheroidal and cow-dung bombs.

Fragments, such as volcanic ash, lapillus, block and bombs are also referred to collectively as pyroclastic materials or pyroclasts ("fire frag-ments", pyro, meaning fire, and klastos, meaning broken). In this sense, tephra is a generic term for any airborne unconsolidated pyroclastic materials.

The rocks related to the activity of volcano can be classified into two different groups. The first one is extrusive or volcanic (named after *Vulcan*, the Roman name for the god of fire) rock, which is produced due to the cooling of lava flowing on the surface. Another type is pyroclastic rock, which is formed by accumulation and cementation (胶结) of tephra. Tephra is something different from pyroclastic rock. If tepha conglomerate into solid rock when they land and cool, then they become a pyroclastic rock. If they do not then they are still called tephra.

There are three types of pyroclastic rocks.

- **Pyroclastic Agglomerate**(集块岩)

Agglomerate is a term for a chaotic assemblage of coarse angular or rounded material. Pyroclastic agglomerate is a pyroclastic deposit whose average pyroclast (火山碎屑) size exceeds 64 mm and contain at least 50% blocks and bombs held together by mineral cement or a fine matrix.

- **Pyroclastic Breccias**(火山角砾岩)

Pyroclastic breccias are pyroclastic rock whose average pyroclast sizes range from 2 to 64 mm.

- **Tuff**(凝灰岩)

Tuff is a rock made of ash-sized fragments (<2 mm).

3.2 Types of Volcanoes

The nature of volcanic eruptions is determined by the chemical composition, viscosity and gas content of magma. An active volcano(活火山) is that erupts either continually or periodically in the past 10 000 years. Dormant volcano (休眠火山) or sleeping volcano, a volcano which is presently inactive but which may erupt again in the future. Extinct volcanoes (死火山) are those that scientists consider unlikely to erupt again. Whether a volcano is truly extinct is often difficult to determine.

Volcanoes can classify by their erupted materials and shapes (Figure 3. 8). In fact, the shape itself is determined by the composition of the erupted materials.

- **Shield Volcano(盾状火山)**

They are large volcanic forms with broad summit areas and slight-sloping sides (shield shape). Their materials mainly consist of low viscosity lava. The basal diameters of this broad, low profile shield volcano vary from a few kilometers to over 100 kilometers. The volcanoes of the Hawaiian Islands and Iceland are typical of this type.

- **Cinder Cones(火山渣锥)**

It is the simplest type of volcano. A volcanic cone is a cone-shaped structure with steep-sided slopes of 30°～40°. It is composed of loosely materials resulted from eruptions that throw out mostly small pieces of rock that finally build up around the vent. Most volcanic cones have a bowl-shaped crater at the summit.

- **Composite Volcanoes(复合火山)**

They are steep-sided, symmetrical cones of large dimension composed of alternating layers of lava flow and ejected materials, which form the strata which give rise to the name. Slopes of a typical composite cone are steepest near the summit and decrease outward to the base. The summit area typically contains a surprisingly small summit crater. Classic example of composite volcano is Mt. Fuji (富士山) in Japan.

3. 3 Worldwide Distribution of Active Volcanoes

Today, there are many active volcanoes worldwide. Active volcanoes are not scattered over the Earth randomly; instead, most occur in belts (Figure 3. 9):

- **Circum-Pacific Volcanic Belt(环太平洋火山带)**

Circum-Pacific belt passes around the Pacific Ocean and includes New Zealand, New Guinea, Japan, the Aleutian Islands, Alaska, and the west coasts of North and South America. About 60 percent of all volcanic eruptions take place in this belt. The ring of volcanoes around the Pacific Ocean is called the"Pacific Rim of Fire".

- **Mediterranean-Indonesia belt (地中海-印度尼西亚带)**

This belt stretches from the Mediterranean, eastward Iranian, Himalayan, the is-

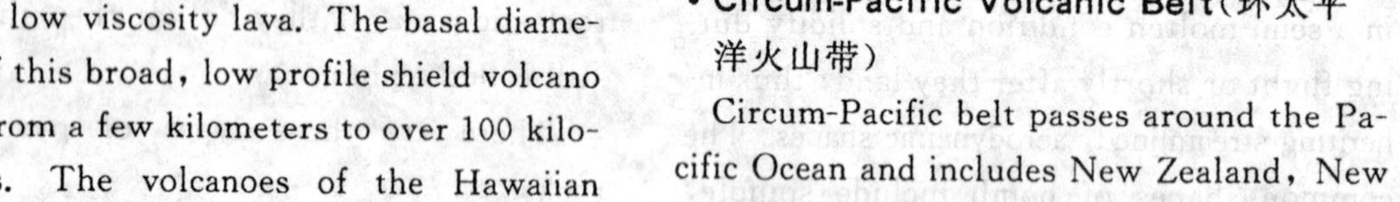

Figure 3. 8 The shapes of different volcanoes

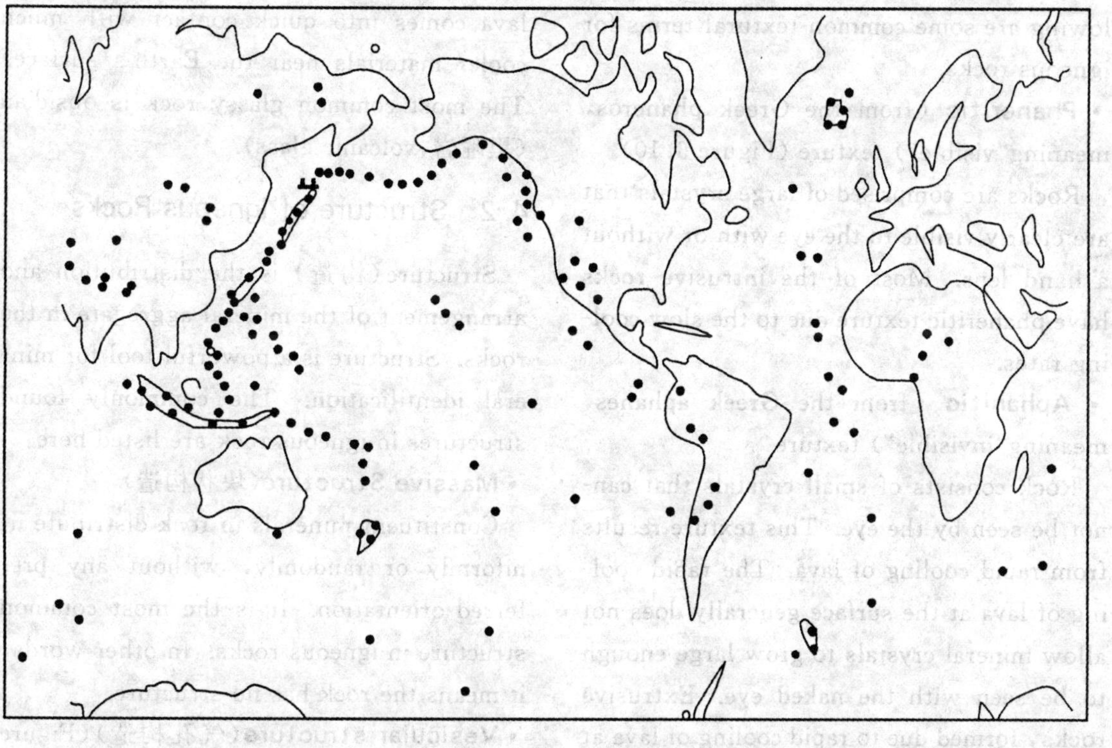

Figure 3.9 World wide distribution of volcanoes (the black dots stand for volcanoes)

land chains of Indonesia and New Guinea, and to the Solomon Islands. About 20% of all active volcanoes are in the Mediterranean-Indonesia belt.

• **Mid-ocean Ridge Belt**(大洋中脊带)

This belt includes the submarine mountains underlying the center of the Atlantic, Indian, and South Pacific Oceans.

Although most volcanoes occur in above well-defined three belts, but some isolated volcanoes also exist. Some of the isolated volcanoes like those in Hawaii or Yellowstone are referred to as hot spot (热点) volcanoes.

4 Igneous Rocks

4.1 Texture and Structure

When we discuss the texture (结构) of rock, we are describing and using terms which refer to the degree of crystallinity(结晶度), grain size and shape, and the geometric configuration between the grains.

The most important, but not only, factor that controls the grain size of a rock is its cooling history. When a large magma body remains underground, it is insulated from heat loss and cools and solidifies very slowly. That is why intrusive igneous rocks will have a texture consisting of coarse-grained crystals. In general, it will take thousands of years to millions of years for constituent minerals in rocks to grow large enough to be easily seen. Conversely, since lava cools quickly at the Earth's surface, volcanic rocks will have a texture consisting of fine-grained crystals or glass.

There are a large number of terms used to describe the texture of igneous rocks. Following are some common textural terms for igneous rocks.

• **Phaneritic** (from the Greek phaneros, meaning "visible") texture (Figure 3.10)

Rocks are comprised of large crystals that are clearly visible to the eye with or without a hand lens. Most of the intrusive rocks have phaneritic texture due to the slow cooling rates.

• **Aphanitic** (from the Greek aphanes, meaning "invisible") texture

Rock consists of small crystals that cannot be seen by the eye. This texture results from rapid cooling of lava. The rapid cooling of lava at the surface generally does not allow mineral crystals to grow large enough to be seen with the naked eye. Extrusive rocks, formed due to rapid cooling of lava at Earth's surface, are composed of tiny crystals (fine-grained) because crystallization was completed within a few seconds, hours, or perhaps days.

• **Glassy texture**

Glassy texture (玻璃质结构) is actually a non-crystalline (glassy) texture of igneous rock, in which no minerals are present. Glass results from cooling that is so fast that minerals do not have a chance to crystallize. This may happen when magma or lava comes into quick contact with much cooler materials near the Earth's surface. The most common glassy rock is obsidian (黑曜岩, volcanic glass).

4.2 Structure of Igneous Rocks

Structure (构造) is the distribution and arrangement of the mineral aggregate in the rocks. Structure is a powerful tool for mineral identification. The commonly found structures in igneous rock are listed here:

• **Massive Structure**(块状构造)

Constituent minerals in rock distribute uniformly or randomly, without any preferred orientation. It is the most common structure in igneous rocks. In other words, it means the rock has no structure.

• **Vesicular structure**(气孔构造)(Figure 3.11)

It describes rocks with numerous cavities. It is a typical feature of extrusive rock, such as basalt and andesite. Vesicular structure is caused by dissolved gases emerging as bubbles on eruption. When the rock cools down and the former gas filled cavities simply solidify. Highly vesicular volcanic rock is also known as pumice. Due to its air chambers, pumice (浮岩) has a very low density and can float on the water.

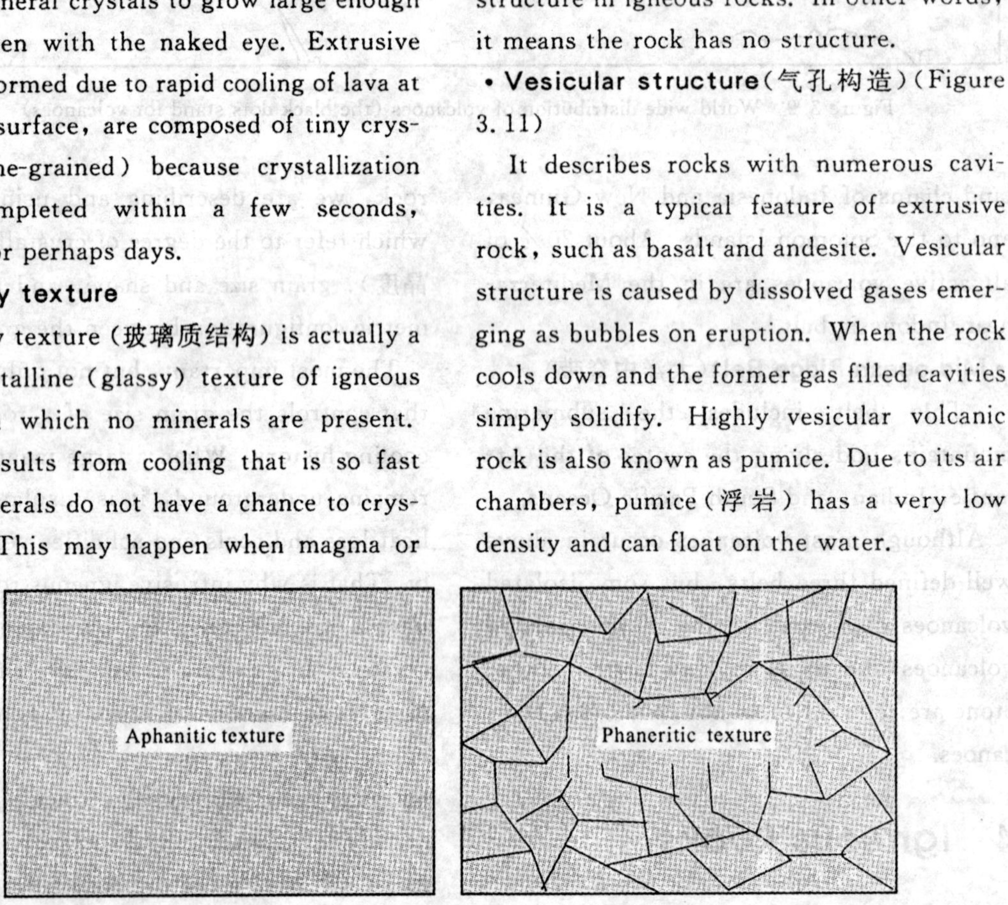

Figure 3.10 Phaneritic and aphanitic texture

Figure 3. 11 Vesicular structure

Image Credit: From Courtesy of the U. S. Geological Survey and the Mineral Information Institute. Public Domain

Figure 3. 12 Left: Columnar structure

Right: Pillow structure

- **Amygdaloidal Structure(杏仁构造)**

If the cavities in vesicular rock are filled with younger new minerals, it is called amygdaloidal structure. The filling is called an amygdule and is usually composed of quartz and calcite. Amygdule(杏仁) forms when fluids containing dissolved minerals flow through the rocks and deposit the minerals as solids in the vesicles(气孔).

- **Flow Structure(流动构造)**

A structure of igneous rocks, generally but not necessarily volcanic rocks, in which the streamlines or flow lines of the magma are revealed.

- **Pillow Structure(枕状构造)**

Pillow structure is characterized by discontinuous pillow-shaped masses ranging in size from a few centimeters to a meter or more in greatest dimension (commonly between 30 cm and 60 cm).

The pillow structure (Figure 3. 12) is a very common structure of basalts that have formed at the ocean floor. They are formed by repeated oozing and quenching of the hot lava underwater. First, when lava erupts underwater or flows into the sea, the cold water quenches the surface and the lava forms a distinctive pillow shape. So the grain sizes within a pillow tend to increase toward the interior and the pillows are rimmed by chilled glass selvedges. Next, the newly erupted basalt will breaks the old one and extrudes like toothpaste, forming another pillow.

The pillow structure can be used to indicate the"up direction"of layered basalt. Pillow tops are usually convex upward, whereas their undersurfaces may be flat, concave upward or project downward between the underlying pillows.

• **Columnar Structure(柱状构造)**

Columnar structure, or columnar jointing (柱状节理)(Figure 3. 13), consists of a 3-D network of fractures that occurs in cooling lava that forms parallel, prismatic columns. They are typically hexagonal in cross section, but 5-sided, 4-sided, and many other shapes also occur. Most columns are straight with parallel sides and diameters from a few centimeters to 3 meters.

The columnar structure is formed due to the shrinkage during cooling of erupted lava. When the lava is erupted onto the Earth's surface in flows, they usually shrink. The tensile stresses(张应力) produced by shrinkage will result in the formation of cracks in rocks. The cracks will grow at right angles to the surface and each of them is at 120° to the other two. Finally six-sided columns will be developed.

The cliffs along the coast region in National Geopark in Zhangzhou, Fujian (福建漳州国家地质公园) exhibits extremely well-developed vertical columns of basalt.

4.3 Classification of Igneous Rocks

Igneous rocks are classified by the composition and texture (Table 3. 1). Igneous rocks are first classified by their chemical compositions. And then they are subdivided by texture and occurrence.

Chemical composition of igneous rock is determined mainly by the source of the magma and any interactions between magma and the country rock. Rock compositions are usually expressed as combinations of the

Figure 3. 13 Basalt columns.

Courtesy of Yellowstone Digital Slide File. Public domain

Table 3.1 Classification of igneous rocks

Rock types			Ultrabasic	Basic	Intermediate	Acidic
Content of silica(%)			<45	45~53	53~66	>66
Content of quartz(%)			No	Few	<5	>20
Feldspar			No	Plagioclase	Plagioclase	Orthoclase
Dark-colored minerals(%)			Olivine, Pyroxene, >90	Pyroxene, amphibole, olivine, <90	Amphibole, pyroxene, biotite, olivine, 90	Biotite, amphibole 10~15
Occurrence		Texture	Rock name			
Intrusive	Plutonic	Phaneritic	Peridotite	Gabbro	Diorite	Granite
Intrusive	Hypabyssal	Porphyritic	…	Diabase	Diorite porphyry	Granite porphyry
Extrusive	Lava	Aphanitic, glassy	…	Basalt	Andesite	Rhyolite

[1] "Porphyry" follows any of the above names except for pyroclastic rocks whenever there are appreciable phenocrysts (斑晶). Such as granite porphyry(花岗斑岩), diorite porphyry (闪长玢岩) and rhyolite phorphyry(斑状流纹岩). Some authorities consider the expression "porphyritic rock"(斑岩) has a better usage than porphyry, since the term refers only to the texture of the rock — not its chemical, physical, or mineralogical composition or color. Porphyritic texture(斑状结构) is especially common in extrusions, e.g., in lava.

[2] The term "lava" is the most commonly applied to streams of liquid rock that flow from a crater or fissure. It also refers to cooled and solidified rock.

oxides rather than as separate elements. For example, the content of K-feldspar ($KAlSi_3O_8$) can be written as 1/2(K_2O + Al_2O_3 + $3SiO_2$). Since oxygen and silicon are by far the two most abundant elements in the crust, so the content of SiO_2 (Silica) can be used as the main basis of chemical classification of all the igneous rocks.

In term of ranges of silica percentages, four major categories can be defined:

• **Ultrabasic Rocks**(超基性岩)

Silica content is less than 45%. They are rich in magnesium and iron with less silica. They are the main component of mantle.

• **Basic Rocks**(基性岩)

Silica content is 45% to 53%. Ca-rich plagioclase is the dominant feldspar. These igneous rocks are mainly in oceanic crust.

• **Intermediate Rocks**(中性岩)

Silica content ranges from about 53% to 66%. Plagioclase feldspars make up more than two-thirds of total feldspars.

• **Acidic Rocks**(酸性岩)

Silica content is greater than 66%. Potassium feldspar makes up more than one-third of total feldspars. These igneous rocks are mainly in continental crust.

In terms of modes of occurrence and texture, igneous rocks can be classified into intrusive or extrusive. The intrusive rocks can be further subdivided into plutonic and hypabyssal rocks. Plutonic rocks (深成岩) are those formed at deeper within the Earth. They often occur as batholith and stock. They are featured by coarse-grained texture. Hypabyssal igneous (浅成岩) rock crystallizes at shallower depths. They often occur as dikes and sills and have a medium to fine grain size.

Major rock types of igneous rock are listed here:

• **Granite**(花岗岩)

Granite is a light-colored acidic intrusive igneous rock. The word granite comes from the Latin granum, a grain, in reference to the coarse-grained structure of such a crystalline rock. It is composed mostly of quartz and potassium feldspar(钾长石). The black sheetlike biotite mica is often a minor constituent of granite. Granite is the most abundant plutonic rock in the crust.

• **Diorite** (闪长岩)

Diorite is an intermediate plutonic rock. The predominant minerals of diorite include plagioclase (light-colored) and amphibole (dark-colored). Quartz is only a minor constituent in diorite. This leads to a characteristic salt-and-pepper appearance (describes hair that is a mixture of dark hairs and gray or white hairs).

• **Gabbro**(辉长岩)

Gabbro is a dark green to black basic intrusive igneous rock. It is composed mainly of plagioclase and pyroxene.

• **Rhyolite** (流纹岩)

Rhyolite is a light-colored volcanic rock. It is generally fine-grained. Rhyolite is chemically the equivalent of granite and thus composed primarily of quartz and K-feldspar with minor amounts of biotite. Rhyolite frequently exhibits a typical flow structure of altering layers that differ slightly in grain size or color. Rhyolites were formed in profusion in the Yandang(雁荡山) area and throughout the southeastern portion of the China.

• **Andesite**(安山岩)

The name andesite is derived from the Andes mountain range. Andesite is a gray fined-grained volcanic rock. The major constituents in andesite include hornblende and plagioclase. Andesite is chemically the extrusive equivalent of diorite.

• **Basalt**(玄武岩)

Basalt is one of the most common rock types on Earth. Some basalts have vesicles (gas bubbles), some don't. It is a black fine-grained volcanic rock. They generally contain the minerals feldspar, pyroxene, and olivine. Basalts in the field usually show columnar, vesicular and pillow structures. Weathered basalt is often red to yellowish or reddish-brown. Basalt erupts on the ocean floor at ocean ridges. This rock type makes up most of the ocean floor.

Basalt can be considered as the extrusive equivalent to plutonic gabbros.

1. Answer the Following Questions

(1) Describe the major features of magma.

(2) What is dike? What is sill?

(3) Describe in detail about the different types of tephra.

(4) How to classify the volcanoes?

(5) Describe the distribution of active volcanoes.

(6) Explain the formation of pillow and columnar structures.

(7) How to distinguish the phaneritic and aphanitic textures? Why some rocks exhibit glassy texture?

(8) How to classify igneous rock by chemical composition.

(9) How to distinguish the granite and basalt?

(10) Describe the classifications of pyroclastic rocks.

2. Multiple Choice

(1) The smallest pyroclastic material (volcanic ejecta) is called ______.

A: Ash　　B: Lapilli

C: Bombs　　D: Blocks

(2) Which of the following types of volcanoes are the largest on Earth?

A: Shield volcanoes

B: Composite volcanoes

C: Volcanic cone

D: None of above

(3) Which of the following is an example of an extrusive, basic igneous rock?

A: Andesite　　B: Gabbro

C: Granite　　D: Basalt

(4) The area of igneous activity commonly called the Ring of Fire surrounds the ______.

A: Indian Ocean

B: Pacific Ocean

C: Atlantic Ocean

D: South China sea

(5) An igneous rock which is commonly black, has a glassy texture and has a conchoidal fracture is which of the following?

A: Andesite　　B: Gabbro

C: Obsidian　　D: Basalt

(6) The word"volcano"comes from Vulcan, the Roman god of what?

A: Fire　　B: Love

C: War　　D: Lightning

(7) This term describes solid or molten rock fragments of any size that are ejected from a volcano.

A: Lapilli　　B: Ash

C: Bombs　　D: Tephra

(8) Gases under pressure in a magma make it ______.

A: Explosive

B: Low viscosity

C: Noncrystalline

D: All of above

(9) The major difference between the extrusive and intrusive rocks is ______.

A: Structure

B: Mineral composition

C: Chemical composition

D: Where they crystallize

(10) A rock with mineral crystals too small to be seen is ______.

A: Granite　　B: Basalt

C: Gabbro　　D: None of above

(11) "Basalt can be considered as the extrusive equivalent to plutonic gabbro" means they have the same ______.

A: color　　B: composition

C: texture　　D: structure

(12) By definition, stocks differ from batholiths in ______.

A: shape　　B: size

C: chemical composition　　D: None of above

(13) Which of these rock textures was created by a period of slow cooling?

A: Aphanitic　　B: Phaneritic

C: Glassy　　D: Massive

(14) Which rock has a phaneritic texture?

A: Basalt　　B: Obsidian

C: Granite　　D: Rhyolite

(15) The grain size of igneous rock is controlled by ______.

A: Temperature

B: Pressure

C: Cooling rate

D: All of above

(16) Granite can be considered as the intrusive equivalent to ______.

A: Rhyolite B: Basalt

C: Gabbro D: Andesite

(17) Which rock is more likely to have vesicular structure?

A: Basalt B: Obsidian

C: Granite D: Tuff

(18) Which rock is extrusive rock?

A: Basalt B: Tephra

C: Granite D: Tuff

(19) Pyroclastic rock is composed of tephra.

A: True B: False

(20) It is impossible for a dike to be made up of the rock ______.

A: Basalt B: Obsidian

C: Tuff D: All of above

(21) Volcanic bombs are ejected in ______ state.

A: Solid B: Plastic

(22) The roof of the laccolith is ______.

A: Flat

B: Convex upward

C: Convex downward

D: None of above

(23) The materials discharged by a volcano include ______.

A: Early solidified lava

B: fragments of country rock

C: Gas

D: All of above

(24) Which mineral has the lowest melting point?

A: Olivine B: Quartz

C: Plagioclase D: Hornblende

(25) Tephra includes the ______.

A: Liquid lava B: Solid rock

C: Ash D: All of above

Chapter 4 External Geological Process and Sedimentary Rocks

Key words

Atmosphere	Troposphere	Hydrologic cycle	Sorting	Clast	Matrix
大气圈	对流层	水文循环	分选	碎屑	基质
Sediment	Clastic texture	Cross bedding	Cement	Bedding	Bed
沉积物	碎屑结构	交错层理	胶结物	层理	层、岩层
Shale	Conglomerates	Mud crack	Sandstone	Limestone	Roundness
页岩	砾岩	泥裂	砂岩	石灰岩	磨圆度
Weathering	Erosion	Diagenesis	Stratosphere	Clastic sedimentary rocks	
风化	侵蚀	成岩作用	平流层	(陆源)碎屑(沉积)岩	

Sedimentary (from Latin, sedimentum, which means a settling) rocks are those rocks which form at or near the Earth's surface at relatively low temperatures and pressures primarily by either deposition by water, wind or ice, or precipitation from solution. Only 5 percent of the Earth's crust consists of sedimentary rocks, yet 75 percent of the rocks exposed at the Earth's surface is sedimentary rock.

Sedimentary rocks are of great economic importance as sources of many important metal and nonmetal ores, such as rock salt, cement and iron ores. In addition, petroleum and natural gas can be found in the porous and permeable reservoirs formed by sedimentary rocks in sedimentary basin.

Sedimentary rocks are formed at or near the surface of the Earth by the deposition and accumulation of particles of rocks, minerals, and fossils or of minerals precipitated from seawater or other aqueous solutions. The major factors or agents responsible of external geological process include wind (atmosphere), water and organism. So, the sedimentary rock is a direct result of interaction between atmosphere, hydrosphere and biosphere. The formation of sedimentary rock is closely related to external geological process. In order to understand the formation of sedimentary rocks, it is necessary to understand the movement of atmosphere, hydrosphere and biosphere.

1 Atmosphere

The gaseous layer surrounding the Earth is called Earth's atmosphere. This blanket of gas starts at ground level and stretches 965.61km into the sky. Atmosphere is composed of nitrogen (78%), oxygen (21%), argon (0.9%), carbon dioxide (0.03%) and other trace gases. The atmosphere is of vital importance for life on Earth. Nitrogen can be used by living things to make proteins. Oxygen is essential for respiration by all living things. Carbon dioxide traps heat being radiated from the Earth. So, without this atmosphere life on Earth isn't possible.

The atmosphere of the Earth may be divided into several distinct layers according to major changes in temperature (Figure 4.1).

• **The Troposphere**

Troposphere is the lowest layer of the Earth's atmosphere. The average thickness of troposphere is about 11 km. The height of the tropopause (top of the troposphere) varies from the equator to the poles. At the equator it is around 18～20 km high, and at the poles just under 6 km high.

The troposphere is the layer where most of the world's weather occurs. Weather is the state of the atmosphere at a specific time or place. Weather (天气) states include rain, snow, sunshine, wind, temperature and thunderstorms. Climate(气候)describes surface and atmospheric conditions over a longer time period or over a larger geographical area. In other words, the term climate refers to long-term weather statistics (mean values, variances, probabilities of extreme values, etc.). Weather conditions change suddenly and climate changes slowly.

The temperature of the troposphere varies vertically. It is normally warmest at ground level and cools higher up where it reaches its top. The average temperature is 15℃ near the surface and －57℃ at the tropopause. The temperature of the troposphere also varies laterally. The tropics receive more heat from the Sun than the poles. Therefore, the temperature is greater at the equator than at the poles. Warm, moist air rising at Equator and cold and dry air sinking at poles lead to the movement of

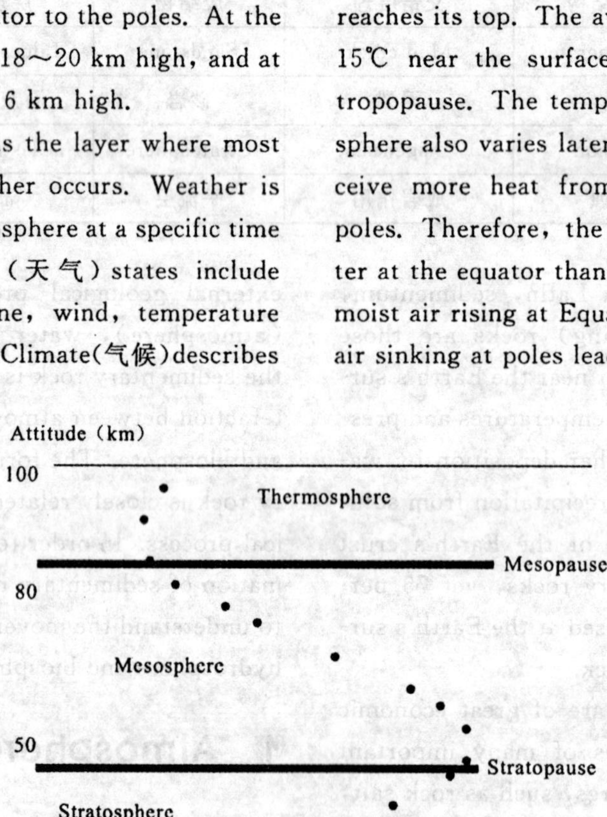

Figure 4.1 Structure of the Atmosphere

atmosphere on global scale. This is atmosphere circulation(大气环流). This convection of air generates clouds and ultimately rain giving rise to the change in weather.

The Figure 4.2 is just an idealized model. The actual atmosphere circulation is far more complex. The rotation of the Earth will cause the air to be deflected toward the right in the northern hemisphere and toward the left in the southern hemisphere.

- **The Stratosphere** (平流层)

The stratosphere extends from the tropopause up to the stratopause(平流层顶部), 50 km above the Earth's surface. The stratosphere is characterized by a slight temperature increase with altitude and by the absence of clouds. Many jet aircrafts fly in the stratosphere because it is very stable (Figure 4.3).

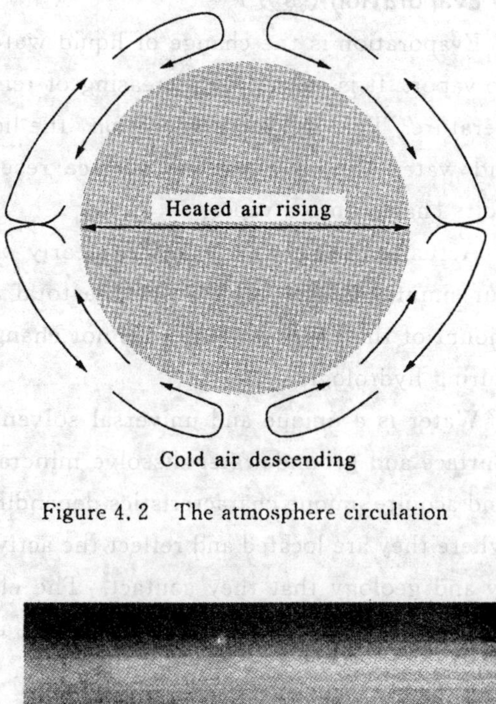

Figure 4.2 The atmosphere circulation

Ozone (O_3) is a relatively unstable molecule made up of three atoms of oxygen (O). It is formed from O_2 by action of high-energy ultraviolet radiation. Most ozone resides in the stratosphere, where it acts as a shield to protect Earth's surface from the sun's harmful ultraviolet radiation.

- **The Mesosphere**(中层)

The mesosphere extends from stratopause up to 85 km. Within this layer, temperature decreases with increasing altitude. This is where we see"falling stars",meteors that fall to the Earth and burn up in the atmosphere.

- **The Thermosphere**(热层)

The thermosphere extends from the top of mesosphere up to 300 km above the Earth's surface. The temperature within thermosphere increases with altitude.

Figure 4.3 Jet flying in the stratosphere, Photo by Chen Fuxing(谌福兴)

2 Hydrologic Sphere

Water is of extremely importance to the external geological process.

The Earth is surrounded by a blanket of water. This is hydrosphere. It (Greek hydro- means"water") embraces all the water on the Earth, including the surface water (stored in the oceans, seas, lakes, ponds, rivers and streams), the water stored in the atmosphere as well as groundwater (Table 4.1). Water will be stored in solid state in the deep glaciers, liquid state in the surface water reservoir, and vapor state in clouds. The hydrosphere covers about 70% of the surface of the Earth.

The water on Earth is in constant movement. It is continuously moving from one reservoir to another. This recycling movement of water among atmosphere, lithosphere and biosphere is called hydrologic cycle. The hydrologic cycle is perhaps the most important natural phenomenon on Earth; it is the driving force behind external geological processes.

Water cycles among the hydrosphere, atmosphere and biosphere through the processes of evaporation, condensation, precipitation, runoff and infiltration (Figure 4.4).

• **Condensation(冷凝)**

Condensation is the process through which the water vapor in the air condenses into clouds. Condensation occurs due to the temperature decreasing.

• **Precipitation(降水)**

Precipitation is the releasing of water held in clouds as moisture. The forms of precipitation can be snow, rain and hail.

• **Infiltration(渗透)**

Infiltration is the process through which the precipitation seeps into the ground.

• **Runoff (地表径流)**

Runoff is the water running on the Earth's surface.

• **Evaporation (蒸发)**

Evaporation is the change of liquid water to vapor. It is caused by increasing of temperature. Through the evaporation, the liquid water stored in various surface reservoirs change into a vapor.

Actually, the above processes nearly occur simultaneously. However, the total amount of the Earth's water does not change during hydrologic cycle.

Water is a unique and universal solvent. Surface and ground water dissolve minerals and acquire unique characteristics depending where they are located and reflect the activity and geology that they contact. The elements dissolved in the lakes and oceans are the important sources for the formation of many minerals and rocks. Movement water can break up rocks and transport the rock

Table 4.1 Inventory of water at the Earth's surface

Reservoir	Percent of Total	Reservoir	Percent of Total
Oceans	97.25	Ice Caps and Glaciers	2.15
Groundwater	0.68	River + Lakes	0.000 1±0.017
Others: Atmosphere, soil moisture and biosphere			

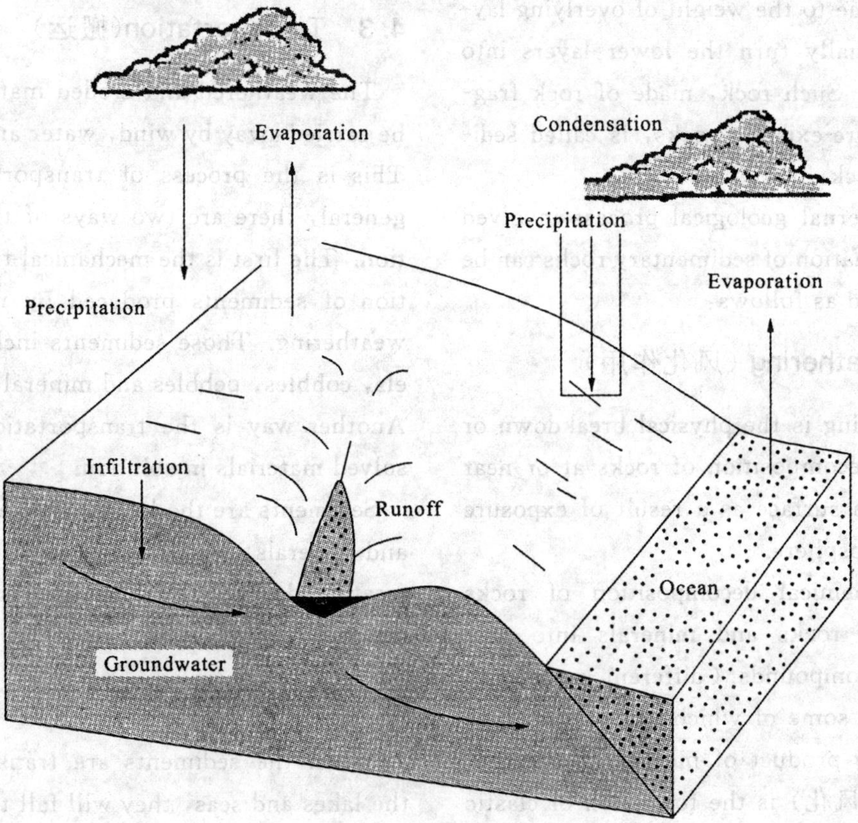

Figure 4. 4 Hydrologic cycle

fragments to place where they deposit.

3 Biosphere

From the broadest point of view, biosphere can be defined as an outer layer of Earth in which life exists. The thickness of biosphere is about 20 km.

The Earth is one system that involves interactions between the different "spheres" of the Earth: biosphere, lithosphere, atmosphere and hydrosphere. The lithosphere is the thin crust between the mantle and the atmosphere. Although the lithosphere is around 100 km thick, only 1 km of it can be considered in interaction with the biosphere. Some important rocks are formed by organism. For example, chert (燧石) is composed of radiolarian(放射虫) skeletons. Organism activity can break up the rocks. For example, plant roots can split rocks. Cracks in rocks can promote plant growth.

4 Formation of Sedimentary Rocks

Due to the interaction of biosphere, lithosphere and atmosphere, the rocks exposed on the Earth's surface will be broken down into small pieces. And the water and wind will carry the rock fragments away and eventually deposit them in some places, such as lakes and seas. As time goes by, perhaps over millions of years, layers of different rock bits settle at the bottom of lakes and oceans and eventually pile up. The huge pressure due to the weight of overlying layers will finally turn the lower layers into solid rock. Such rock, made of rock fragments of pre-existing rocks, is called sedimentary rock.

The external geological process involved in the formation of sedimentary rocks can be summarized as follows:

4.1 Weathering (风化作用)

Weathering is the physical breakdown or chemical decomposition of rocks at or near the Earth's surface as a result of exposure to the atmosphere.

The chemical decomposition of rocks transforms rocks and minerals into new chemical compounds (different rocks and minerals), some of which can be dissolved away. The product of mechanical weathering (物理风化) is the formation of clastic materials (碎屑物质).

4.2 Erosion(侵蚀)

Erosion is the process through which the rocks and minerals are detached from their original positions. Weathering and erosion are different geological processes. Both of them occur in place, but no movement is involved in weathering process. As soon as a rock particle (loosened by weathering processes) moves, we call it erosion.

Wind and water are two major agents responsible for erosion. Gravity is another key factor of erosion. The rock bodies will move downward due to gravity. When it rains, the rocks and soils are washed down a mountain or down a stream. Soils are washed away. The stream strikes against its bank and breaks it apart. They are all examples of erosion.

4.3 Transportation(搬运)

The weathered and eroded materials will be carried away by wind, water and glacier. This is the process of transportation. In general, there are two ways of transportation. The first is the mechanical transportation of sediments produced by mechanical weathering. Those sediments include gravels, cobbles, pebbles and mineral particles. Another way is the transportation of dissolved materials in solution.

Sediments are the loose particles of rocks and minerals, which have been, or is being weathered, eroded, transported and finally deposited.

4.4 Deposition(沉积)

When the sediments are transported to the lakes and seas, they will fall to the bottoms of lakes and seas. This laying down of sediments by wind, water or glacier is called deposition. At this stage, the settled materials are still called sediments. They need to be changed into sedimentary rocks by a series of processes collectively called diagenesis.

4.5 Diagenesis(成岩作用)

Diagenesis is all of the chemical and physical changes, which turn the sediments into sedimentary rocks after deposition. Diagenesis occurs at relatively low temperatures and pressures.

• **Compaction (压实)**

Compaction is the packing together of sediment grains due to the great weight of overlying layers of rock. Compaction can lead to the reduction of volume and thickness of original sediment layers. During compaction, the pore spaces around the sediment grains are also reduced and water filling the free spaces will be squeezed out of pores.

- **Cementation(胶结作用)**

During the compaction, the ions dissolved in the solution will precipitate. The precipitated minerals will fill the free spaces around the sediment grains, resulting in binding together the sediments. The common cementing minerals include calcite, silica and iron oxide.

- **Recrystallization(重结晶作用)**

With the increasing of temperature and pressure, new mineral grain will grow around the old one. This is the recrystallization. The newly formed interlocking crystals will connect the other sediment grains together into a solid rock.

5 Minerals of Sedimentary Rock

Minerals in sedimentary rock include quartz, feldspar, muscovite, clay minerals, calcite, gypsum and so on. Among which, quartz, feldspar and muscovite are also the common minerals in igneous rocks. But some common minerals in igneous rock, such as olivine, pyroxene, hornblende, biotite and basic plagioclase (基性斜长石) are rarely found in sedimentary rocks. That is because those dark-colored minerals and plagioclase are formed at depth and unstable on the Earth's surface. When they are exposed to Earth's surface, the chemical weathering will decompose them to new stable minerals such as clay minerals(黏土矿物).

Calcite, gypsum and clay mineral are the typical sedimentary minerals formed under the surface condition. They can precipitate directly from the solution.

Besides the mineral particles, there are lots of fragments of pre-existing rocks. Those rock fragments are the result of incomplete weathering.

6 Texture of Sedimentary Rock

The term "texture" used in sedimentary rock has a different sense than for igneous rocks. Texture in sedimentary rocks refers to size, shape and arrangement of particles. Texture provides a clue to the history of the sediment and its lithification(石化,成岩作用) into rock. In general, sedimentary rocks exhibit either a clastic or non-clastic texture (chemical, and biologic)(非碎屑结构).

Clastic texture (陆源碎屑结构) means that the sedimentary rocks are composed of grains compacted together with spaces (pores or voids) between the grains. Those rocks with clastic texture are called clastic sedimentary rocks. In other words, clastic sedimentary rocks are made up of pieces (clasts) of pre-existing rocks.

Clastic texture is composed of three diffe-

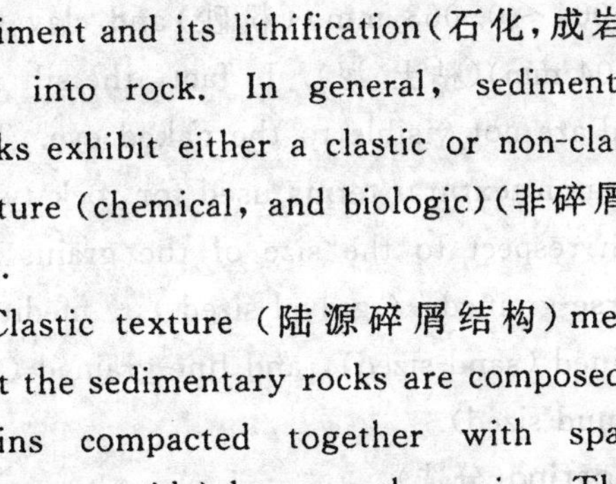

Figure 4.5 The clastic texture

rent components: clast, matrix and cement (Figure 4.5).

6.1 Clast（碎屑、陆源碎屑）

Clasts are larger particles in rocks. There are the products of mechanical weathering and erosion of pre-existing igneous, metamorphic, or sedimentary rocks. All of the clasts are derived from land.

Texture of clasts can be described in terms of grain size, roundness and sorting.

- **Grain size**(粒级)

Particle size is a most important feature of clastic rocks. Usually we classify the clasts into four different groups: gravel (> 2 mm)(砾), sand (2～0.063 mm)(砂), silt (0.004～0.063 mm)(粉砂) and clay (< 0.004 mm)(粘土、泥). In fact, the silt and mud are not visible to the naked eye. The common textural terms used for rock types with respect to the size of the grains are coarse-grained (gravel-sized), medium-grained (sand-sized), and fine-grained (silt or mud-sized).

- **Sorting**(分选)

Sorting is the degree to which clastic particle sizes are similar. It refers to the distribution of grain sizes in a rock. The degree of grain size similarity can be described as poorly-sorted(分选差) and well-sorted(分选好)(Figure 4.6).

Well-sorted sediments mean that all grains in rock are approximately the same size. The poorly sorted sediment or rocks are a mixture of various particle sizes. The sediments and particles can be sorted by grain size and density during transportation. The larger and heavier particles will settle down first and smaller and lighter sediments will be transported far away. Therefore, the sorting can be used to indicate the energy of the transporting medium. The larger the size of grains or the heavier the particle, the more energy it takes to move that particle to the place of deposition.

- **Roundness**(圆度、磨圆度)

The shape of clastic sediments is described by the term "roundness". Roundness indicates the extent to which the corners of individual grains have been rounded off. Rounding can be described by relative terms, such as angular(棱角状), subangular(次棱角状) to subrounded (次圆状) and rounded(圆状)(Figure 4.7).

During the transportation process, rand-

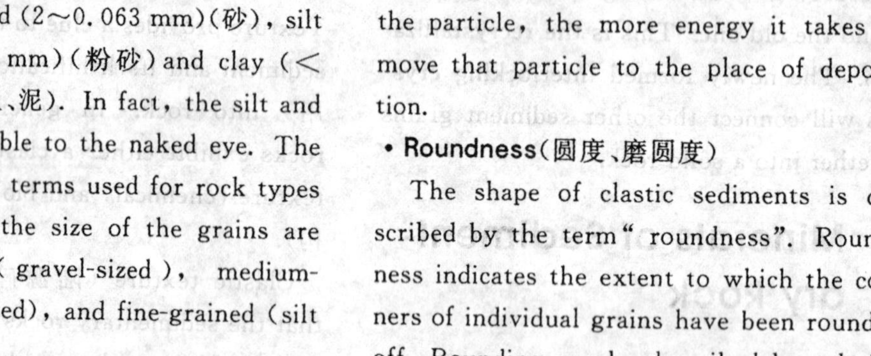

Figure 4.6 Sorting. Left: well sorted
Right: poorly sorted

Figure 4.7 The roundness of clasts

om abrasion between different particles results in the eventual rounding off of the sharp corners and edges of grains. Thus, the roundness is an indicator of transporting distance. The rounder a grain is, the greater distance it has been transported.

6.2 Matrix(基质)

Matrix is the finer material in which larger particles are embedded. Like clasts, they are the products of mechanical weathering (机械风化) of pre-existing rocks and transported to depositional basin along with clasts. The only difference between clast and matrix is the grain size. Matrix is commonly composed of clay with other silt-sized particles, such as quartz and feldspar.

6.3 Cement(胶结物)

Cement is a chemically precipitated material that glue the clasts all together. During the chemical breakdown of rocks, some elements will dissolve in the water. Under suitable conditions, the dissolved materials will precipitate from the water and fill the spaces between clasts. Thus they will hold or"cement"the clasts together. Many common minerals, such as calcite, iron oxide and silica, can precipitate directly from solution. They can serve as"cementing minerals"(胶结矿物) in clastic rocks. Clastic rock cemented by iron oxide (铁质胶结)often takes a red color. And silica-cemented (硅质胶结) rock is very hard. Rocks cemented by calcite (钙质胶结) can effervesce weakly with dilute hydrochloric acid.

Clast, matrix and cement are the essential components of clastic texture. But there are other sedimentary rocks which show no clastic texture. Those rocks formed by chemical and biological ways often exhibit totally different textures. Those textures are collectively termed as non-clastic textures. One of the important non-clastic textures is crystalline texture, in which the grains are interlocked and fit together like the pieces of a jigsaw puzzle. Crystalline texture (结晶结构) is very similar to the crystalline texture of igneous rocks. It is formed by combination effects of chemical precipitation (化学沉淀) and recrystallization (重结晶).

7 Structure of Sedimentary Rock

The structure in sedimentary rocks refers to the general distribution and arrangement of rock components. Like textures, the structures found in sedimentary rocks not only can be used to distinguish different rock types, but also to deduce the environment of deposition.

7.1 Bedding(层理)

Bedding (Figure 4.8) is the most obvious and important feature of sedimentary rock. It is the layering indicated by the variations of composition, grain size, roundness and color. The original bedding is usually horizontal or sub-horizontal. The individual layer is called bed(层). The original beds are usually parallel to each other. The texture (grain size, sorting), composition and color with a bed are relatively uniform. The upp-

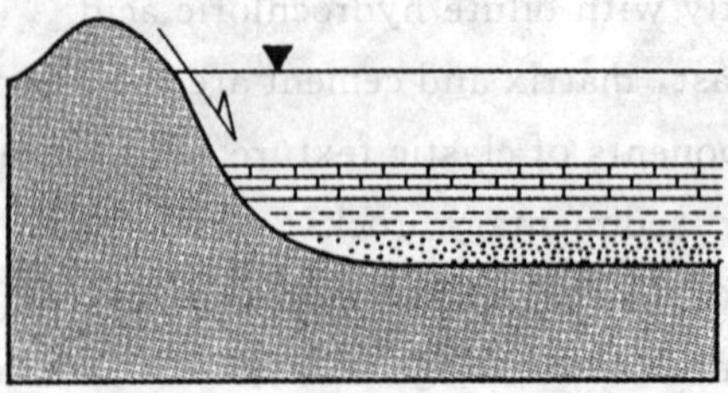

Figure 4.8 The formation of bedding

er and lower surfaces of these beds are called bedding planes(层面). The bedding plane stands for a cessation of deposition(沉积间断)at that location for a short period of time. These surfaces are probably the easiest physical features of sedimentary rocks to identify in outcrop (Figure 4.9).

Based on thickness, beds can be described as massively-bedded (>0.5 m)(巨厚层), thickly-bedded (0.1~0.5 m)(厚层), thinly-bedded (0.01~0.1 m)(薄层) and laminated (<0.01 m). Bed thinner than 1 cm are called lamination(纹层).

7.2 Cross Bedding(交错层理)

Cross bedding (Figure 4.10) is a term for special bedding structure in which the layering with a bed is inclined to the main bedding plane. Cross-beds usually become tangent to the lower bed surface and angular to the overlying bedding plane. Hence, cross beds can indicate the"up"direction in tilted rocks (Figure 4.11). Cross bedding mostly forms in response to the migration of sand dunes, sand-waves and ripples in the direction of current flow, down-stream or down-wind under conditions of erosion and deposition.

7.3 Mud Cracks(泥裂)

Mud crack is a polygonal pattern of cracks produced by the desiccation (干裂) of mud surface. The V-shaped cracks can be used to indicate the top and bottom of a bed (Figure 4.12). The concave upward direction of crack is the up direction.

7.4 Ripples(波痕)

Ripples are undulations of the sediment surface produced by water or wind movement. Ripples can be either asymmetrical or symmetrical. Asymmetrical ripples can be

Figure 4.9 Horizontal beds are easily seen in the cliffs of Grand Canyon, Arizona. Photo by Dr. Yu Shan(俞姗)

Figure 4.10 Cross bedding in Walnut Canyon, United State, Photo by Dr. Yu Shan(俞姗)

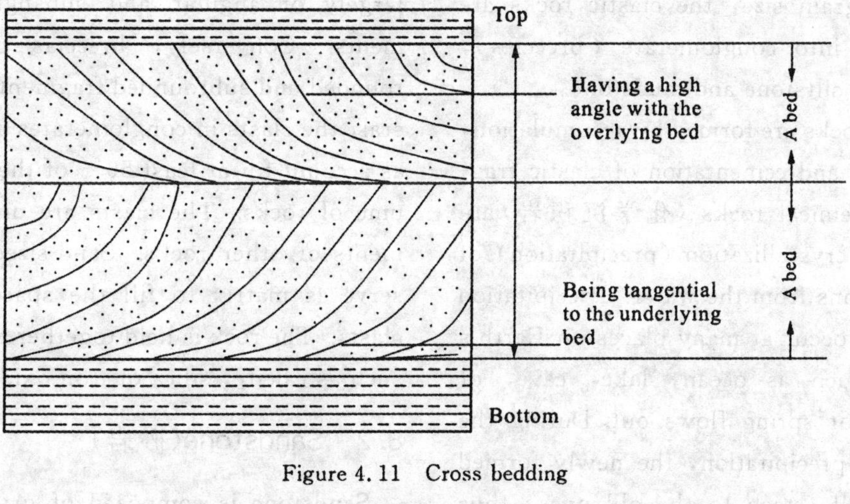

Figure 4.11 Cross bedding

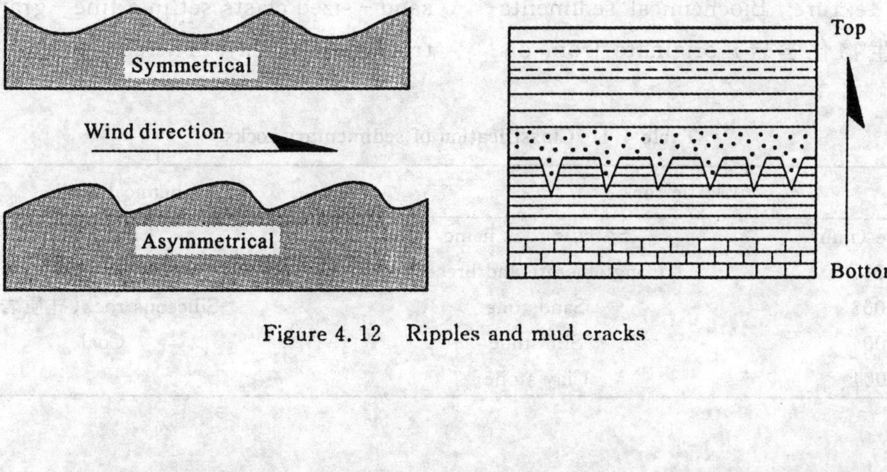

Figure 4.12 Ripples and mud cracks

used to indicate the direction of wind or current. Steep side of an asymmetrical ripple points in the direction of current flow or wind movement. Symmetrical ripples form by back and forth movement of wave in shallow water.

8 Classification of Sedimentary Rocks

Sedimentary rocks are classified by origin into two major types, clastic and chemical. Further subdivision of these two major sedimentary rock types is made based on their composition and texture. For example, based on grain size, the clastic rocks are subdivided into conglomerate (breccias), sandstone, siltstone and mudstone.

Clastic rocks are formed by accumulation, compaction and cementation of clastic fragments. Chemical rocks (化学沉积岩) are formed by crystallization (precipitation) of dissolved ions from the water. Precipitation (沉淀) can occur at many places on Earth's surface, such as ocean, lake, cave, or where a hot spring flows out. During the process of precipitation, the newly formed crystals will attach to the old ones, thus forming a typical interlocking (镶嵌状) crystalline texture. Biochemical sedimentary rocks (生物化学岩) originate from dissolved material that is precipitated through the activities of organisms. But it is really difficult to distinguish chemical and biochemical rocks in the field. So, in this textbook, both the traditional chemical rocks and biochemical rocks are grouped together as chemical rocks in a broad sense.

Here are the descriptions of some common sedimentary rocks (Figure 4. 13).

8. 1 Conglomerate and Breccia (砾岩和角砾岩)

Conglomerates (砾岩) and breccias (角砾岩) are the coarsest clastic rock with grains 2 to 256 mm. Conglomerate are composed largely of angular and subangular fragments. Conversely, breccias consist of rounded and subrounded fragments. In general, the clasts in conglomerates and breccias account for at least 50% of the total volume of rocks. The clasts are mainly fragments of other rocks. Sand-sized particles serve as matrix to fill the space between clasts. The rock is held together by cements such as calcite, silica or iron oxide.

8. 2 Sandstone (砂岩)

Sandstone is composed of grains in the sand size range (1/16～2 mm size). The sand - sized clasts set in a fine - grained matrix of silt or clay. The cement may be calc-

Table 4.1 Classification of sedimentary rocks

Clastic rock		Chemical rock
Grain size (mm)	Specific rock name	Limestone
>2	Conglomerate and breccia	Evaporite(蒸发岩)
>0. 063	Sandstone	Siliceous rock(硅质岩)
>0. 004	Silt stone	Coal
<0. 004	Clay stone	

Figure 4.13 Some sedimentary rocks

Image credit: Courtesy of the U.S. Geological Survey and the Mineral Information Institute. Public Domain

ite, silica or iron oxide. Sandstone comes in many colors, including gray, red or tan.

The clasts in sandstone are dominated by quartz. Other components include feldspar, muscovite and fragments of other rocks. Quartz sandstone (石英砂岩) contains 90% or more of quartz. Pure quartz sandstones have a white or gray color. A feldspar-rich sandstone is called an arkose(长石砂岩). Arkose is usually red or brown in color because of large amount of feldspar and iron oxide cementing. Sandstone forms at desert deposits and shallow seas. Most of the world's oil and natural gas is found in sandstones due to their porous texture.

8.3 Siltstone(粉砂岩)

Siltstone is composed of particles between 0.004 mm and 0.063 mm. The major mineral component in siltstone is quartz.

8.4 Mudstone(泥岩)

Mudstone is the finest clastic stone, consisting mostly of clay minerals. The individual grains are too small to see with the naked eye or a hand lens, and the rock feels smooth to the touch. Usually the mudstone with well-developed laminations (纹理、页理) is called shale(页岩). Shale can be easily broken along parallel surfaces due to the distribution of the flat clay minerals on the surface of lamination.

8.5 Limestone (灰岩)

The dominant mineral in limestone is calcite. Limestones are generally gray in color.

Limestones are formed through both chemical and biochemical processes. But the exact mechanism for the formation of limestone is very complex. They can be formed by the accumulation of organism remains, by direct chemical precipitation from the water, or by compaction and cementation of eroded pre-existing limestones. Thus the texture and structure in limestone are described with many terms.

8.6 Evaporites (蒸发岩)

Evaporites are formed by the evaporation of water. Common evaporite rocks include rock salt, gypsum and soda. Evaporites generally have a crystalline texture because they crystallize directly from solution.

1. Answer the Following Questions

(1) What is the clastic texture? How to distinguish clast, matrix and cement?

(2) Describing the main features of diagenesis.

(3) How to classify sedimentary rocks? How to classify clasitic rocks?

(4) Describing the major characteristics of the troposphere.

(5) Differentiate between weathering and erosion.

(6) What is the difference between climate and weather?

(7) What is cross bedding? How to use cross bedding to determine the up direction of beds?

(8) Are there any differences of mineral composition between igneous and sedimentary rocks? Why?

(9) What is sorting? How are the fragments of rocks and minerals sorted in the nature?

(10) List and briefly describe the various processes involved in the diagenesis.

2. Multiple Choices

(1) What rock is composed of angular-shaped clasts?

A: Sandstone B: Limestone
C: Breccia D: Conglomerate

(2) Which sedimentary rock is made of clay minerals?

A: Sandstone B: Limestone
C: Breccia D: Conglomerate

(3) What is a sandstone with a high feldspar content?

A: Arkose B: Limestone
C: Breccia D: Evaporite

(4) What ripple can be used to indicate the direction of wind or current.

A: Asymmetrical B: Symemetrical

(5) The thickness of a lamination bed is ______.

A: >1 m B: <1 cm
C: >10 cm D: <0.5 m

(6) Which rock has a clastic texture?

A: Arkose B: Limestone
C: Evaporite D: None of above

(7) What term is used to describe the degree of similarity of grain sizes?

A: Roundness B: Sorting
C: Shape D: Clastic

(8) What mineral is not a cement in clastic rock?

A: Silica B: Calcite
C: Iron oxide D: Feldspar

(9) Falling stars occur within the ______.

A: Troposphere B: Stratosphere
C: Mesosphere D: Thermosphere

(10) Which rock has well-developed laminations?

A: Sandstone B: Limestone
C: Shale D: Conglomerate

(11) Once the particle starts moving, we call it ______.

A: Weathering B: Erosion
C: Cementation D: None of above

(12) Weathering occurs ______.

A: at depth
B: near-surface
C: higher temperature
D: higher pressure

(13) Which mineral is rarely found in sedimentary rock?

A: Quarts B: Feldspar
C: Muscovite D: Biotite

(14) Clasts are formed by ______ weathering.

A: Mechanical B: Chemical
C: Biochemical D: All of above

(15) Ozone is found in ______.

A: Troposphere
B: Stratosphere
C: Mesosphere
D: Thermosphere

(16) The rock on the cliff is knocked down by the ocean current, this is ______.

A: Weathering B: Erosion
C: Cementation D: Deposition

(17) The second most abundant gas in atmosphere is ______.

A: Nitrogen B: Oxygen
C: Carbon dioxide D: Water vapor

(18) Which gas can protect us from high-energy ultraviolet radiation?

A: Nitrogen B: Oxygen

C: Carbon dioxide D: Ozone

(19) Compaction during digenesis can lead to the ______.

A: Reduction of volume

B: Increasing of density

C: Expulsion of water

D: All of above

(20) Which mineral can precipitate directly from the solution?

A: Olivine B: Feldspar

C: Calcite D: Biotite

(21) Which rock can be easily broken along parallel surfaces?

A: Sandstone B: Limestone

C: Shale D: Conglomerate

(22) Which rock may have a interlocking crystalline texture.

A: Sandstone B: Limestone

C: Shale D: Conglomerate

(23) Most of the world's oil and natural gas is found in sandstone due to its ______.

A: Composition B: Porous texture

C: Hardness D: None of above

(24) The red - colored clastic rock is probably cemented by ______.

A: Silica B: Calcite

C: Iron oxide D: K-feldspar

(25) Clouds can only occur in ______.

A: Troposphere B: Stratosphere

C: Mesosphere D: Thermosphere

(26) Clay is the major component of ____.

A: Sandstone B: Limestone

C: Shale D: Conglomerate

(27) Which rock is formed by biochemical precipitation?

A: Sandstone B: Chert

C: Shale D: Conglomerate

(28) The temperature will increase in ______.

A: Troposphere B: Stratosphere

C: Mesosphere D: None of above

Chapter 5 Metamorphic Rocks

Key words

Gneiss	Metamorphism	Contact metamorphism	Country rocks	Foliation	Marble
片麻岩	变质作用	接触变质作用	围岩	叶理	大理岩
Schist	Phyllite	Regional metamorphism	Foliated rocks	Slate	Quartzite
片岩	千枚岩	区域变质作用	叶理化岩石	板岩	石英岩
Hornfels	Aureole	Dynamic metamorphism	Chemically active fluid		Protolith
角岩	(变质)晕	动力变质作用	化学活动性流体		原岩

Metamorphic rocks are one of the three main groups of rocks, making up about 27.4% of the Earth's total volume. Metamorphic rock results from the transformation of a pre-existing rock type in a process called metamorphism. Metamorphism is a process where an existing rock type (igneous, sedimentary or metamorphic) is transformed in the solid - state to a new rock type.

Metamorphic rocks are formed deep beneath the Earth's surface due to high pressure and temperature. But now most metamorphic rocks older than 200 million years are mainly exposed in the orogenic (mountain building) zones and continental shields.

Metamorphic rocks are forming today in the cores of active mountain ranges and in other geologically active areas. The study of metamorphic rocks provides us with valuable information about the temperatures and pressures that occur at great depths within the Earth's crust.

1 Intruduction

Metamorphism literally means to"change form" (meta = change, morphic = form). Metamorphism is a geological process which means the changing of pre-existing rocks in the mineral composition and/or texture that occurs in the solid state (i. e., without melting) as a response to the environment. Or the metamorphism takes place when rock is subjected to conditions unlike those in which it originally formed.

Firstly, metamorphic process is that which acts on pre-existing rocks. The pre-existing rocks, otherwise known as protolith(原岩), may be sedimentary rock, igneous rock or another older metamorphic rock. Once the protolith is subjected to new environment, a profound physical and/or chemical change will take place.

Secondly, the mineral changes will take place during metamorphism. The resulted metamorphic rocks are composed of new minerals, which are stable under new condition.

New minerals or mineral assemblages stable under the new conditions of temperature and pressure are created by chemical reactions that take place during metamorphism. In the process atoms are exchanged between the minerals, and thus new minerals are formed. There are lots of metamorphic reactions(变质反应).

One of the commonly found metamorphic reaction is dehydration action(脱水反应). A metamorphic reaction giving off water is called a dehydration reaction. For example, the commonly found clay minerals in sedimentary rocks can change into andalusite (红柱石)and quartz during dehydration reaction:

$Al_2Si_4O_{10}(OH)_2 \Leftrightarrow Al_2SiO_5 + 3SiO_2 + H_2O$

Kaolinite (高岭石) ⇔ Andalusite + Quartz + H_2O

In this case, the new mineral andalusite is produced. Actually, andalusite is a typical metamorphic mineral. Metamorphic minerals are those that form only at the high temperatures and pressures associated with the process of metamorphism. These minerals include kyanite(蓝晶石), staurolite(十字石), sillimanite(矽线石), andalusite, and some garnets(石榴子石). Other minerals, such as micas, feldspars, and quartz, may be found in metamorphic rocks, but are not necessarily the result of the process of metamorphism.

Another example of metamorphic reaction is the reaction of muscovite and quartz.

$KAl_3Si_3O_{10}(OH)_2 + SiO_2 \Leftrightarrow (Al_2SiO_5 + KAlSi_3O_8 + H_2O$

Muscovite + Quartz ⇔ Andalusite + Kfeldspar + H_2O

Recrystallization(重结晶) of the original rock can also take place during metamorphism. Recrystallization is the growth of new mineral grain in a rock at the expense of old grains which are same kind. For example, during metamorphism, a pure limestone will recrystallise (重结晶) to a marble. Both the limestone and marble are composed of calcite. But the calcite grains are far larger than that of limestone.

Thirdly, textural transformation will occur during metamorphism. The most remarkable textural change occurred during metamorphism is the increasing or decreasing of grain size. Grain size increases with increasing temperature, but decreases as directed pressure increases. Directed pressure can lead to the development of foliation(叶理), where mineral grains are arranged into layers. Some of the textures produced during metamorphism have little resemblance to the original protolith. One of the typical metamorphic textures is granoblastic texture (花岗变晶结构), which is the result of mutual adjustment of grain boundaries in the solid state.

Finally, it is important to note that metamorphism is a process where protolith is transformed in the solid state to a new rock type. Solid state indicates the change of mineral identity without melting. All ion migration occurs while the rock is still solid. Once a rock is melted, the new product will be igneous. The process of metamorphism does not melt the rocks, but instead transforms them into denser, more compact rocks. The lower limit of metamorphism is diagenesis. The upper limit of metamorphism is melting. It is important that you realize that the rock remains in its solid state.

Metamorphism is a two way street, however. Rocks that are slowly buried and, therefore, progressively metamorphosed at higher and higher temperatures and pressures are said to undergo prograde metamorphism(进变质). But, if the temperature and pressure decline then the rock undergoes retrograde metamorphism(退变质)as the rocks and minerals adjust to the reduced temperature and pressure.

2 Factors Controlling Metamorphism

Factors controlling metamorphism include heat, pressure, and chemically active fluids. During metamorphism, rocks are often subjected to all three metamorphic agents simultaneously. However, the degree of metamorphism and the contribution of each agent vary greatly from one environment to another.

2.1 Temperature

Temperature is the most important factor contributing to metamorphism. An increase in temperature leads to the loss of water or carbon dioxide, forming one or more new minerals in the process. Increasing temperature can also result in faster chemical reactions. Generally, geologists consider the temperature range of metamorphism to be approximately 150℃ up to the point where a rock begins to melt (700 ～1 000℃).

There are three basic sources of heat:

- **Geothermal Gradient**(地温梯度)

The temperature will increase naturally with depth in the Earth. That is because the Earth is a hot planet, with temperature reaching several thousand degrees Celsius at the center of the Earth. The deeper the rock is buried, the hotter it gets. Geothermal gradient is the rate of increase in temperature per unit depth in the Earth. Although the geothermal gradient varies from place to place, the average geothermal gradient of the Earth is about 25℃/km. In regions where geothermal gradients are higher (such as parts of Tibet, island arcs around the Pacific), temperature increases more rapidly with depth.

- **Intrusive Bodies of Magma**(岩浆热)

Another way for heat to form metamorphic rocks is through the introduction of underground magma into an area of pre-existing solid rock. When the magma (molten rock) moves from deep within the Earth toward the surface, there is a significant temperature difference between the magma and surrounding rock. As these magma bodies cool they release heat to the surrounding country rock, leading to metamorphism.

- **Crust Fracture**(地壳断裂)

Another source of heat is friction among bodies of rock as they grind against each other.

2.2 Pressure

Pressure can be classified into two different types: confining pressure and directed pressure (Figure 5.1).

- **Confining Pressure**(围压)

Rocks buried deep beneath the Earth are exposed to confining pressure, or the lithostatic (Greek lithos, "rock"; statikos, "in place") pressure(静岩压力). The confining pressure is caused by the material that sits above a particular location. Within the

Earth, an object is under pressure from all directions due to the weight of the column of rock above the object. So confining pressure is equal in all directions. Because confining pressure is the push from all directions, it tends to confine material to the smallest possible volume (space). New metamorphic minerals crystallize under high confining pressure and tend to be denser than their low-pressure counterparts.

• **Directed Pressure(定向压力)**

Directed pressure is not equal in all directions. It is caused by tectonic movement(构造运动).

One of the most important textures associated with directed pressure is foliation(叶理)(Figure 5.2). Foliation is defined as a planar structure that results from the nearly parallel alignment of sheet minerals such as mica and/or compositional layering in the rock. Under great directed pressure, platy mineral (especially micas) grains may be aligned into bands perpendicular to the direction of greatest pressure.

2.3 Chemically Active Fluid(化学活动性流体)

Rock masses contain a fluid phase, except in areas where high temperatures or impermeable rocks prevail. Chemically active fluid is largely water, but also contains salts and volatiles and other elements, such as CO_2, CH_4, N_2, Cl, S, B, Na and K. Sources of fluids include surface infiltration, pore spaces between minerals, magma and heated minerals.

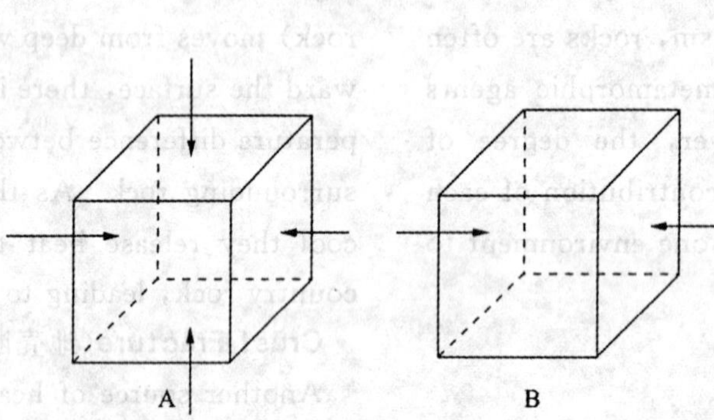

Figure 5.1 Confining pressure and directed pressure

A: Confining Pressure ; B: Directed Pressure

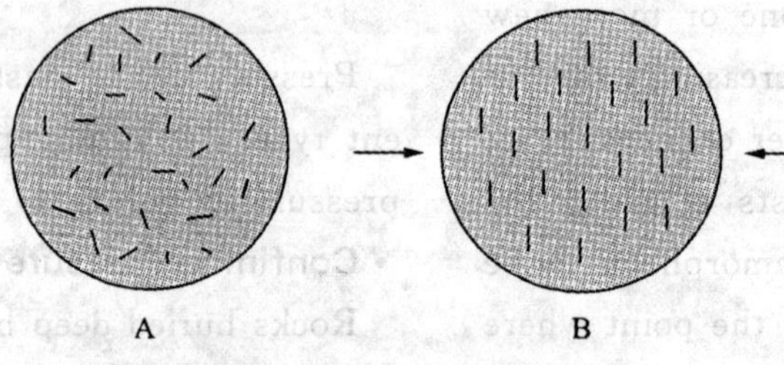

Figure 5.2 Formation of foliation

A: Randomly oriented particles;

B: Parallel alignment of minerals due to directed pressure

Chemically active fluids play an important role during metamorphism. Chemically active fluids accelerate and enhance metamorphic reactions. Fluids can also transport dissolved solids to and from rock masses and thus are instrumental in the formation of new minerals. Over time, these fluids can change the chemical makeup of the parent rock. Fluids tend to wander away from their birthplaces, interacting with rocks elsewhere. That process, the interaction of rock with chemically active fluids, is called metasomatism(交代作用).

Time is also very important for the metamorphic reaction. The mineral reactions are exceptionally slow. Generally, therefore, metamorphic rocks that have undergone higher pressures and temperatures for longer periods of time tend to be coarser grained than their lower grade counterparts. Metamorphism can take millions of years as in the slow cooling of magma buried deep under the surface of the Earth.

3 Types of Metamorphism

Metamorphism occurs under a variety of different conditions that controls the geographic distribution of metamorphic rocks and their significance to Earth surface features. Based on the general geological setting, metamorphism can be classified into three types: contact metamorphism, regional metamorphism and dynamic metamorphism. The three different types of metamorphism have different settings and metamorphic processes. It should be noted that the general types grade into one another. The distinctions between the types of metamorphism are sometimes unclear and subjective.

3.1 Contact Metamorphism(接触变质作用)

Contact metamorphism occurs when local rocks are metamorphosed by the heat from an igneous intrusion, such as limestone turning to marble along the contact zone. Contact metamorphism accounts for less than 2% of the metamorphic rocks on the Earth.

When magma invades cooler rock, the country rock will be heated and metamorphoses. Contact metamorphism is characterized by high temperature, low pressure conditions, and so is also known as thermal metamorphism(热变质). Temperatures of metamorphism vary widely from 400℃ to 1 000℃. They usually occur at relatively shallow depths (low pressure) because it is only at shallow depths where there will be a large contrast in temperature between the intruding magma and the surrounding country rock.

Contact metamorphism is restricted to the contact zones adjacent to large igneous intrusive bodies (Figure 5.3). As the metamorphism envelopes an igneous intrusion, the zone of metamorphism is called a metamorphic aureole(变质晕). Outside of the contact aureole, the rocks are not affected by the intrusive event. The aureole ranges in width from several centimeters to several kilometers. Small intrusive bodies like dikes and sills have aureoles only a few centimeters thick. By contrast, batholiths may create zones of metamorphic rock several kilometers thick.

This type of metamorphism produces a

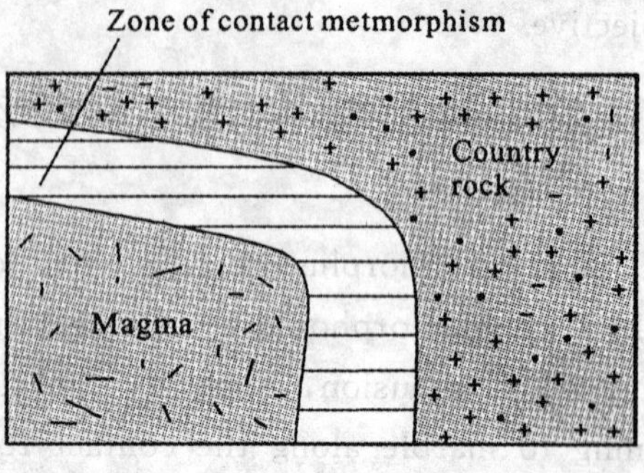

Figure 5.3 Contact Metamorphism

recrystallisation with no preferred mineral elongation direction, i. e. foliation, resulting in a massive, hard rock of fused, interlocking grains. The rock is called a hornfels. During contact metamorphism, a quartz sandstone will recrystallise to a quartzite and a pure limestone to a marble(大理岩).

Hydrothermal fluids(热液、水热流体)circulating through the surrounding rock are also important in metamorphosing the rock as they transport heat. Fluids are particularly important in the metamorphism of carbonate rocks(碳酸盐岩).

3.2 Regional Metamorphism

Regional Metamorphism occurs over a very large area in response to increased temperature and pressure. Regional metamorphism is associated with the plate tectonics (板块构造)(Figure 5.4).

Pressure plays a key role in the development of regional metamorphism. The pressure results from the tectonic movement, such as continental collision(碰撞)and subduction(俯冲). As tectonic plates collide and converge, intense pressure alters sedimentary and igneous rocks already buried in the Earth. Thus, regional metamorphic rocks usually occur in the cores of fold/thrust mountain belts or in eroded mountain ranges. Tectonic movements exert high pressure to the rocks and result in folding of rock and thickening of the crust. That is why most regional metamorphic rocks are strongly deformed. These kinds of strongly deformed metamorphic rocks are most commonly exposed in old mountain chains(山链). The Qinlin－Dabie(秦岭－大别山) mountains are an excellent example of where this has taken place. The general pattern of folds(褶皱) in the regionally metamorphosed rocks can be used to indicate the direction of the intense pressure.

Heat during metamorphism may come from large bodies of molten rock rising under a wide geographic area.

Due to the high directed pressure, the regionally metamorphosed rocks are usually strongly foliated (页理化). Most metamorphic rocks are the production of regional metamorphism. Rocks are foliated with mineral assemblage characteristic of varying metamorphic temperatures and pressures. Please keep in mind that not all rocks that undergo regional metamorphism will form a foliation. Some rocks that lack tabular minerals such as mica will not generate foliation. That is because the foliation is the parallel alignment of tabular minerals. For example, quartzite and marble can also be produced during regional metamorphism. Because they are composed mainly of quartz grains and calcite grains respectively, neither quartzite (石英岩)nor marble contains a foliation.

A variation on regional metamorphism is burial metamorphism, the latter occurs solely in response to burial. Burial metamorp-

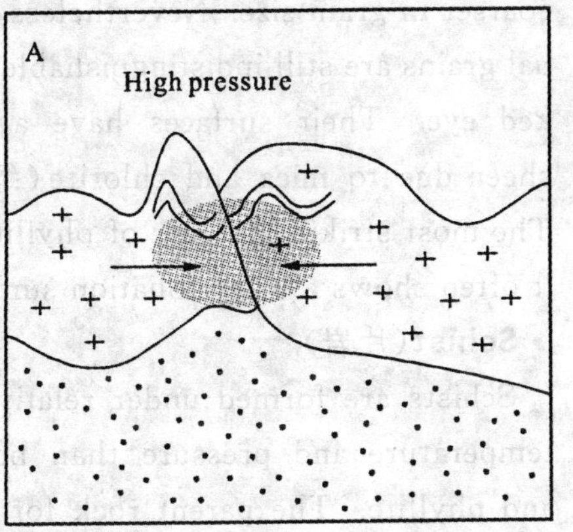

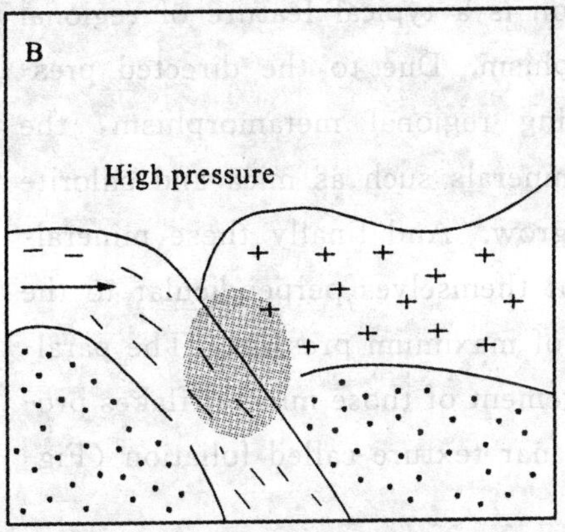

Figure 5.4 Rocks are altered by regional metamorphism by continent collide (A) or ocean crust subducts (B)

hism occurs in areas that have not experienced significant deformation. It is generally restricted to large, relatively undisturbed sedimentary piles away from active plate margins.

3.3 Dynamic Metamorphism

Due to the rupturing and movement of rocks along fault planes (断层面), rocks near large faults are often intensely deformed and reconstituted. This is dynamic metamorphism. Dynamic metamorphism usually is caused by high directed pressures. Frictional heat (摩擦热) due to the relative movement of rock bodies also contribute much to the dynamic metamorphism. Because fluid can move readily along the fault plane, dynamic metamorphism is also influenced by chemically active fluid.

Dynamic metamorphism causes only the structural alteration of rock through pressure. At shallow depths in the Earth, rock behaves in a brittle manner. Under this condition, the rocks are mechanically pulverized along faults. At deeper levels, the rocks behave in plastic (塑性) or ductile (韧性) manner due to high temperature. In this case, recrystallization occurs and results in the formation of mylonite. Mylonite (糜棱岩) is a strongly-foliated rock. It is produced in the ductile deformation due to fault movement.

4 Classification of Metamorphic Rocks

Based on the different foliation textures, metamorphic rocks can be classified into two broad divisions: foliated and nonfoliated. This is not a rigid classification. Some common nonfoliated metamorphic rocks can develop foliations. For example, the quartzite and marble may be either foliated or nonfoliated.

4.1 Foliated Metamorphic Rocks

Foliation (from the Latin *folium* — leaf) is a planar structure that results from the nearly parallel alignment of sheet minerals such as mica and/or compositional layering in the rock. Foliated metamorphic rocks are those in which the minerals have been flattened and/or oriented into parallel layers. They generally exhibit a layered look.

Foliation is a typical feature of regional metamorphism. Due to the directed pressure during regional metamorphism, the tabular minerals such as mica and chlorite start to grow. And finally these minerals will orient themselves perpendicular to the direction of maximum pressure. The parallel arrangement of those mineral flakes produce a planar texture called foliation (Figure 5.5).

Here are some common foliated rocks.

• **Slate(板岩)**

Slate is one of the most common foliated metamorphic rocks. It forms at low temperature and pressure. Slate is generally fine-grained and minerals are too small to be seen with the naked eye. Slate can be easily to break into a set of even parallel sheets when you strike it with a rock hammer. Slate is formed when shales and clays are put under great pressure and heated inside the Earth for millions of years.

• **Phyllite(千枚岩)**

Phyllites are similar to slates but are coarser in grain size. Nevertheless, individual grains are still indistinguishable with naked eye. Their surfaces have a lustrous sheen due to mica and chlorite(绿泥石). The most striking feature of phyllite is that it often shows a wavy foliation surface.

• **Schist(片岩)**

Schists are formed under relative higher temperature and pressure than both slate and phyllite. The parent rock for schist is phyllite. They have a high luster than phyllites. Schists are characterized by a preferred alignment of moderately coarse grains, each grain generally being visible to the naked eye. Because it has been squeezed harder than slate and phyllite, it is often found folded and crumpled. Schists are usually named by the main minerals contained in the rocks. Bitotite schist(黑云母片岩), garnet mica schist(石榴云母片岩) and quartz schist(石英片岩) are some examples of this.

• **Gneiss(片麻岩)**

Gneiss has been subjected to more heat

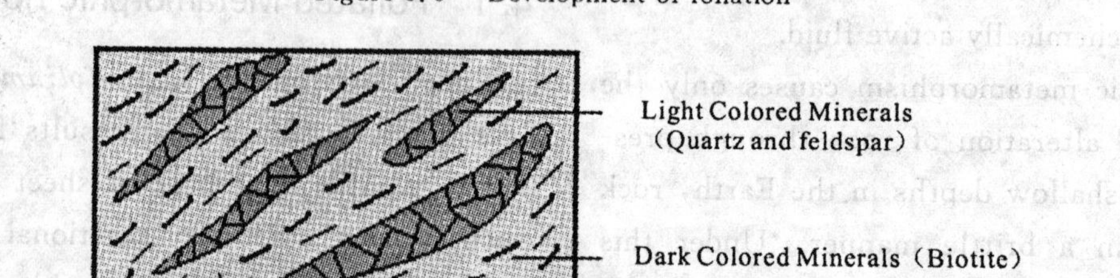

Figure 5.5 Development of foliation

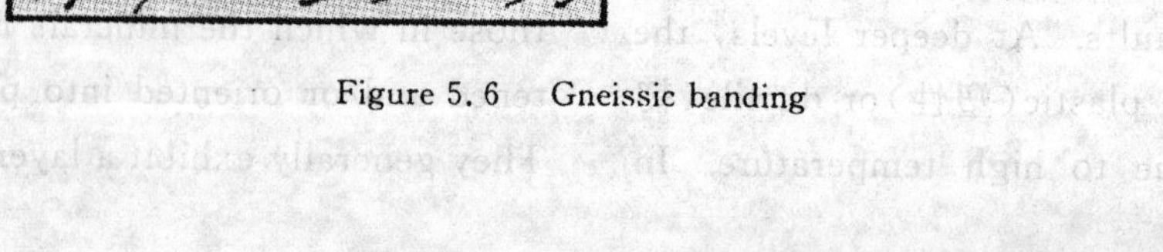

Figure 5.6 Gneissic banding

and pressure than schist. Gneisses are coarser than schists and commonly display parallel compositional layers(成分层). Unlike schist, which is more strongly aligned, gneiss doesn't fracture along the planes of the mineral bandings(矿物条带). The foliation in gneiss is actually distinct mineral banding in which layers rich in quartz and feldspar often alternate with layers rich in mica. The banding give gneiss a striking feature of alternating layers of dark and light colors.

The minerals that compose gneiss are the same as granite. They are mainly composed of quartz, feldspars, micas, and amphiboles. Under higher pressure and heat, the granite can be changed into gneiss. Gneiss can also be formed from sandstone.

4.2 Nonfoliated Metamorphic Rocks

Common nonfoliated (无叶理构造) rocks are hornfels, quartzite, and marble. In general, they don't have the layered look because they have neither preferred mineral orientation (矿物优选方位) nor compositional banding. Nonfoliated metamorphic rocks are apt to form in a contact metamorphic environment where pressure is unimportant.

Figure 5.7 Foliated metamorphic rocks
Image Credit: Courtesy of the U.S. Geological Survey and the Mineral Information Institute. Public Domain

• **Quartzite** (石英岩)

Quartzite is a metamorphic rock composed predominantly of quartz. The protolith (原岩) is typically sandstone. It forms from sandstone that has come into contact with deeply buried magmas. Quartzite looks similar to its parent rock. Pure quartzite is usually white to grey. Quartzites often occur in various shades of pink and red due to varying amounts of iron oxide. The best way to tell quartzite from sandstone is to break the rocks. The quartzite fractures through the grains (rather than between the grains as it does in sandstone).

Quartzite can be produced either by regional metamorphism or by contact metamorphism (Figure 5. 8).

• **Marble** (大理岩)

Marble is composed predominantly of calcite or dolomite. Most marbles have no foliation texture. The protolith is typically limestone. Both marble and limestone may have the same composition but marble typically has larger grains. Marble has many color variances due to the impurities (杂质) present at formation. Marble is much harder than its parent rock. That is why it is extensively used for sculpture and building material.

• **Hornfels** (角岩)

Hornfels (German, meaning "hornstone") is typically very fine-grained and compact, and occurs in contact aureoles. Hornfels results from the contact metamorphism of shales or other clay rich rocks in the vicinity of an igneous intrusion. Hornfels has no foliation.

5 Rock Cycle

It is not always true that "once a rock, always that rock". Rocks are constantly forming and being destroyed within and at the surface of Earth's crust. Rocks take different forms at different times. Igneous rock can change into sedimentary rock or into metamorphic rock . Sedimentary rock

Figure 5. 8 Nonfoliated metamorphic rocks

Image from "Minerals in Your World", a cooperative project between the U. S. Geological Survey and the Mineral Information Institute. Public Domain

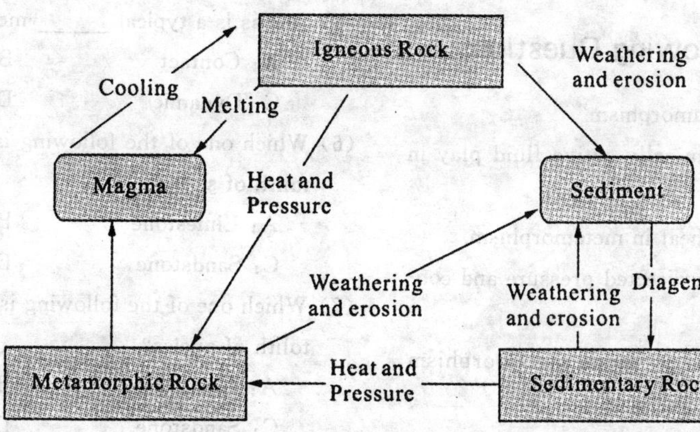

Figure 5.9 Rock cycle

can change into metamorphic rock or igneous rock. Metamorphic rock can change into igneous or sedimentary rock.

The rock cycle is a fundamental concept in geology that describes the dynamic transitions through geologic time between the three main rock types: sedimentary, metamorphic and igneous. The concept of the rock cycle was proposed by James Hutton (1727 — 1797), an eminent 18th century gentleman farmer and founder of modern geosciences. Rocks are always on the move through the rock cycle (Figure 5.9)!

Magma is molten rock below the Earth's surface. When it reaches to the Earth's surface, it is called lava. The origin of all rocks can be ultimately traced back to the solidification of molten magma.

Igneous rocks are created when magma and lava cool and become solid. Igneous rocks that form below the Earth's surface are called intrusive igneous rocks. Igneous rocks that solidify above the Earth's surface are called extrusive igneous rocks. When igneous rocks are forced deep underground, it will eventually melt if temperature rises enough, returning the material to the state of magma.

When the igneous rocks are exposed on Earth's surface, the various weathering agents, such as wind and water can break them into pieces. The rock pieces, called sediments, will be picked up and transported to some places where they are compressed into new sedimentary rocks.

When the sedimentary rocks are subjected to higher temperature and pressure, the metamorphism will take places. The compositional and textural changes can finally result in the formation of metamorphic rocks. Of course, the opposite can occur due to tectonic uplift of the crust. And if the metamorphic rock is forced into deep interior of the Earth, it will melt and form magma. At this stage, the rocks have accomplished a complete cycle.

Above demonstration is just a very simple and ideal way to show the rock cycle. Actually, the cycle may be interrupted at any time and stage. For example, after the formation of sedimentary rocks, if the tectonic movement forces them quickly into mantle (as in the case of subduction), the sedimentary rocks will be changed into magma directly. Then the process will begin all over again. The rock cycle never stops!

1. Answer the Following Questions

(1) Define the term "metamorphism".
(2) What role does chemically active fluid play in metamorphism?
(3) Describe the role of heat in metamorphism.
(4) Differentiate between directed pressure and confining pressure.
(5) Differentiate between prograde metamorphism and retrograde metamorphism.
(6) How do foliated and nonfoliated metamorphic rocks differ?
(7) Define the term "rock cycle".
(8) Contrast quartzite and sandstone.
(9) Differentiate marble and limestone.
(10) Contrast the slate and phyllite.
(11) Define the term "foliation". How does foliation form?
(12) List and describe briefly the types of metamorphism.
(13) Contrast the bedding and foliation.
(14) List the sources of heat responsible for metamorphism.
(15) What is dehydration action?
(16) Define the metamorphic aureole.

2. Multiple Choice

(1) Which one of the following is not a foliated rock?
A: Slate B: Phyllite
C: Hornfels D: Gneiss

(2) Which one of the following has a coarser grain size?
A: Slate B: Gneiss
C: Schist D: Phyllite

(3) Which one of the following is formed at a higher pressure?
A: Slate B: Gneiss
C: Schist D: Phyllite

(4) Foliation is formed due to ______.
A: Confining pressure
B: High temperature
C: Directed pressure
D: High temperature and pressure

(5) Gneiss is a typical ______ metamorphism rock.
A: Contact B: Regional
C: Dynamic D: All of above

(6) Which one of the following is probably the protolith of slate?
A: Limestone B: Shale
C: Sandstone D: All of above

(7) Which one of the following is not a possible protolith of gneiss?
A: Limestone B: Schist
C: Sandstone D: Granite

(8) Which rock can be easily to break into a set of even parallel sheets?
A: Slate B: Gneiss
C: Schist D: Phyllite

(9) The major component in chemically active fluid is ______.
A: Water B: CO_2
C: S D: CH_4

(10) The intrusion of magma can result in ______ metamorphism.
A: Contact B: Regional
C: Dynamic D: All of above

(11) Burial metamorphism is a kind of ______.
A: Contact B: Regional
C: Dynamic D: All of above

(12) With the increasing of temperature, the limestone will turn into ______.
A: Slate B: Phyllite
C: Marble D: Gneiss

(13) With the increasing of temperature, the limestone will turn into ______.
A: Slate B: Phyllite
C: Marble D: Gneiss

(14) Temperature is extremely important for ______ metamorphism.
A: Contact B: Regional
C: Dynamic D: All of above

(15) Pressure is not important for ______ metamorphism.
A: Contact B: Regional
C: Dynamic D: All of above

(16) Which mineral can only be found in metamorphic rocks?

A: Quartz　　B: Feldspar
C: Calcite　　D: Garnet

(17) The average geothermal gradient of the Earth is about ______ ℃ /km.

A: 25　　B: 35
C: 45　　D: 55

(18) The lower limit of metamorphism is ______.

A: Diagenesis　　B: Melting
C: Precipitation　　D: All of above

(19) The upper limit of metamorphism is ______.

A: Diagenesis　　B: Melting
C: Precipitation　　D: All of above

(20) Which one of the following rocks will fizz upon the application of a drop of acid?

A: Slate　　B: Phyllite
C: Marble　　D: Gneiss

(21) Which one of the following rocks can only be found in contact metamorphism?

A: Slate　　B: Hornfels
C: Marble　　D: Gneiss

Chapter 6 Geologic Time Scale

Key words

Relative age	Principle of original horizontality	Stratigraphic principles	Half-life
相对年代	原始地层水平律	地层层序律	半衰期
Absolute age	The principle of superposition	Radiometric dating	Unconformity
绝对年代	叠置律	同位素测年	不整合
Index fossil	The Principle of biologic succession	Geologic time scale	Angular unconformity
标准化石	生物层序律	地质年代表	角度不整合
Geological time units	Rock-stratigraphic unit (Local)	Time-stratigraphic unit	Disconformity
地质年代单位	岩石地层单位、地方性地层单位	年代地层单位	假整合
Principle of cross-cutting relationships		Conformable contact	Nonconformity
切割律		整合接触	非整合

Two of the most often asked questions are"how old is the Earth?" or " how has the geology of someplace varied throughout geological time?". In order to answer above questions, we need to know how the geologists unravel the complexities of the past.

The only way or at least the basic way for geologists to work out the Earth evolution history is to study the rocks, especially sedimentary rocks, and geological events in which they are involved. That is because the geological history was recorded in the layers of rocks. Geological events are those events occurs on or within the Earth, such as earthquake and mountain-building (造山运动). However, the geological events by themselves have little meaning until they are put into a time perspective. In other words, the geological events should be listed in sequential order. That means that we need a geological calendar to unravel the Earth's history. Forturenately, by studying the rocks exposed on the Earth's surface and using some simply principles revealed by the rock records, geologists have been able to construct a complete calendar that outlines the periods of time when rocks were formed. Such a calendar is called geological time scale.

Geological time is often discussed in two forms: relative and absolute dating.

Relative dating doesn't assign a precise age, such as in years or millions years to rocks or geologic events. Relative age refers to the age of rocks in relation to other rocks. So relative age simply means something is old and something is young. We don't know exactly how old something is and all we know is the sequence of geological events. The problem is that how geologists know one rock is old than other rocks? The answer is simple, that is studying the rock itself!

Absolute dating assigns specific dates to

rock formations and geologic events. Absolute age is determining the actual number of years that have passed since an event occurred. With the absolute age, we can say exactly how old something is. For example, The limestone is 450 million years old. Absolute dates are obtained from natural geologic clocks such as radiometric dating. With the development of technology, the absolute dates are pinpointed more exactly in recent years.

By studying the relative and absolute ages of various rocks and geological events, we can determine the geological history of an area. By piecing together the histories of enough areas we can unravel the geological evolution of Earth.

1 Stratigraphic Principles

Stratigraphic principles are basic principles that all geologists use in deciphering the spatial and temporal relationships of rock layers.

Layers of sedimentary rocks are called strata (singular stratum). Casually, bed is sometimes used for "stratum" or "layer". Stratigraphy(地层学) is the studying and description of strata.

The beds or strata are the result of natural depositional event in geological history. As time goes, the sediment will pile up layer upon layer, and eventually lead to the formation of layered sedimentary rocks. All of the information of geological evolution are preserved in the beds as geological records(地质记录). For example, the entire history of Grand Canyon was written into the cliffs and its horizontal beds. The only way to unravel its mysterious past is to study the spatial and temporal relationships of those magnificent layers. The sequence of stratification (地层层序) preserved in the rocks can be served as a clock which we can use to measure time and events.

- **Principle of Original Horizontality （原始地层水平线）**

Principle of original horizontality states that all sediments strata are originally deposited in horizontal or nearly horizontal layers (Figure 6.1).

When they are transported to seas or lakes, the sediments will settle down and accumulate on the lake bottom or sea floor due to the influence of gravity. As a result, a flat, horizontal layer of sediment will be formed.

Actually the originally deposited horizontal layers can be tilted, folded and overturned due to the tectonic movement (Figure 6.2). So in order to decipher the correct sequences of deformed strata, some of the sedimentary structures, such as cross bedding and mud cracks can be used to find out the"up directions".

So, in the field, if you find some tilted beds, you can immediately conclude that some unusual geological events may occur. Actually, it is this kind of simple clue given us by sedimentary strata that help us to interpret geological history.

- **Principle of Superposition （叠置律）**

The principle of superposition states that in an undeformed sedimentary sequence of sedimentary rocks, younger rock units are deposited on top of older rock units. That means that the original deposited layers have the oldest layer on the bottom because the bottom layer is the first one to form.

Figure 6.1 This photograph illustrates the horizontal beds exposed in Grand Canyon, Arizona. Note that the individual bed layers are horizontal.

Photo by Dr. Yu Shan (俞姗)

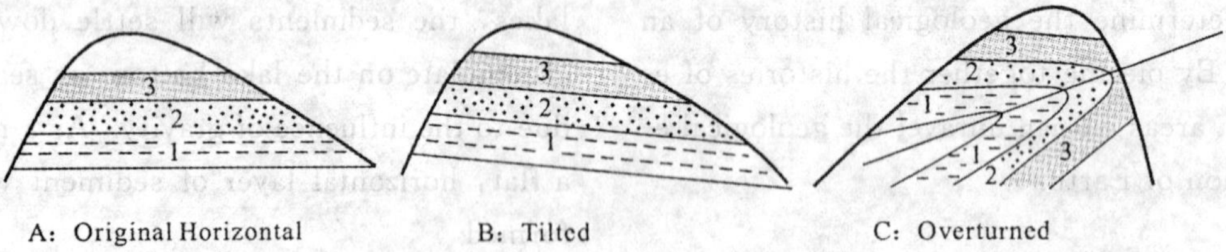

Figure 6.2 Tilted and overturned strata

This principle allows spatial relations between layers of rock to be translated into temporal relations.

Due to the tectonic forces, the original sequences may be totally disturbed. It is first necessary to determine the "up direction" before you can determine the sequence of beds in the field.

How to use this principle to decipher the history of the Earth? It is very simple. For example, if you find a very old outcrop (露头) of sedimentary rock on the Earth's surface, then you may conclude that some crust movements (地壳运动) have taken place, which result in the uplifting or turnover of the originally deep—seated rocks.

• **Principle of Lateral Continuity (水平连续律)**

Principle of lateral continuity states that sedimentary rocks are laterally continuous over large areas. That means that initially deposited sediment layers extend horizontally a considerable distance in all directions just like the case of snowing. They may extend over hundreds of thousands of square kilometers of the Earth's surface. These layers will thin out (尖灭) and end eventually due to the limit of depositional basin or supplies of depositional materials.

Along the Three Gorges (长江三峡) or the Great Canyon (seeing Figure 6.3), some of the horizontal layers extend as long as a few or tens of kilometers. In most cases, it is rarely to find such well-exhibited vast layers due to topography or tectonic movement.

The principle of lateral continuity is of extremely importance for field works (野外

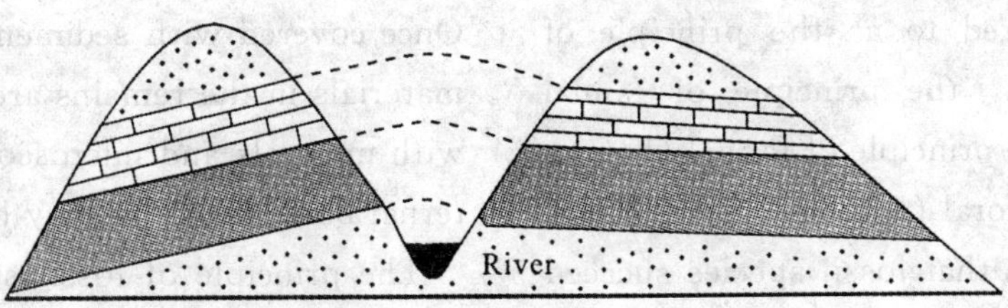

Figure 6.3 Rock layers in isolated hills were once continuous

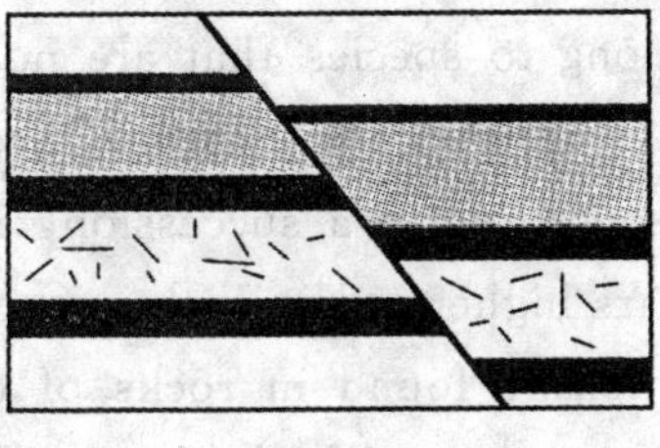

(a)

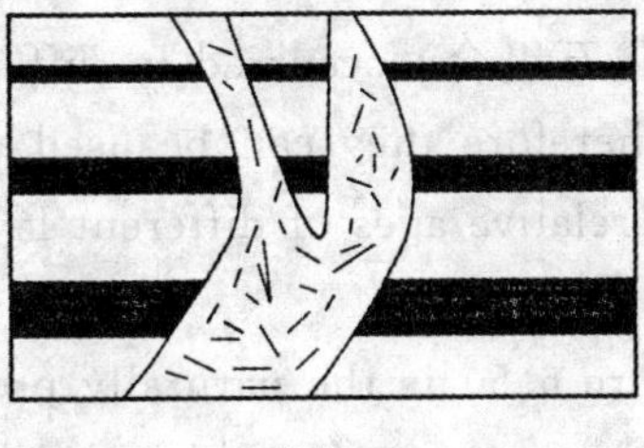

(b)

Figure 6.4 Principle of cross-cutting relationship

A: Fault is younger than the rock; B: Dyke is younger than the country rock

工作). One of the applications of this principle is to correlate(对比) rocks exposed at different places. The physical properties and chemical composition with a layer are relatively uniform. Isolated rocks layers with similar chemical and physical features may once belong to same layer. In Figure 6.3, according to the principle of lateral continuity, similar layers exposed at different hills were once continuous. It is the erosion that separates them. Another application is to infer the existence of fault. Once the layers disappear abruptly, faulting (断层作用) can be inferred.

2 Principle of Cross-Cutting Relationship

This principle states that younger features cut across older feature, or the thing being cut is older than the thing doing the cutting.

This principle is typically used to determine the relative age of rocks that have been intruded by igneous bodies or rock bodies that have been displaced by faults (Figure 6.4). An intrusive rock is always younger than the rock that surrounds it. The sedimentary rocks are older than the fault which cuts them, because they had to be there first, before they could be faulted.

This principle is very useful to determine the relative age of faults and intrusion in the field. For example, you find a volcano crater surrounded by the sand rocks, and then the immediate conclusion is that the crater is younger than sand rocks. A specific section of Great Wall built in Ming Dynasty was destroyed by an unknown earthquake, then this earthquake must have occurred after 1368 (洪武元年, the first year of Ming dynasty). You can't destroy a building that is not there yet!

3 Principle of Biologic Succession

Principle of biologic succession is also

sometimes referred to as the principle of fossil succession, the principle of faunal succession or the principle of faunal (belong to animal) and floral (belong to plant) succession. It states that fossil species succeed on another in a definite and recognizable order. In general, different species or fossil assemblage (化石组合) occurred in different layers. Therefore they can be used to determine the relative ages of different layers .

Fossil (Figure 6. 5) is the naturally preserved remains or traces of prehistoric organisms. Some fossils only consist of skeletal remains or teeth; other fossils contain traces of skin, feathers or even soft tissues. In order for an organism to be fossilized, the remains need to be covered by sediment as soon as possible. Or the key first step in the process of fossilization is the burying or otherwise entombing of an organism prior to that organism's complete decomposition. Once covered with sediments, the chemical materials in the remains are slowly replaced with minerals and microscopic details of internal bone structure may be preserved.

The principle of fossil succession can be summarized as followings: ①Fossils are the remains of extinct organisms; that is, they belong to species that are no longer living anywhere on the Earth. Once a fossil has disappeared in a succession, it never reappears higher in the succession. ② The kinds of fossils found in rocks of different ages differ because life on the Earth has changed through time. In general, the earlier fossil life forms are simpler, and the recent forms more similar to present life forms. ③ Each species is restricted to a specific interval of geologic time.

This principle allows geologists to identify rocks of the same age in widely separated regions and made it possible to correlate and divide layers based on fossil assemblages.

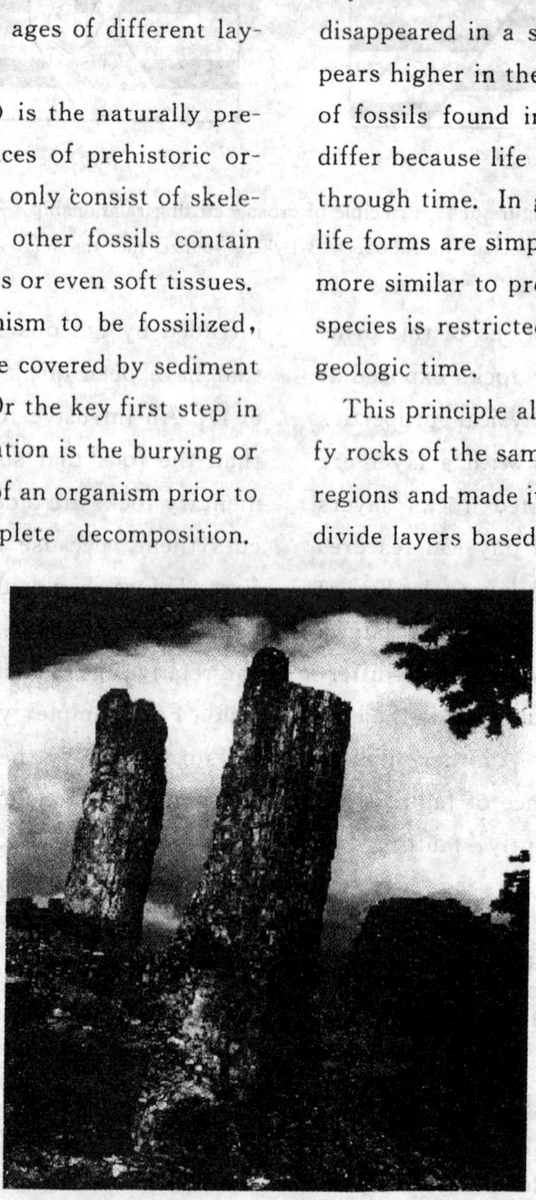

Figure 6. 5 Petrified tree (硅化木)

Courtesy of Schmidt; 1977, from Yellowstone Digital Slide File. Public domain

For example, when we find the same kinds of fossils in rocks from different places, we know that the rocks are of the same age.

A variety of fossils from each layer of strata have been chosen to be what are called"index fossils". An index fossil is a fossil that can be used to identify and date the strata in which it is found. An index fossil has the following characteristics:

• The species existed for a rather short period of time.

• The fossil has distinct characteristics that make it relatively easy to recognize.

• The species lived over a wide geographic area.

4 Unconformity

If the depositional basin keeps in stable for a long time, beds deposited continuously from the bottom up will contact parallel. This kind of contact relationship between beds is called conformable contact(整合接触). And the beds are said to be conformable to each other as they lie as parallel beddings one above the other.

The Earth's crust is in constant movement. Sometimes, the early deposited sediments or sedimentary rocks will be uplifted and exposed to the Earth's surface due to the tectonic movement. Once the rocks are out of water, erosion will take place. At this stage, the continuous record of sedimentation will be broken. The early deposited sediments will be taken away and an erosion surface will form. When the rocks or sediments are putted under water again, the new sediments will deposit on the erosional surface and thus the eroded surface will be preserved. This surface is unconformity.

So an unconformity is a surface of erosion or non-deposition separating older from younger rocks units. It is a gap in the sedimentary rock record. It represents a period of time in which deposition stopped and erosion occurred due to tectonic activity in the form of uplift of land. Or in other words, it represents a time when no rocks were formed, or a time when rocks were formed but then eroded away.

Three types of unconformities are recognized:

• **Angular Unconformity**(角度不整合)

Angular unconformity is an erosional or non—deposition surface where the beds above and below lie at different angles. Usually it separates the underlying tilted older strata and overlying horizontal younger layers. The angular relationship between different layers makes it very easy to recognize angular unconformity in the field.

The Figure 6.6 shows us the entire procedure of formation of angular unconformity. At first stage, the continuously deposed beds pile up layer by layer. The parallel layers are conformable to each other. Then due to the geological movement, the layers may undergo strong deformation and original parallel layers may be tilted. At this moment, erosion may also take place during this phase if the beds are exposed to atmosphere due to the tectonic uplift(构造抬升, uplifting caused by crustal movement). After a long period of time, maybe millions of years, the entire region may become submerged. At this stage, a set of new conformable beds may deposit on the eroded surface. And the angular contact between the new horizontal layers and old tilted layers may form. Of course, the eroded surface will be preserved in the sequences as angular unconformity.

In the sense of geological evolution, an unconformity always imply regional-scale (区域规模) or global-scale(全球规模) tectonic movement. So, if you find an unconformity in the field, it means that region may experience some very important geological events long long time ago. That unconformity may be the result of plate collision, like Himalayas, who knows?

• **Disconformity** (假整合)

A common, but less conspicuous unconformity is disconformity in which the layers on either side are essentially parallel. It represents a considerable period of non-deposition or erosion and marked by visible, irregular or uneven erosional surface (Figure 6.7).

It can be difficult to distinguish within a series of parallel conformable strata. The uneven erosional surface may help us to recognize the disconformity. Another way is to correlate from other place and find if any strata are missing.

The formation of disconformity is similar to that of angular unconformity, except that the layers in the case of disconformity are not folded or tilted.

• **Nonconformity** (非整合)

Nonconformity is the erosional surface separating younger overlying sedimentary rocks from underlying eroded igneous or metamorphic rocks.

When the deep-seated igneous or metamorphic rocks are exposed to the Earth's surface, they will be subjected to erosion. All of the rocks overlying the metamorphic or intrusive igneous rocks will be removed by erosion. After a long period of time, the eroded rocks may get submerged and then covered by the new sediments. When this is

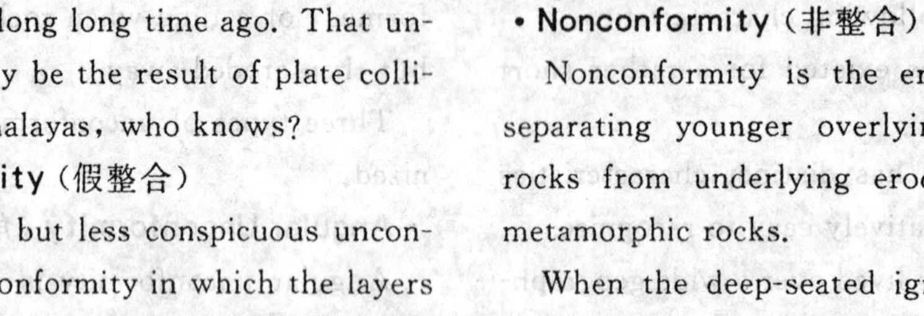

Figure 6.6 Development of angular unconformity

(a) Deposition; (b) Uplift and tilting; (c) Subsidence

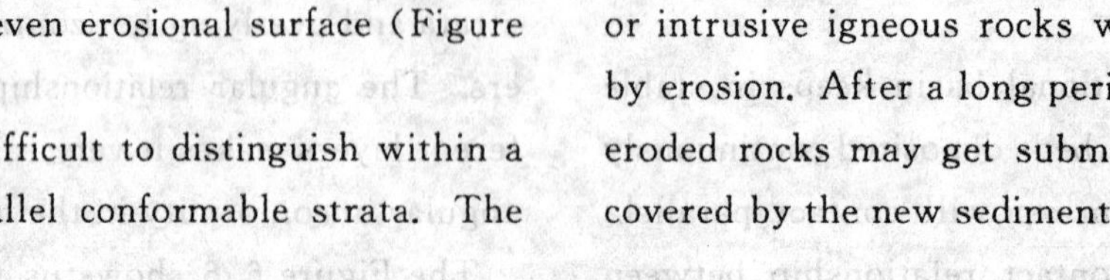

Figure 6.7 Development of disconformity

(a) Deposition; (b) Uplift; (c) Subsidence

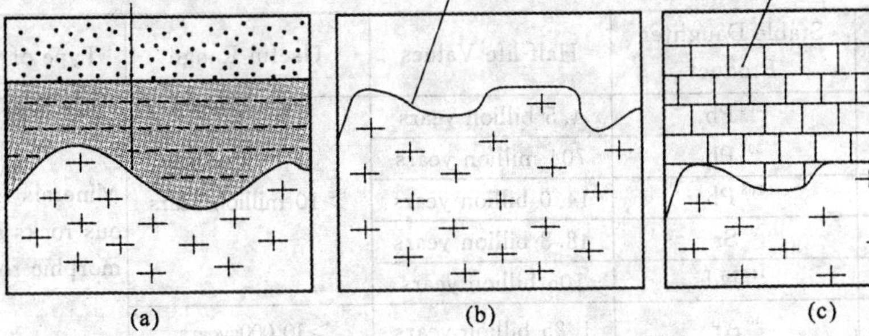

Figure 6.8 Development of nonconformity
(a) Intrusion; (b) Uplift; (c) Subsidence

the case, the term nonconformity is applied to refer to the preserved erosional surface (Figure 6.8).

No matter what the geologic situation, all of the above-introduced principles can be applied reliably to reconstruct the history of various geological events, including depositional, erosional, deformational, and others, for the geology of a region. And by piecing together the reconstructed histories of different regions, or different continents, as much as possible, a regional or global "relative geological time scale" can be constructed. With this relative time scale, we can tell that one event occurred prior to another, regardless of the actual duration of time between them.

5 Radiometric Dating

Radioactive decay is the process in which naturally occurring radioactive materials break down into other materials at known rates.

Dating rocks by radiometric way is simple in theory. Rock is made up of minerals, and mineral is usually made up of at least several different chemical elements. Most of the elements in nature are stable. However, some elements, called radioactive elements, are not completely stable in their natural state. They can decay gradually into other elements at a constant exponential or geometric rate. The original element is called the parent element, and the result of the decay process is called the daughter element. For example, the Carbon 14 is radioactive and decays to Nitrogen 14. In this case, Carbon 14 is the parent and the stable Nitrogen 14 is the daughter. The rate of decay is proportional to the number of parent atoms present. By measuring the ratio of daughter to parent, we can measure how old the sample is. These ages are usually called radiometric ages. The rate of decay is conveniently expressed in terms of an isotope's half-life. A half-life is the time it takes for half of the parent radioactive element to decay to a daughter product. Most radioactive isotopes have rapid rates of decaying (that is, short half-lives) and losing their radioactivity within a few days or years. The half-lives of isotopes used in geology range from thousands of years (^{14}C) to billions of years (U).

The mathematical expression that relates radioactive decay to geologic time is called the age equation and is:

Table 6. 1 Isotopes used to date geologic time

Parent Isotope	Stable Daughter Product	Half-life Values	Useful Range	Type of Material
^{238}U	^{206}Pb	4. 5 billion years	>10 million years	Minerals from igneous rocks and metamorphic rocks
^{235}U	^{207}Pb	704 million years		
^{232}Th	^{208}Pb	14. 0 billion years		
^{87}Rb	^{87}Sr	48. 8 billion years		
^{147}Se	^{143}Nd	106 billion years		
^{40}K	^{40}Ar	1. 25 billion years	>10 000 years	
^{14}C	^{14}N	5 730 years	100~50 000 years	Organic material Cloth

$$t = \frac{1}{\lambda}\ln(1 + \frac{D}{P})$$

where t is the age of the rock or mineral specimen, D is the number of atoms of a daughter product today, P is the number of atoms of the parent isotope today, ln is the natural logarithm, and λ is the appropriate decay constant.

The potassium-argon method and uranium-lead method can be used on older rocks due to their long half-life values (Table 6. 1). Potassium can be found in most rock-forming minerals, such as feldspar and mica. The carbon-nitrogen is a useful and efficient tool in dating events that have taken place only within the past 50 000 years because of the relatively short half-life of ^{14}C.

Radiometric dating of rocks tells how much time has passed since some events occurred. For igneous rocks the event is usually its cooling and hardening from magma or lava. To date a sedimentary rock is relatively very complex because the minerals found in sedimentary rocks are derived from the weathering of igneous rocks. If the sedimentary rocks were dated, the age date would be the time of cooling of the magma that formed the igneous rock. The date would not tell anything about when the sedimentary rock formed.

Radiometric dating provides the absolute age of an appropriate rock, usually expressed in millions or billions of years. It can be used to calibrate the relative time scale constructed by various geological principles. Therefore by combination of relative and absolute time scale, an integrated geological time scale can be constructed.

6 Geologic Time Scale

Geologists measure the age of the Earth and the history of life in ages of millions and even billions of years. The entire history of humankind is but a blink of an eye next to the vastness of geological time. For this reason a special sort of "calendar" is required, one that measures not days, weeks, months or years, but millions and tens of millions of years. This is the Geological Time-Scale.

Think of relative time as physical subdivisions of the rock found in the Earth's stratigraphy, and absolute time as the measurements taken upon those to determine the actual time which has expired. The final time scale is actually an integrated geologic time scale.

6.1 Geologic Time Unit

Geologic time units are divided into eon (宙), era(代)and period(纪). The periods are in turn divided into epochs(世).

The Geological time scale (Table 6.2) is usually represented as a vertical table with oldest at the bottom and youngest at the top — the present day is at the zero mark. The reason for this strange convention is due to the table being a rough and rather symbolic representation of the layers of sedimentary rocks that make up the Earth's crust. The earlier layers were deposited first, the younger ones were on top of them. The geological time scale can also be used to define the major stages in the history of life on the Earth. Often each era ends with a major extinction, which eliminates the dominant life forms of the time and paves the way for newcomers.

Unfortunately for us, the names of the various subdivisions within the scale are a bizarre mix of Greek roots and geographic terms that few of us will recognize. This makes memorization difficult, but memorize it we must. You should also know the age boundaries between the eras.

Periods are the most basic unit of geologic time. Many of their names are derived from Western Europe where much of the initial time scale development was done. It is very helpful to know the origin of the terms used in geological time scale:

- **The Phanerozoic Eon** "revealed life".

Table 6.2 Simplified geological time scale

Eon 宙	Era 代	Period 纪		Date (Ma)
Phanerozoic 显生宙	Cenozoic Kz 新生代	Quaternary 第四纪 Q	Mammals 哺乳动物	2.6
		Neogene 新近纪 N (Late Tertiary,晚第三纪)		23.5
		Paleogene 古近纪 E (Early Tertiary,早第三纪)		65
	Mesozoic 中生代 Mz	Cretaceous 白垩纪 K	Reptiles 爬行类	137
		Jurassic 侏罗纪 J		205
		Triassic 三叠纪 T		250
	Paleozoic 古生代 Pz	Permian 二叠纪 P	Amphibians 两栖类	295
		Carboniferous 石炭纪 C		354
		Devonian 泥盆纪 D	Fishes 鱼类	410
		Silurian 志留纪 S		438
		Ordovician 奥陶纪 O	Invertebrates 无脊椎动物	490
		Cambrian 寒武纪 ∈		543
Sinian 震旦纪 Z				
Precambrian (87% of geologic time scale) 前寒武纪,Algae and simple bacteria.				

* The term"Tertiary"was previously used to refer to the period from the end of the Cretaceous period to the start of the Quaternary period, however the International Commission on Stratigraphy (国际地层委员会) no longer endorses this term as part of the formal stratigraphic nomenclature.

* The age in the table is the bottom boundary for each of the related periods.

• **Paleozoic era** The era of"ancient life"is characterized by fossils of invertebrates, primitive tetrapods, etc; This era witnessed the age of invertebrates, of fish, of tetrapods, reptiles.

• **Mesozoic era** the era of"middle life", by fossils of dinosaurs etc. Lasting little more than half the duration of the Paleozoic, this was a spectacular time.

• **Cenozoic era** the era of"recent life"by mammals and modern plants and invertebrates. This period begins with the extinction of the dinosaurs and the end of the Age of Reptiles. It is sometimes called the Age of Mammals and is divided into the Tertiary and Quaternary.

• **Sinian Period** Sina was an old spelling for China. This term is only used in our country.

• **Cambrian Period** Actually, it is the biggest break point in all of geology. It marks the beginning of the Phanerozoic Eon and the Paleozoic Era. The Cambrian Period was named after the region of Cambria in North Wales, where rocks of this age were first found. The name "Cambria" is a version of Cumbria, a latinisation the Welsh Cymry (= countryman). Precambrian simply means"before Cambrian".

• **Ordovician Period** The name"Ordovician" comes from an ancient tribe that once inhabited the region in Wales where rock strata of this period occur.

• **Silurian Period** after the Silures, an ancient tribe that lived along what is now the Welsh-English border.

• **Devonian Period** The term Devonian was used to describe a sequence of rocks in Devonshire, England.

• **Carboniferous Period** coal bearing, was proposed by the English geologist to designate coal-bearing strata in north-central England.

• **Permian Period** Named after the ancient kingdom of Permia and the present city of Perm near the Ural Mountains.

• **Triassic Period** Originally refers to a set of layers consisting of three distinctive formations in western Germany.

• **Jurassic Period** Originally described massive limestone formations of the Jura Mountains in Switzerland.

• **Cretaceous Period**"chalk-bearing"(from Creta, the Latin word for chalk). The Belgian D'Omalius D'Halloy named the chalk deposits, the"Terrain Cretace", found in the Paris Basin and traceable from England eastward into Sweden and Poland.

• **Tertiary Period** The several geologic eras were originally named Primary, Secondary, Tertiary, and Quaternary. The first two names are no longer used. Tertiary (Third) and Quaternary (Fourth) have been retained but used as period designations.

• **Quaternary Period** See above.

6.2 Rock Stratigraphic Units

Stratigraphy is the study of the stratified rocks (sediment and volcanic) especially their sequence in time, the character of the rocks and the correlation of beds in different localities. Two types of stratigraphic classification are used, one based on physical characteristics or material properties of the rocks — Rock Stratigraphic Units, and the other based on the time over which the material was formed — time stratigraphic units.

A rock-stratigraphic unit, also called Local Stratigraphic Unit, is any distinctive stratum that differs physically from the strata above and below. They are identified by composition, texture, color, or some other observable physical characteristics of rock without considering fossils or relative time.

In stratigraphy, the fundamental unit of rock units is the Formation. Formations can be subdivided into Members, which are in turn divided into Beds, and they may be combined into groups. All formations are named after the geographic location of the best outcrop or a prominent named geographic feature of their typical sections, such as Bantou Formation(板头组), Shaxian Formation (沙县组). The names of rock-stratigraphic units have capital letters, even when they do not have the unit type used in their name: for example, Qixia Limestone (栖霞灰岩), though limestone does not have capital letters when we are just writing about a rock type, it does when we are using it in the name of a stratigraphic unit.

- **Group** (群) The largest unit of rock-stratigraphic sequence, such as Dabie Group (大别群) and Taishan Group (泰山群). Group consists of two or more continuous formations that are genetically related.

- **Formation**(组) As a fundamental unit of a rock-stratigraphic sequence, formation is a collection of similar strata that are sufficiently different from adjacent groups of strata. They are defined in the field, distinguished on the basis of its distinctive physical characteristics, and named for a geographic locality. They are recognizable units with distinctive upper and lower bounds that can be used for geologic mapping (地质填图) over a wide area. Example is Qixia Formation (栖霞组) in Nanjing. The formation itself is without any time implications.

- **Member** (段) Member is a distinctive bed or set of beds within a formation. It is a distinct part of a formation. They can be any thickness or extent.

- **Bed**(层) The smallest rock stratigraphic unit is called a bed. It is typically a few centimeters to few meters thick and must be distinguishable from layer above and below it.

6.3 Time Stratigraphic Unit

The time-stratigraphic unit, also called time-rock units, is the classification of sediments or rocks deposited during a particular time interval. The boundaries of time-stratigraphic unit may be considered on the basis of fossils and other features. These units may or may not coincide with rock units, such as Formations, since they are dependent on the presence of unique clusters of fossils being contained in the units in question. Rock stratigraphic unit may not have been deposited at the same time everywhere it is found. That is why we need a second unit, the time-stratigraphic unit.

The classification scheme used terminology is as follows:

The fundamental unit of the sequence of rocks deposited within a specified period of time is a system. The interval of absolute time during which a system of rocks was deposited is known as a period; it is the fundamental unit of geologic time.

For example, the Jurassic Period is a time unit, but the rocks deposited during that time are called the Jurassic System, a time-

Table 6.3 Time Units & Time-rock Units

Time Unit	Example	Time-rock Unit	Example
Eon(宙)	Phanerozoic	Eothem(宇)	Phanerozoic
Era(代)	Mesozoic	Erathem(界)	Mesozoic
Periods(纪)	Cretaceous	System(系)	Cretaceous
Epoch(世)	Early Cretaceous	Series(统)	Lower Cretaceous

rock unit. The Cambrian Period was the time during which the Cambrian time stratigraphic units were laid down. When referring to rocks, we use time-rock units, such as Lower, Middle and Upper or informally, lower, middle, and upper. When referring to geologic time, we use geologic time units — Early, Middle and Late or informally, early, middle and late.

Although rock layers and geologic time are obviously linked (the former is the basis for the later), it should be remembered that rocks are real, physical entities whereas time is an abstract concept. This is why we have two different systems of units: Geologic time units (divisions of time) and time-stratigraphic units (divisions of rock layers).

1. Answer the Following Questions

(1) You must be very familiar with the following table in Chinese.

全球性		地方性
地质年代单位	年代地层单位	岩石地层单位
宙	宇	
代	界	群
纪	系	组
世	统	段
期	阶	

(2) What is the principle of superposition?
(3) What is the principle of cross—cutting relationships?
(4) What is the principle of original horizontality?
(5) What is an unconformity?
(6) Memorize the Geological Time Scale.
(7) Differentiate the Formation, Period and System.
(8) What is the rock stratigraphic unit?
(9) What is the time—stratigraphic unit?
(10) What is the geologic time unit?

2. Select the Best Answer from the Choices Given

(1) Outcrops exposed at different locations belong to same layers.
A: Principle of superposition
B: Principle of lateral continuity
C: Principle of crosscutting relationships
D: Principle of original horizontality

(2) The tilted layers are the result of deformation?
A: Principle of superposition
B: Principle of lateral continuity
C: Principle of crosscutting relationships
D: Principle of original horizontality

(3) The displaced rocks remind you of the ______.
A: Principle of superposition
B: Principle of lateral continuity
C: Principle of crosscutting relationships
D: Principle of original horizontality

(4) An index fossil has the characteristics of ______.
A: Existing for a rather short period of time
B: distinct appearance
C: Large geographic distribution
D: All of above

(5) The names Cenozoic, Mesozoic and Paleozoic refer to geological ______.
A: Period B: Formation
C: Era D: Eon

(6) Placing rocks in their proper sequence of formation is called ______.
A: Relative dating
B: Absolute dating
C: Principle of superposition
D: None of above

(7) Formation is a ______.
A: Time—stratigraphic unit
B: Rock Stratigraphic Unit
C: Geologic time unit
D: None of above

(8) The smallest rock stratigraphic unit is ______.
A: Period B: Formation
C: Bed D: Member

(9) The ______ was an era dominated by the dinosaurs.
A: Cenozoic B: Mesozoic
C: Paleozoic D: Precambrian

(10) The first 85% of the history of the Earth is referred to as ______.
A: Cenozoic B: Mesozoic
C: Paleozoic D: Precambrian

(11) The word "Jurassic" in the phrase "The Jurassic Park" is used as ______.
A: Time—stratigraphic unit
B: Rock stratigraphic unit
C: Geologic time unit
D: None of above

(12) The youngest period in the geological time scale is ______.
A: Quaternary B: Tertiary
C: Cambrian D: Jurassic

(13) The erosional surface separating granite from overlying shales is called ______.
A: Angular unconformity
B: Disconformity
C: Nonconformity
D: None of above

(14) Which period marks the beginning of the Phan-

erozoic Eon?

A: Quaternary B: Tertiary

C: Cambrian D: Jurassic

(15) The ______ Period marked the beginning of major changes that were to take place throughout the Mesozoic Era.

A: Triassic B: Tertiary

C: Cambrian D: Jurassic

(16) Group is divided into ______.

A: Period B: Formation

C: Bed D: System

(17) Which isotope is frequently used to date archaeological artifacts?

A: ^{40}K B: ^{14}C

C: ^{232}Th D: ^{235}U

(18) Which isotope has the shortest half-life?

A: ^{40}K B: ^{14}C

C: ^{235}U D: ^{238}U

(19) Which one is the correct name?

A: Bantou Formation

B: Bantou formation

C: The bantou formation

D: All of above

(20) The eroded surface in angular unconformity is parallel to ______.

A: Underlying beds

B: Overlying beds

C: None of above

(21) Fossils are the most common in which rock types?

A: Igneous B: Metamorphic

C: Sedimentary D: All of above

Chapter 7 Earthquakes

Key words

Epicenter	Collapse Earthquakes	Tsunami	Surface Wave	Aftershocks	Seimogram
震中	塌陷地震	海啸	面波	余震	地震波谱
Focus	Seismograph	Longitudinal wave, compressional waves, P wave			
震源	地震仪	纵波、压缩波、P波			
Intensity	Seismic waves	Transverse or shear waves, S wave			
烈度	地震波	横波、剪切波、S波			
Richter Scale	Volcanic Earthquakes	Isoseismal map		Travel-time graph	
里氏震级	火山地震	等震线图		(地震)走时线	
Magnitude	Tectonic Earthquake	Elastic rebound theory		Intraplate earthquakes	
震级	构造地震	弹性回跳说		板内地震	

Earthquake is a kind of internal geological process. The seismic waves produced by earthquake can destroy the ground buildings and kill human beings. On the other hand, because the seismic waves can reflect and refract on the various geological boundaries with the Earth's interior, it can provide us valuable information about the internal architecture of the Earth, thus being an extremely important way to study the Earth's deep structure and mineral composition. Study of the earthquake is also of great importance for constructing dams and tunnels.

1 What is Earthquake?

Earthquake is a trembling or shaking of the ground caused by a sudden release of energy stored in the rocks beneath the Earth's surface.

An earthquake can be likened to the effect observed when a stone is thrown into water. After the stone hits the water, a series of concentric waves will move outwards from the center. The same events occur in an earthquake. There is a sudden movement within the crust or mantle, and concentric shock waves move out from that point. Geologists and geographers call the origin of the earthquake the focus (Figure 7.1). The focus is the place within the Earth where the rock breaks, producing an earthquake. This is the actual location where fault movement begins. Almost all of the time, the focus of an earthquake is below the surface. Since this is often deep below the surface and difficult to map, the location of the earthquake is often referred to as the point on the Earth's surface directly above the focus. This point is called the epicenter. The epicenter of an earthquake is the point on the ground's surface directly above the focus and is the location normally reported in the news or shown on maps. Or the place

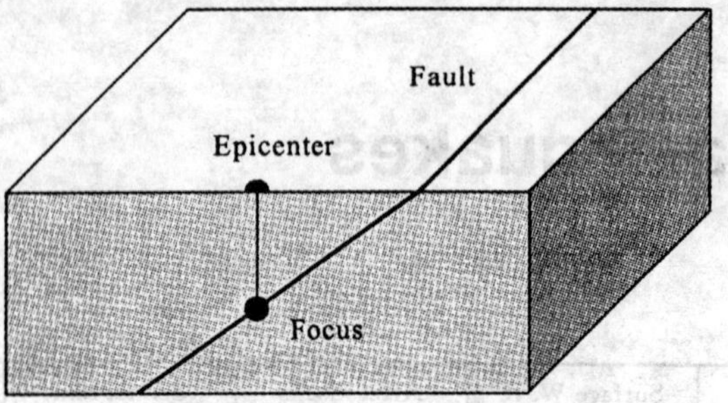

Figure 7.1 Epicenter and focus

on the Earth closest to the focus is the Epicenter.

Among the natural calamities, earthquakes are the most destructive, in terms of loss of life and destruction of property (Table 7.1). The hazard of earthquakes varies from place to place, dependent upon the regional and local geology.

It is estimated that there are 5 million earthquakes in the world each year. Many of these earthquakes go undetected because they occur in remote areas or have very small magnitudes.

Here is a table (Table 7.2) of frequency of occurrence of earthquakes based on observations from 1900 to 1946. From this table, you can find that the most destructive earthquakes (higher than 8 magnitude) will occur once a year.

2 Types of Earthquake

An earthquake is the shaking of the Earth. The shaking of the Earth can be caused by many reasons. According to the cause of shaking, earthquake can be classified into following types:

Table 7.1 Some destructive earthquakes since 1899

Date	Location	Magnitude	Approximate Toll In Human Lives
May, 1960	Chile	8.9	5 700
September, 1923	Kwanto, Japan	8.3	27 000
December, 1920	Gansu, China	8.5	100 000
July 1976	Tangshan, China	7.8	200 000

Table 7.2 Frequency of occurrence of earthquakes from 1900 to 1946

(Modified After Bangdong Xia)

Descriptor	Magnitude	Annual average
Great	7.7～8.6	2
Major	7.0～7.7	12
Strong	6.0～7.0	108
Moderate	5.0～6.0	800
Light	4.0～5.0	6 200
Minor	3.0～4.0	49 000
Very minor	2.5～3.0	100 000

2.1 Volcanic Earthquakes

Volcanic earthquakes are those directly associated with volcanic activity, which occur in conjunction with tectonic movement.

Most of Hawaii's earthquakes are the direct result of volcanic activity. Magma moving beneath the Earth's surface can exert extra forces against the country rocks, and thus results in the shaking or fracturing of ground surface. Earthquakes may occur before or during an eruption, or they may result from the underground movement of magma that comes close to the surface but does not erupt.

Volcanic earthquakes account for about 7% of total earthquakes.

2.2 Collapse Earthquakes

In a limestone region, large caverns will form due to the dissolution of groundwater. Once the roof of the cavity cannot support itself, collapse occurs. Sometimes, large-scale collapse can result in the vibration of neighboring regions.

2.3 Tectonic Earthquake

Tectonics means large-scale deformation of the outer part of the Earth resulting from forces in the Earth. Tectonic earthquake is that occurs when the Earth's crust breaks due to geological forces on rocks and adjoining plates that cause physical and chemical changes.

In 1911 Harry Fielding Reid, after studying the effects of the 1906 California earthquake, proposed the elastic rebound theory to explain the generation of certain earthquakes. This theory states that when the force applied to rocks becomes too great then the rocks separate and an earthquake results. So, the earthquake only happens in rocks that have been stressed. This stress is stored until the strength of the rock is exceeded. The actual break then releases the energy which travels in the form of waves. The figure 7.2 shows how the rock is gradually distorted in response to forces, eventually leading to sudden slip or displacement along the fault — what we call an earthqu -

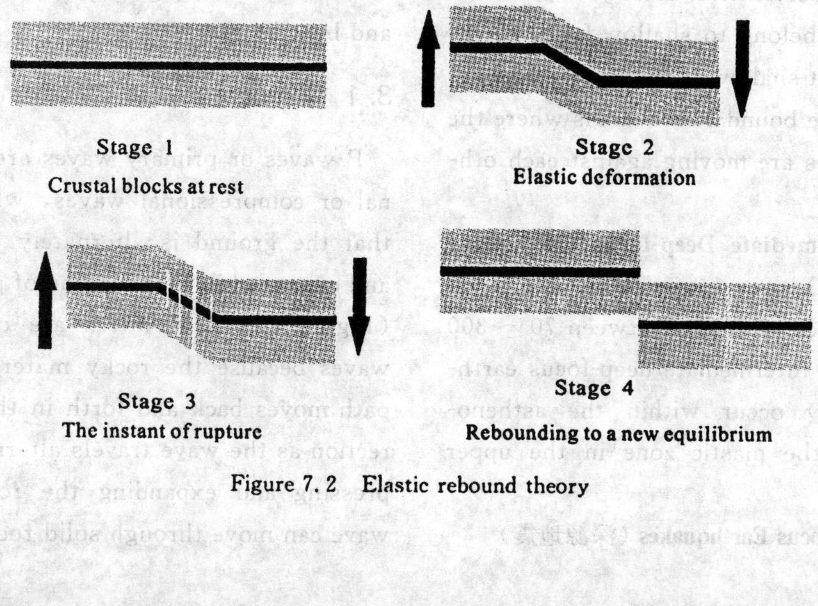

Figure 7.2 Elastic rebound theory

ake.

Stage 1: the original undeformed rocks.

Stage 2: rocks are gradually deformed.

Stage 3: rupture starts at weak point on the fault and then spreads quickly to other points of fault. The fault slip radiate waves out along the rupture surface, and thus causes vibrations of surrounding rocks. Aftershocks (smaller earthquakes that occur after the initial earthquake) are the result of further slips of the main fault. The "focus" of the earthquake is actually the point on the fault at which the displacement begins.

Stage 4: After the release of energy, rocks rebound to a new equilibrium, similar to stage 1.

The earthquakes can be classified by depth of focus:

(1) Shallow-focus Earthquakes (浅源地震)

They are earthquakes with depths of focus between 0～70 km. Thus they originate within the lithosphere. Shallow-focus earthquakes are much more common than deep-focus ones, and they are usually larger, which in turn makes them more dangerous and destructive. Nearly 90% of all earthquakes belong to shallow-focus earthquakes. Most shallow-focus earthquakes occur near plate boundaries — areas where the Earth's plates are moving against each other.

(2) Intermediate Deep-focus Earthquake (中深源地震)

Earthquakes with foci between 70 ～300 km deep are intermediate deep-focus earthquake. They occur within the asthenosphere, or the plastic zone in the upper mantle.

(3) Deep-focus Earthquakes (深源地震)

The foci of deep-focus earthquakes will be deeper than 300 km. Deep-focus earthquakes constitute only about 3% of all recorded quakes. The greatest depth of focus known is 700 km. Deep-focus earthquakes originate from the mantle lying below the plastic zone.

3 Earthquake Wave

The sudden movement of rocks along a fault causes vibrations that transmit energy through the Earth in the form of waves. A seismic wave is an elastic wave generated by an impulse such as an earthquake or an explosion. The two main types of seismic waves are body waves and surface waves. Body waves can travel through the Earth's inner layers, but surface waves can only move along the surface of the planet like ripples on water. Body waves can be classified into P waves and S waves. Both P and S waves help seismologists to locate the focus and epicenter of an earthquake. As P and S waves move through the interior of the Earth, they are reflected and refracted, or bent, just as light waves are reflected and bent by glass.

3.1 P wave

P waves or primary waves are longitudinal or compressional waves, which means that the ground is alternately compressed and expanded in the direction of propagation (Figure 7.3). P waves are compression waves because the rocky material in their path moves back and forth in the same direction as the wave travels alternately compressing and expanding the rock. The P wave can move through solid rock and flui-

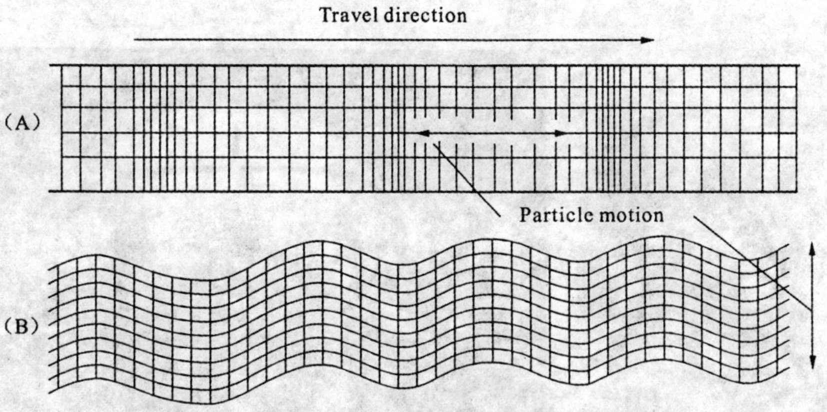

Figure 7. 3 A: P wave or shear wave
B: S wave or compressional wave

ds, like water or the liquid layers of the Earth. It pushes and pulls the rock. It moves through just like sound waves push and pull the air. This is the fastest kind of seismic wave. These waves generally travel twice as fast as S waves and can travel through any type of material. Typical speeds are 0. 330 km/s in air, 1. 45 km/s in water and about 5 km/s in granite.

3. 2 S wave

S waves or secondary waves are the second waves you feel in an earthquake. Once the P waves have begun radiating out from the focus of the rupturing fault, S waves follow. They are transverse or shear waves, which means that the ground is displaced perpendicularly to the direction of propagation. The S waves sheer or twist the rocks by moving up and down or side-to-side, perpendicular to the direction that the waves are traveling in, rather than compress it. The side-to-side motion of S waves causes the most damage to structures during an earthquake. S waves can travel only through solids. Their speed is about 58% of that of P waves in a given material.

3. 3 Surface Wave

Surface Waves, often referred to as L waves or long waves, are seismic waves traveling along the outer surface of the Earth at slower speeds, which are analogous to water waves traveling along the surface of a body of water. They travel more slowly than body waves. Because of their low frequency, they are more likely than body waves to stimulate resonance in buildings, and are therefore the most destructive type of seismic wave.

The various seismic waves travel through the Earth and bounce off different features of the Earth's interior. The patterns they form after bouncing off these features can be used to infer the interior structure of the Earth.

4 Seismograph

Seismograph is a device that automatically detects and records the intensity, direction, and duration of a movement of the ground, especially of an earthquake.

One of the earliest instruments used to

Figure 7.4 Zhang Heng and his seismograph

(From http://www.kepu.com.cn/english/quake/relic/rlc04.html)

measure earthquakes was invented by a Chinese scientist Zhang Heng, in AD 132～1 600 years before anyone in the West did (Figure 7.4). Zhang's invention was a large (2m in diameter) bronze dragon jar. The cylindrical jar had eight dragonheads gazing outward in eight principal directions, each with a ball in its mouth. Around the base of the jar, directly under each dragon, sat a frog with his mouth open to receive a ball from the dragon's mouth. When an earthquake happened, a ball dropped from a dragon's mouth and was caught by the frog's mouth. The direction of the earthquake was indicated by which of the dragon heads had dropped a ball.

A typical modern seismograph works in a very simple way. Basically, a seismograph is a simple mass (pendulum) freely suspended from a frame attached to the ground (Figure 7.5). Thanks to its inertia, the mass is reasonably independent of the frame's motion. This mass is free to swing from side to side if the Earth shakes. One end of pendulum holds a pen. In addition, a sheet of paper is wrapped around a slowly revolving drum. As time passes, the drum rotates so that the paper is marked by moving pen.

If there is no shaking, a straight line will be recorded on the moving paper. If the ground shakes , however , the pendulum

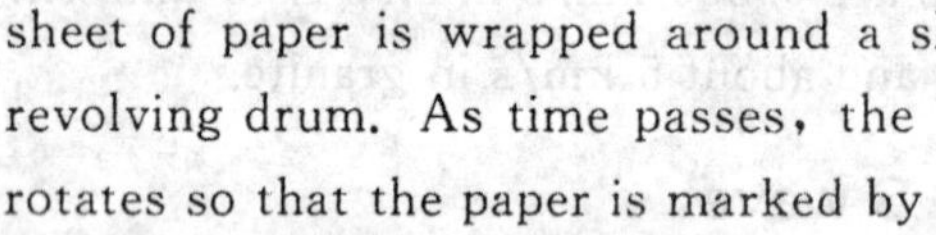

Figure 7.5 Seismographs

swings and so the pen draws a zigzag line as the paper turns.

The zigzag line recorded on the paper during earthquake is called seismogram (Figure 7.6). In other words, a seismogram is the paper record of seismic waves over time. Today, earthquakes are not recorded straight onto paper; they are turned into an electric signal first. Sensitive seismographs, which greatly magnify these ground motions, can detect strong earthquakes from sources anywhere in the world. The location, time and magnitude of an earthquake can be determined from the data recorded by seismic stations.

The P waves will usually be the first ones that are recorded on seismograph records because they are the fastest seismic waves. The next ones will be the S waves. The last seismic waves recorded by seismogram are surface waves. They are generally often larger waves marked on the seismogram.

Seismologist can use the seismogram to determine the location of the epicenter of an earthquake. To figure out just where that earthquake happened, at least three seismographs recorded at different stations for the same earthquake are required. The steps for determining the location of the epicenter of an earthquake are summarized as following:

(1) Measure elapsed time between the arrival of the first P wave and the first S wave.

This can be read directly from the seismogram of a seismograph station.

(2) Use the travel-time graph to determine the distance from a seismograph station to the earthquake epicenter.

Travel-time graph is a graph in which the distance is plot against seismic wave arrival time (Figure 7.7). From the travel-time graph we can find that the time interval between the first arrivals of P and S waves increases with distances from the focus of an earthquake. The greater the difference in time of arrival for the P and S waves, the farther away an earthquake occurred from a seismic station. So, the time interval between the first arrivals of P and S waves can be used to determine the distance from seismograph station to the epicenter of an earthquake.

(3) Draw circle around the station on a map with radius of that distance.

The epicenter can be anywhere on this circles. So in order to determine exactly where the epicenter is, more seismograms from different seismograph station are needed.

(4) Repeat the above steps for station B and station C.

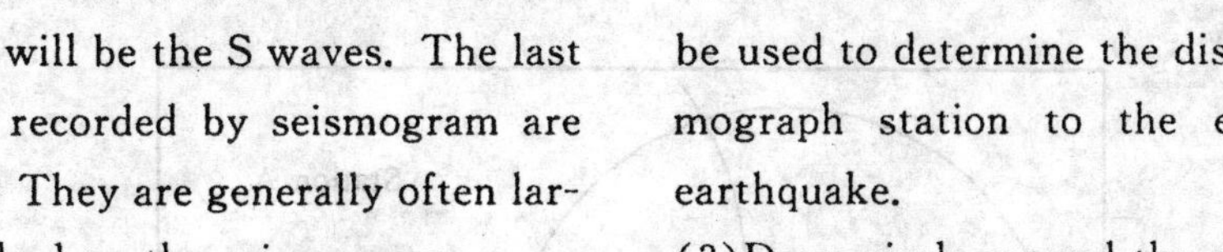

Figure 7.6 Seismogram

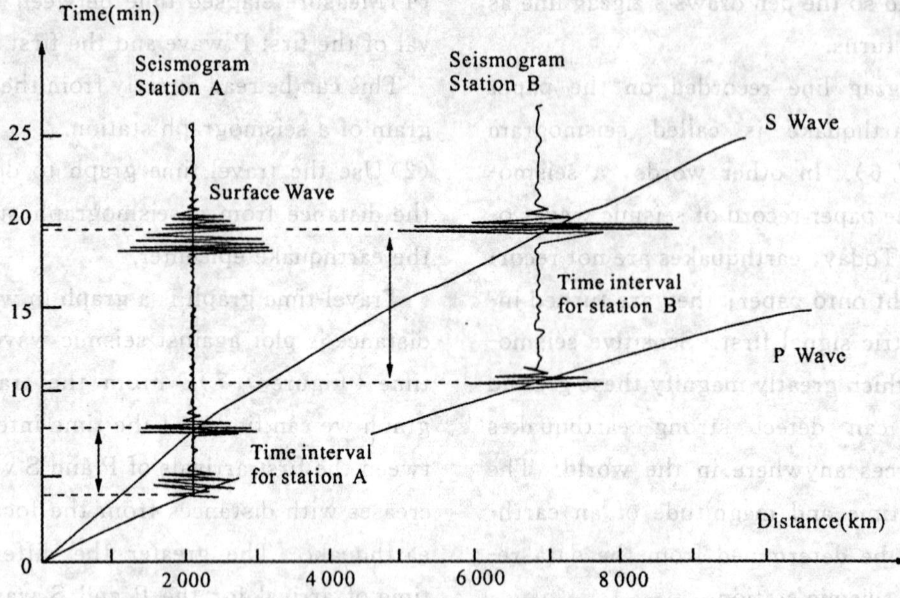

Figure 7. 7 Travel－time graph.
Horizontal axis is the distances from epicenter and vertical axis is the travel time since the earthquake occurred

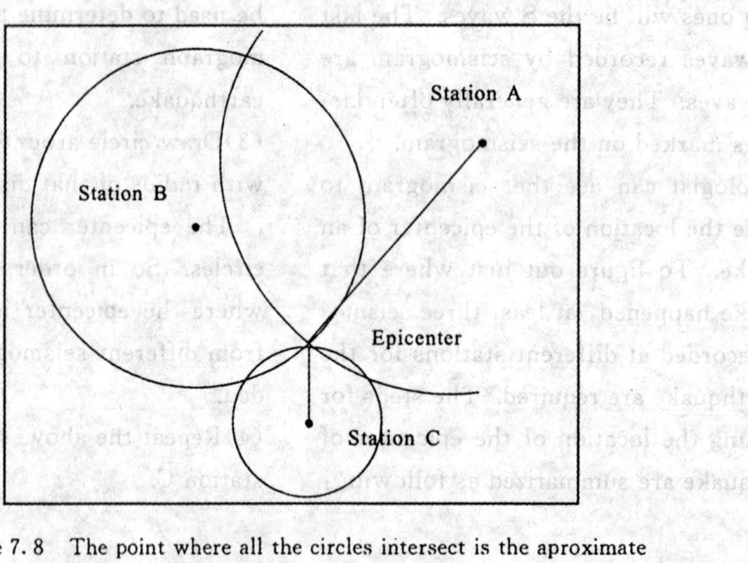

Figure 7. 8 The point where all the circles intersect is the aproximate epicenter of the earthquake

Finally three circles around different stations will be drawn (Figure 7. 8). All of the circles should overlap. The point where three circles meet is the approximate epicenter of the earthquake.

This method does not tell you how deep the earthquake was. To determine the focus requires more information. Accurate determination of the time intervals between arrivals of P, S, and surface waves, enable us to calculate depth of focus for every earthquake. From study of seismograms, seismologists can detect most underground nuclear explosions.

5 Magnitude and Intensity of Earthquake

The severity or the size of an earthquake can be expressed in several ways. The magnitude of an earthquake, usually expressed by the Richter scale, is a measure of the energy released in an earthquake. Each earthquake has a unique magnitude assigned to it. The intensity, as expressed by the modified Mercalli Scale, is a subjective measure that describes how strong a shock was felt at a particular location.

5.1 Magnitude

Magnitude is a measure of the amplitude of the seismic waves. The magnitude of an earthquake is most commonly measured through the use of the Richter Magnitude Scale, which was developed in 1935 by the American seismologist Charles F. Richter. The equation for Richter Magnitude is:

$M = \lg A + (\text{Distance correction factor} = 3\lg 8\Delta t - 2.92)$

M is the magnitude and A is the amplitude of an earthquake, in millimeters, measured directly from the seismograph. Thus after you measure the wave amplitude (Figure 7.9) you have to take its logarithm, and scale it according to the distance of the seismograph from the epicenter of earthquake, estimated by the S-P time difference. The S-P time, in seconds, makes Δt.

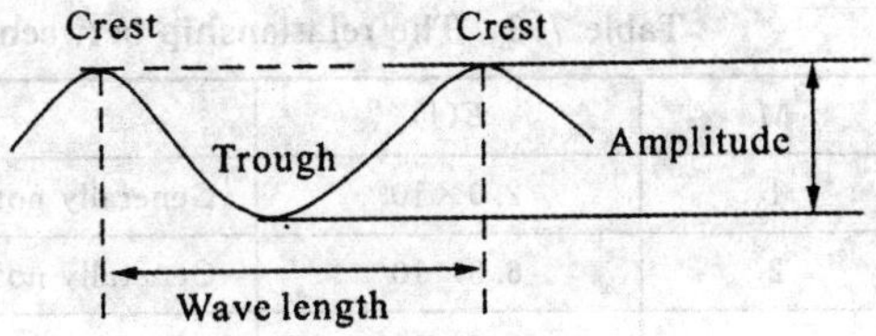

Figure 7.9 Amplitude

The Richter scale is logarithmic, meaning that whole-number jumps indicate a tenfold increase. In this case, the increase is in wave amplitude. That is, the wave amplitude in a level 4 earthquake is 10 times greater than in a level 3 earthquake, and the amplitude increases 100 times between a level 5 earthquake and a level 7 earthquake. Although the Richter scale has no upper limit, the largest known shocks have had magnitudes in the 8.8 to 8.9 ranges.

The Richter scale does not measure energy directly. The relationship between magnitude and energy is:

$$\lg E = 11.8 + 1.5M$$

Giving the energy E in ergs(尔格) from the magnitude M. Note that E is only the amount of energy radiated from the earthquake as seismic waves. For every unit increase in magnitude on the Richter scale, there is roughly a thirty-fold increase in the energy released by an earthquake. That is, each whole number step in the magnitude scale corresponds to the release of about 32 times more energy than the amount associated with the preceding whole number value. Or a tenfold increase in the size of the Earth vibrations is caused by an increase of about 32 times in terms of energy. A magnitude 6 quake is almost 1 000 times more powerful in terms of energy released than a magnitude 4 earthquake. A magnitude 8.6 earthquake releases energy equivalent to about 10 000 atom bombs of the kind that destroyed Hiroshima (日本广岛) in World War Ⅱ (Table 7.3).

Table 7.3 The relationship between magnitude and energy and earthquake severity

M	E(J)	Earthquake Severity
1	2.0×10^{6}	Generally not felt
2	6.3×10^{7}	Generally not felt
3	2.0×10^{9}	Often felt, but recorded. No damage
4	6.3×10^{10}	Often felt, but rarely causes damage
5	2.0×10^{12}	Felt widely, slight damage to buildings near epicenter
6	6.3×10^{13}	Can be destructive
7	2.0×10^{15}	Major earthquake. Can cause serious damage over larger areas
8	6.3×10^{16}	Great earthquake can cause serious damage and loss of life in areas several hundred kilometers across
9	2.0×10^{19}	areas several hundred kilometers across

5.2 Intensity

Although each earthquake has a unique magnitude, its effects will vary greatly according to ground conditions, construction standards, distance, and other factors. Seismologists use a different intensity scale to express the variable effects of an earthquake.

Intensities are measured by means of the degree of damage to structures, the amount of disturbance to the ground surface, and the extent of human and animal reactions to the shaking. Intensity does not have a mathematical basis and is determined only by observed effects. You shouldn't confuse these two terms: intensity is based on the observed effects of an earthquake on its surroundings, whereas magnitude measures the amount of seismic energy released.

One of the widely used intensity scale is the Mercalli Intensity scale, which is a semi-quantitative linear scale to measure the intensity of earthquakes by examining the damage rather than measure their magnitude. It was conceived by Italian volcanologist Giuseppe Mercalli in 1902. Mercalli intensity denotes how strongly an earthquake affects a specific place and has 12 divisions using Roman numerals from I to Ⅻ. In order to better fit modern conditions, some modifications have be made to the original Mercalli intensity scale. Following Table 7.4 is one of the examples of modified Mercalli Intensity scale commonly used in China.

The modified Mercalli Scale is a good way of saying how bad it was at some particular location, but it isn't a good way of saying how strong the earthquake was in general. For example, intensity depends a great deal on the nature of the ground. An earthquake's intensity will be much greater in a town built on fill than in one built on granite. On the other hand, different places has different intensity scales for a particular earthquake. Earthquakes of large magnitude do not necessarily cause the most intense surface effects. The effect in a given region depends to a large degree on local surface and subsurface geologic conditions.

So intensity experienced at or near the epicenter of the earthquake is totally different from that experienced far away from the epicenter. Isoseisms are any line on a map

Table 7. 4 Modified Mercalli intensity scale (modified after Xie Yushou, 1957)

Intensity	Summary	Full Description
Ⅰ	Not felt	Not felt at all. Can only be recorded by instruments
Ⅱ		A few people might notice movement if they are at rest.
Ⅲ	Felt indoors	Felt quite noticeably by persons indoors. Hanging objects swing back and forth
Ⅳ		Most people indoors and a few people outdoors feel movement. At night, some awakened. Dishes, windows, and doors rattle
Ⅴ		Felt by nearly everyone; many sleeping people are awakened. Some dishes, windows and doors are broken. Pendulum clocks may stop. Small objects are turned over. Plaster in walls might crack
Ⅵ	Felt by all	Poor-built buildings suffer slight damage. Some destroyed. Sometimes, ruptures can be found in the wet soil. Occasionally, large-scale landslides occur
Ⅶ		Damage is slight to moderate in well-built buildings; considerable in poorly built buildings. Some pagodas or chimneys slightly damaged
Ⅷ		Well-built buildings suffer slight damage. Poorly built structures suffer severe damage. Tall structures such as towers, chimneys, monuments, walls might twist and fall
Ⅸ		Panic is general. Well-built buildings suffer considerable damage. Great in substantial buildings, with partial collapse. Buildings are shifted off foundations
Ⅹ		Some well-built structures destroyed. Ground fractures can extend a few kilometers. Some solid rocks are fractured
Ⅺ		Most buildings collapse. Landslide and lumps are common in mountainous regions. Large cracks appear in the ground. Underground pipelines, bridges and railroad tracks are badly bent or destroyed
Ⅻ		Virtually total destruction

connecting places having the same value of intensity (Figure 7. 10). We can construct the isoseismal map according to the intensity scales in different places. Due to the different ground conditions, the isoseisms often take a unregular wavy pattern.

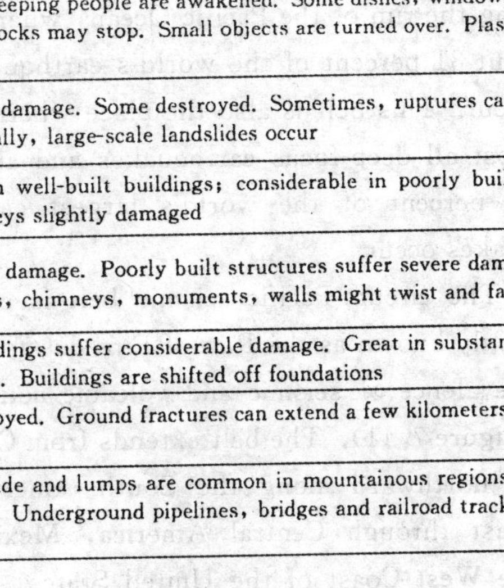

Figure 7. 10 Irregular patterns of isoseisms

6 World Wide Distributions of Earthquake

Although earthquakes can occur anywhere on the Earth, most earthquakes (>90%) occur where tectonic plates move against one another. The largest events usually happen where two plates are colliding — This is where large amounts of stress can build up rapidly. Intraplate earthquakes(板内地震) occur less commonly and do not follow recognizable patterns. They take place in the relatively stable interior of continents, away from plate boundaries. This type of earthquake generally originates at more shallow levels. Less than 10 percent of all earthquakes occur within plate interiors.

6.1 The Circum-Pacific Seismic Belt(环太平洋地震带)

The world's greatest earthquake zone, the circum-Pacific seismic belt, is found along the rim of the Pacific Ocean, where about 81 percent of the world's earthquakes occur. This belt is also the place where almost all deep-focus earthquakes and about 90 percent of the world's largest earthquakes occur.

The circum-Pacific seismic belt is commonly known as the Rim of Fire due to the prevalence of seismic and volcanic activity (Figure 7.11). The belt extends from Chile, northward along the South American coast through Central America, Mexico, the West Coast of the United States, and the southern part of Alaska, through the Aleutian Islands to Japan, the Philippine Islands, New Guinea, the island groups of the Southwest Pacific, and to New Zealand.

6.2 The Mediterranean-Himalayan Belt(地中海—喜玛拉雅地震带)

The next most seismic region, the Mediterranean-Himalayan belt, extends from Mediterranean region, eastward through Turkey, Iran, and Himalayas. This belt accounts for about 17 percent of the world's largest earthquakes. Shallow and intermediate-focus earthquaks are prevalent in this belt.

6.3 The Oceanic Ridges Belt(洋中脊地震带)

The oceanic ridge is the third most likely area where quakes occur, and covers the region of from the north-south axis of the Atlantic to the Arctic Ocean to the southernmost tip of Africa. In such areas the lithosphere is in a state of tensional stress and

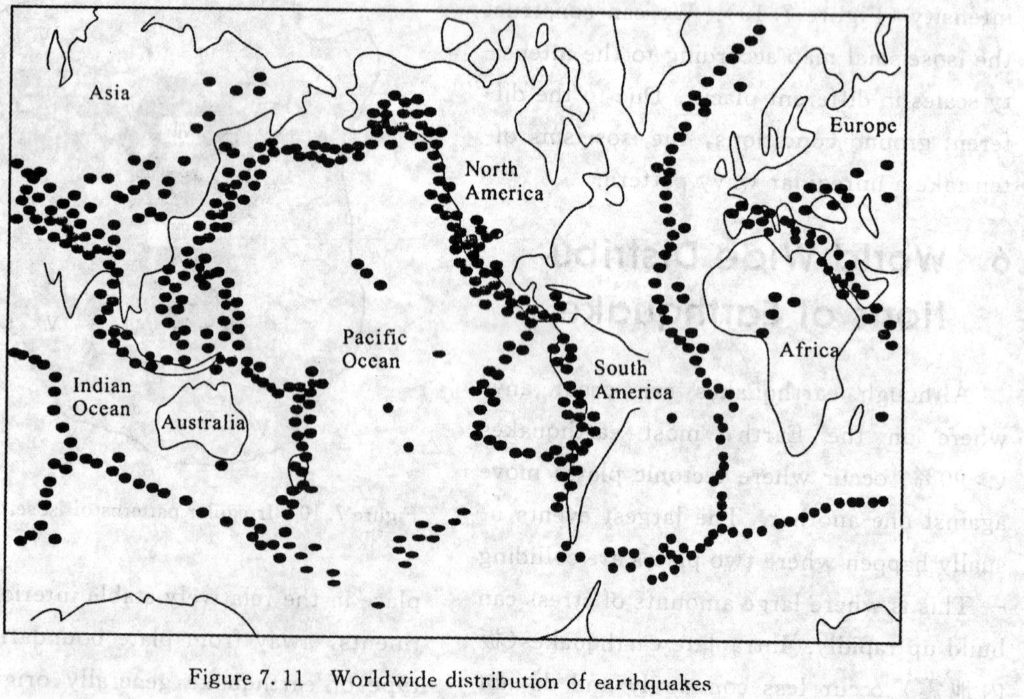

Figure 7.11 Worldwide distribution of earthquakes

thus normal faults and rift valleys occur. Earthquakes that occur along such oceanic ridges show normal fault movement and tend to be shallow focus earthquakes, with focal depths less than about 20 km.

The remaining shocks are scattered in various areas of the world.

7 Effects of Earthquakes

Earthquakes produce various damaging effects to the areas they act upon. The effects of earthquakes at a given location depend upon a number of widely varying factors. These factors include magnitude, distance from focus, building construction and types of underlying soil or rock.

The earthquake's effects can be classified into two types: primary and secondary.

7.1 Primary Effects

Primary effects are permanent features produced by the earthquake. They are the direct results of ground movement. Examples of primary effects include ground ruptures, fault scarps, and offsets of bridges, pipes and human-constructed objects (Figure 7.12, Figure7.13).

Figure 7.12 Devastation of freeway structure
Image credit: U.S. Geological Survey. Public domain

Figure 7.13 Road damage following earthquake
Image credit: NASA Marshall Space Flight Center (NASA-MSFC). Public domain

7. 2 Secondary Effects

Secondary effects result from propagation of seismic waves away from the fault rupture. These effects are not directly caused by fault movement. Examples include fire, landslide and tsunami. Most of the damage done by earthquakes is due to their secondary effects.

A tsunami, or seismic sea wave, is a series of waves generated by any underwater disturbance. Tsunami is from the Japanese word for harbor wave. The possible causes include an earthquake, volcanic eruption, landslide and meteorite impact, of which the undersea earthquake is the most common cause.

Sudden large scale vertical movements of seafloor can displace the overlying water, thus generating tsunamis. Such large vertical movements of the Earth's crust can occur at plate boundaries (Figure 7. 14).

Tsunami usually has much smaller amplitude (typically 30～60 cm) and a very long wavelength (in excess of 100 km). That is why tsunamis are harmless in the open ocean. However, as they run onto shallow water regions and pass into continental coasts, their speed diminishes which results

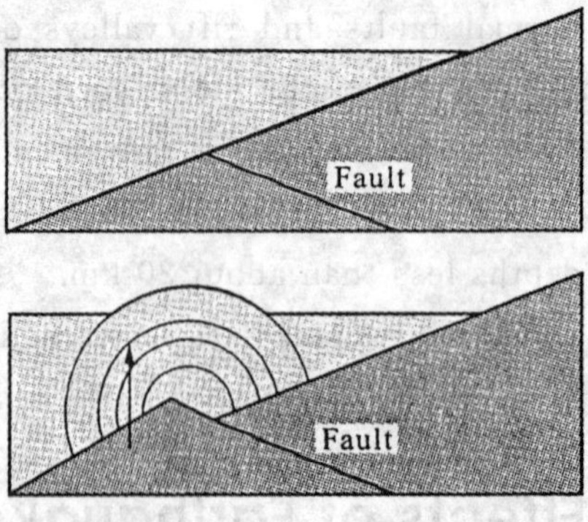

Figure 7. 14 Tsunami is generated by vertical movement of seafloor

in increase in the wave height in order to conserve the total energy. The towering walls of water smash into the shore and can cause a lot of destruction on the shore.

It is a misconception that a tsunami is a normal wave only much bigger. Tsunamis are often no taller than normal wind waves, but they are much more dangerous. Tsunami is more destructive than the normal wind generated waves beause they behave in totally different ways when they come ashore. Wind waves come and go without flooding higher areas. But the tsunami surge over low-lying coastal areas as a wall of water due to its long wavelengths (Figure 7. 15).

On the 26th December 2004, the Indian coastline experienced the most devastating tsunami in recorded history . The tsunami

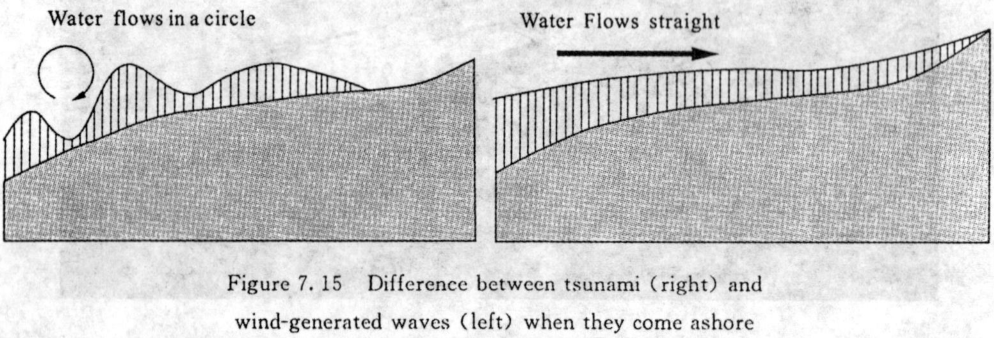

Figure 7. 15 Difference between tsunami (right) and wind-generated waves (left) when they come ashore

was triggered by an earthquake of magnitude 9.0 on the Richter scale off the coast of Sumatra (苏门答腊) in the Indonesian Archepelago. The massive tsunamis obliterated cities, seaside communities and holiday resorts, killing tens of thousands of people in a dozen countries.

8 Prediction and Prevention of Earthquake

Earthquakes are probably the most frightening naturally occurring hazard encountered because earthquakes typically occur with little warning. They may occur at any time of day or on any day of the year. Due to earthquakes's devastating potential, there is great interest in predicting earthquakes. The goal of earthquake prediction (预报) is to give warning of potentially damaging earthquakes early enough to allow appropriate response to the disaster, enabling people to minimize loss of life and property. Although a great deal is known about where earthquakes are likely, there is currently no reliable way to predict the days or months when an event will occur in any specific location.

One well-known successful earthquake prediction was for the Haicheng (海城) earthquake of 1975, when an evacuation warning was issued the day before an earthquake with magnitude 7.3. In the preceding months changes in land elevation and in ground water levels, widespread reports of peculiar animal behavior, and many foreshocks had led to a lower-level warning. An increase in foreshock activity triggered the evacuation warning. Unfortunately, most earthquakes do not have such obvious precursors.

Earthquake prediction is usually defined as the specification of the time, location, and magnitude of a future earthquake within stated limits.

Relatively speaking, it is easy to determine location, and magnitude of a future earthquake. Since earthquakes are usually associated with faulting, any region containing active faults is potentially dangerous. On the other hand, we can decide where and how the large shocks will occur by studying the past frequency of large earthquakes in a specific area.

It is very difficult to predict the exact time when a damaging earthquake will occur. At present, the prediction of earthquake time has three different types: long-term (decades) prediction, intermediate-term (years) prediction, and short-term (days) prediction. The former two is mainly based on the scientific study on the geological condition. The last one, short-term (days) prediction, is done by using the various of precursors as well as geological researches. Phenomena which occur in a characteristic way prior to an earthquake is called precursor. Precursors include an increase in micro-seismicity, which has been credited with causing unusual animal behavior. Foreshocks (前震) are relatively smaller earthquakes that precede the largest earthquake in a series, which is termed the mainshock (主震). Over time, foreshocks generally grow weaker. Not all mainshocks have foreshocks. Well water quality may change, as well as spring discharge.

Earthquakes are part of a global tectonic process that generally occurs well beyond the influence or control of humans. Earthquakes can happen in virtually any region of the world. There is no known way to prevent (预防) earthquakes, but it is possible to lessen the impact and mitigate earthquake hazard. The amount of devastation from an earthquake can be greatly diminished by building structures using earthquake resistant design (so-called "earthquake proof" buildings), making the interiors of buildings safe from falling objects, and educating people about earthquake safety.

Because no one can predict exactly or prevent effectively an earthquake, only thing we can do is to act properly and immediately when an earthquake strikes. Here are some tips on what to do before and during an earthquake.

• Prepare an emergency plan for your family and household.

• Usually it's better to stay inside if you are already inside, and to stay outside if you are already outside.

• Know the safe places to be in your home during an earthquake. If you are inside, try to take cover in such safe places, for example, sturdy desk and bed.

• Avoid dangerous places: windows, heavy furnitures that might fall, chimneys and walls and electrical power lines.

1. Answer the Following Questions

(1) Differentiate epicenter and focus of an earthquake.
(2) Describe in detail how epicenter is located.
(3) Explain the bound elastic theory.
(4) Compare and contrast the concepts of magnitude and intensity of earthquake.
(5) Describe the worldwide distribution of earthquakes.
(6) Describe with a sketch the theory of seismograph.
(7) Name and describe the various types of earthquakes.
(8) Differentiate the P wave and S wave.
(9) Discuss briefly the prediction of earthquake.
(10) Show in a sketch how and why tsunamis can cause much damage.
(11) Differentiate the foreshocks and aftershocks.

2. Multiple Choice

(1) The location of the source of an earthquake is called ______.

A: Epicenter B: Focus
C: Center D: Fault

(2) Short—term prediction of an earthquake predict the time in ______.

A: Hundreds of years B: Tens of years
C: Days D: Hours

(3) A magnitude 7 earthquake releases nearly ______ times more energy than a magnitude 5 earthquake.

A: 20 B: 32
C: 64 D: 1 000

(4) Of the seismic waves produced by an earthquake, which one travels fastest?

A: Surface waves B: S waves
C: P waves D: Shear waves

(5) In a general/global sense, most earthquakes occur ______.

A: On plate boundaries B: On the continent
C: In the oceans D: All of them

(6) Which seismic body wave cannot travel through a liquid?

A: Surface waves B: S waves
C: P waves D: None of above

(7) With increasing of travel time, the time interval between the first arrivals of P and S waves will ______.

A: Increase B: Change to zero
C: Remain the same D: Decrease

(8) Long—term prediction of an earthquake predict the time in ______.

A: Hundreds of years B: Tens of years
C: Days D: Hours

(9) Earthquakes that occur along oceanic ridges tend to be ______ earthquakes.

A: shallow focus B: Deep focus
C: All of them D: None of them

(10) Which of these statements is true?

A: Each increase of 1 in Richter Magnitude represents a tenfold increase in the size of the earthquake.

B: For each increase in 1 in the Richter Magnitude, there is a tenfold increase in amplitude of the wave.

C: Each increase of 1 in Richter Magnitude represents a 10 time increase in the size of the earthquake.

D: For each increase in 1 in the Richter Magnitude, there is a 10% increase in amplitude of the wave.

(11) Approximately what percentages of earthquakes occur along the circum-Pacific seismic belt?

A: 60 B: 70
C: 80 D: 90

(12) The most destructive type of seismic wave is ______.

A: Surface waves B: S waves
C: P waves D: Shear waves

(13) The device used to measure an earthquake's magnitude is called a ______ and the resulting record is called a ______.

A: Seismograph, Seismogram
B: Seismogram, Seismograph
C: Seismogram, Seismometer
D: Seismogram, Seismometer

(14) If the time interval between the first arrivals of P and S is 10 minutes, the distance from the epicenter to the seismograph station will be about ______.

A: 7 000 km B: 7 500 km

C: 8 000 km D: 8 500 km.

(15) How does the P wave travel?

A: Side to side

B: Back and forth parallel to the direction of wave propagation

C: Up and down

D: Back and forth perpendicular to the direction of wave propagation

(16) The location normally reported in the news is ______.

A: Focus B: Source

C: Epicenter D: All of the above

(17) Which one is not the way of S wave propagation?

A: Side to side

B: Side to side parallel to the direction of wave propagation

C: Up and down

D: Up and down perpendicular to the direction of wave propagation

(18) If you find yourself indoors during an earthquake, is it usually better to rush outside or to take cover inside?

A: Stay inside

B: Run outside

(19) Tsunami is caused by undersea ______.

A: Earthquake

B: Volcanic eruption

C: Landslide

D: All of the above

(20) Which one is primary effect of earthquake?

A: Fire B: Tsunami

C: Surface rupture D: Landslide

Chapter 8 Geological Structures

Key words

Tectonic movement		Geologic structure		Neotectonic movement		Orientation
构造运动		地质构造		新构造运动		产状
Bedrock	Outcrop	Horst	Dip	Dip direction	Strike	Trend
基岩	露头	地垒	倾角	倾向	走向	倾伏向
Plunge	True dip	Apparent dip	Bearing	Compass	Inclinometer	Azimuth
倾伏角	真倾角	视倾角	方位角	罗盘	测斜仪	方位角
Hogback	Fault	Fault line	Fault block	Hanging wall	Footwall	Fault plane
猪背岭	断层	断层线	断层盘、断盘	上盘	下盘	断层面
Normal fault	Reverse fault	Left - lateral	Right - lateral	Strike - slip fault		Graben
正断层	逆断层	左行	右行	走滑断层		地堑
Anticline	Syncline	Monocline	Fold core	Fold limb	Axial plane	Hinge
背斜	向斜	单斜	褶皱核部	褶皱翼	褶皱轴面	枢纽
Cuestas	Anticlinal valleys		Synclinal mountains		Inversion of relief	Mesa
单面山	背斜谷		向斜山		地形倒置	桌子山

Tectonic movement is the movement resulted from or causing the crustal deformation. Geological structure refers to the nature, arrangement and orientation of the Earth's materials making up the geological feature being studied. The fault and fold are the two basic geologic structures occurred within the crust.

The Earth's crust is constantly in movement. Mountains, basins and plains are all the products of tectonic movements. Most of the tectonic movement is too slow to be observed by human beings. But the deformation associated with an earthquake takes place very quickly, from a few seconds to several days and months, and produces displacement ranging from a few centimeters to several meters in magnitude.

According to movement direction, the tectonic movement can be divided into two basic types: vertical movement and horizontal movement. Vertical movement is the relatively uplift and subsidence of the adjacent geological blocks, forming the highland and mountain in some regions and basin and plain in other regions respectively. Horizontal movement is the movement of geological blocks in horizontal.

For engineering project, such as the construction of dams and tunnels, we are more interested with those tectonic movements occurred recently. Neotectonic movement is the tectonic movement taking place during the Quaternary period and it is a specific subject in geology.

1 Orientations

Bedrock is a general term for solid rock that lies beneath soil, loose sediments, or other unconsolidated material. Outcrop is the part of a bedrock formation that appears above the surface of the surrounding land. Although the geological structures have different scales, shapes, but from the geometrical viewpoints, all of the geological structures may be reduced to combinations of two basic geometric elements — planes and lines, regardless of their complexity. So, in order to deduce the three - dimensional geometry of geological structure, it is necessary to study carefully the orientation of lines and planes in bedrock outcrop. Orientation is the spatial position of any of the geologic structure.

Planar features within rocks include bedding planes, joints and faults. Linear features within rocks include channels, axes of elongated pebbles, crests of ripples, dunes, intersections of fractures or other planar features. In order to indicate the spatial position of the planar and linear structure, the orientation elements have been introduced.

1.1 The Orientation of Planar Structure

The orientation of planar structure (面状构造) in space can be described with strike and dip.

For an inclined plane, the strike is the bearing (relative to North) (方位角) of the intersection line of the plane measured with the horizontal plane. The angle between the strike line and north is used to describe the strike (Figure 8.1).

The angle that the measured plane makes to the horizontal is called the dip angle. Notice that the bearing of the projection of the dip on a horizontal surface is in a direction at right angles to the strike. This is called the dip direction. The direction of dip is 90° from the strike. The azimuth directions of strike and dip are perpendicular. Azimuth (compass bearing) refers to the direction of a horizontal line contained in a vertical plane, measured by quadrant or (°). Examples: N90°E, N90°W, S90°W, 270°. We often record the orientation of planar structure in the form of dip-direction∠dip angle, for example: 123° ∠34°, in which 123° and 34° are dip direction and dip angle

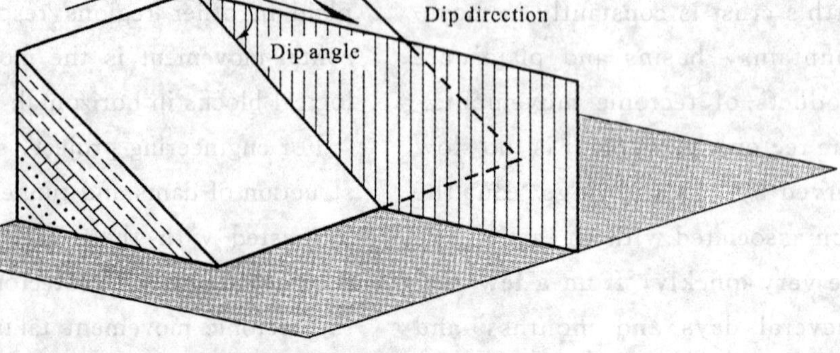

Figure 8.1 Strike and dip of a plane

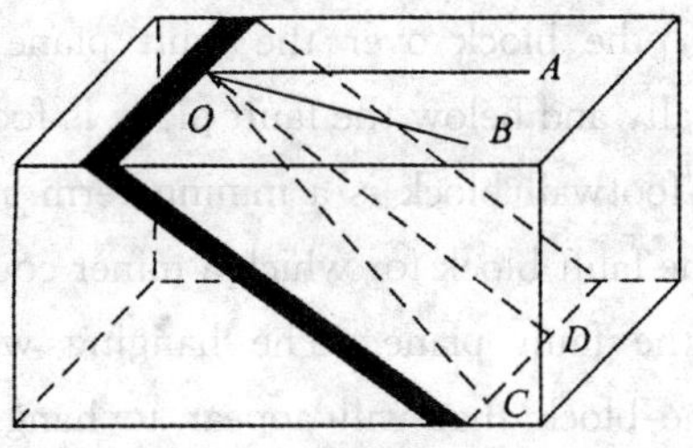

Figure 8.2 True dip and apparent dip

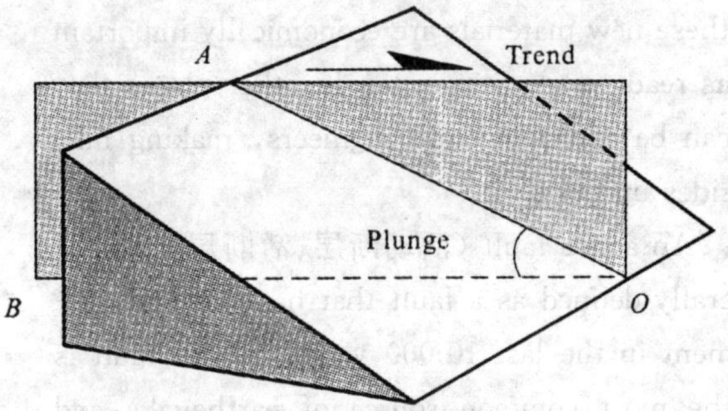

Figure 8.3 Orientation of a line

Figure 8.4 Symbols denoting attitudes of plane and line on geological map

respectively.

Geological structures are often visualized in the field in vertical sections. In such sections we can see dipping beds. However, the amount of dip we see in a section may not be so called true dip. True dip ($\angle AOD$) of a bed is the dip observed in a vertical plane perpendicular to the strike of the bed (Figure 8.2). In all other sections, we get apparent dips ($\angle BOC$) which are less than the true dip angle.

1.2 Orientation of Linear Structure (线状构造)

In structural geology, the orientations of linear features are also important. The orientation of lines in space can be described using two distinct angles referred to as the trend and plunge (Figure 8.3).

Trend is the direction (azimuth) of a vertical plane containing the line of interest. This is generally recorded in the direction pointing downwards and measured in degrees clockwise from North (through 360°). Plunge is the angle of inclination of a line.

Traditionaly, we use the following symbles to denote attitude of plane and line on geological map(地质图)(Figure 8.4).

All of the orientation data can be measured by compass. Actually, the strike and trend are measured with a compass , and the dip and plunge are measured using an inclinometer(测斜仪).

2 Fault

Fault is a kind of crack in the crust of the Earth along which there has been apparent movement of the rocks on either side of the crack. Fault is one of the most important structures occurred in the Earth's crust. Large scale faults, often forming the regional geological framework, not only control the evolution and structure of regional geology, but also exert a strong influence on the distribution of mineralization (矿化)and the subsurface accumulations of hydrocarbons. And they can have a major influence on the shaping of the landscape. Active faults have a close relation with the earthquake. Movement on faults, with earthquakes, shatters rocks. In some places these new materials are economically important as ready-made aggregate! In other places they can be a problem for engineers, making hillsides unstable.

An active fault (活动断层、活断层) is generally defined as a fault that has had displacement in the last 10 000 years. Active fault is the most common source of earthquake and likely to have another earthquake sometime in the future.

Fractures (断裂、破裂) are surfaces along which rocks or minerals have broken. Fractures that show an offset are called faults. Joints(节理) are the fractures without apparent displacement in rocks.

2.1 Geometrical Elements of Faults

Fault is a planar structure. Fault plane is fracture surfaces, which divide the rock body into two parts. The spatial orientation of fault plane is decided by its strike, dip direction and dip angle. Fault plane is not always a stable plane, often varying along its strike or dip direction.

Fault line is the intersection of fault plane and the Earth's surface. Fault line is controlled by the curvature and orientation of fault plane and topography.

Fault blocks(断盘) are the displaced rock bodies along the fault plane. If the fault plane is inclined, the block over the fault plane is hanging wall, and below the fault plane is footwall. The footwall block is a mining term and refers to the fault block for which a miner could stand on the fault plane. The hanging wall block is the block that will appear to hang if the other block is removed.

2.2 Types of Faults

We classify faults by how the two rocky blocks on either side of a fault move relative to each other(Figure 8.5).

- **Normal Fault** (正断层)

Normal fault is a fault where the hanging wall side appears to have moved downwards relative to the footwall. Normal fault often has a steep fault plane with dip angles steeper than 50°. In recent years study has shown that some of the normal faults often have a very low degree of dip, especially some of the large - scale normal faults with steep fault plane can become gently dipping at depth.

Normal faults may dip at a variety of angles, but they most typically dip between about 40° and 70°. Some normal faults dip at angles as low as 10° or less.

A kind of special mountains, called block - fault mountain (断块山),is produced by nor-

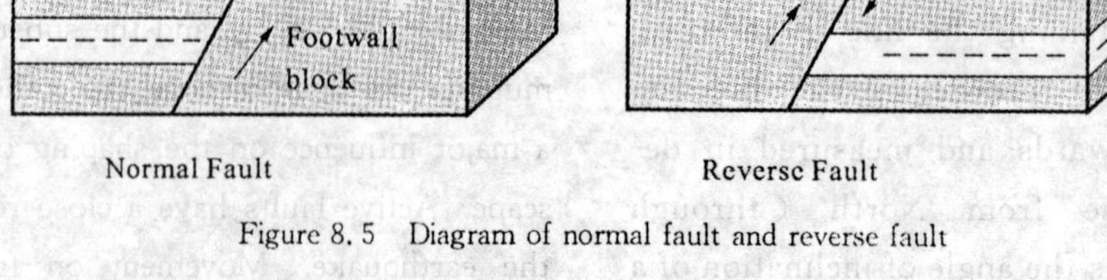

Figure 8.5 Diagram of normal fault and reverse fault

mal faulting（正断层作用）in areas subjected to tensional forces. In these areas, normal faults often occur in pairs or series, with adjacent faults dipping in opposite directions. A fault block that moves down is called a graben and one that moves up is called a horst (Figure 8.6). A horst is bounded by normal faults dip away from each other.

The African Rift Valley is the largest graben in the Earth's surface. It extends for 4 800 km from Syria to the Zambesi River. Weihe（渭河）valley in Shannxi Province（陕西省）and Fenghe（汾河）valley in Shanxi Province（山西省）also are examples of grabens. The famous horsts in our country include Taishan（泰山）, Lushan（庐山）and Huashan（华山）. Those mountains are also called block-fault mountains. Actually, horsts and grabens often occur in pairs, for example, the Lushan horst is immediately adjacent to the Poyanghu（鄱阳湖）graben. The uplifting of Huashan（华山）is accompanied by the relative depression of Weihe（渭河）valley in the North. The basin and range province of the western U.S.（美国西部盆岭省, including Nevada, Utah, and Idaho）is perhaps the most famous example of horst and graben system.

• **Reverse Fault**（逆断层）

If the hanging wall side appears to have moved relatively upwards, the faults are reverse faults. This fault motion is caused by compressional forces (Figure 8.7).

Low angle reverse faults(less than 45°), also known as thrusts or thrust faults(逆冲断层), are noteworthy because they are common in mountain belts.

• **Strike-Slip Fault**（走滑断层）

In strike-slip faulting, the two blocks move horizontally past one another and the rock layers beneath the surface haven't been moved up or down on either side of the fault. Strike slip faults can be of two varieties, depending on the sense of displacement. To an observer standing on one side of the fault and looking across the fault, if the block on the other side has moved to the left, we say that the fault is a left-lateral strike-slip fault. Right-lateral fault is a strike-slip fault on which the displacement of the far block is to the right when viewed from either side (Figure 8.8).

The San Andreas fault in California is probably the best known active strike - slip fault in the world (Figure 8.9). A total displacement of about 500 km is estimated on this fault. It has been responsible for destructive earthquakes occurring in San Francisco.

2.3 Recognition of fault in the field

Fault in the field can be recognized in many ways, among which the special topography,

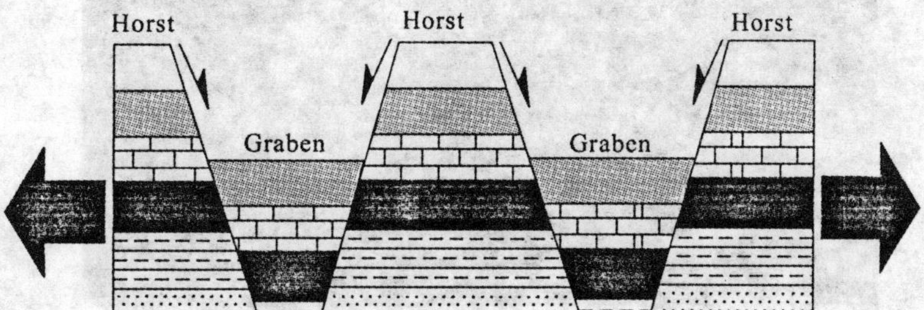

Figure 8.6 Schematic cross section of horst and graben pairs

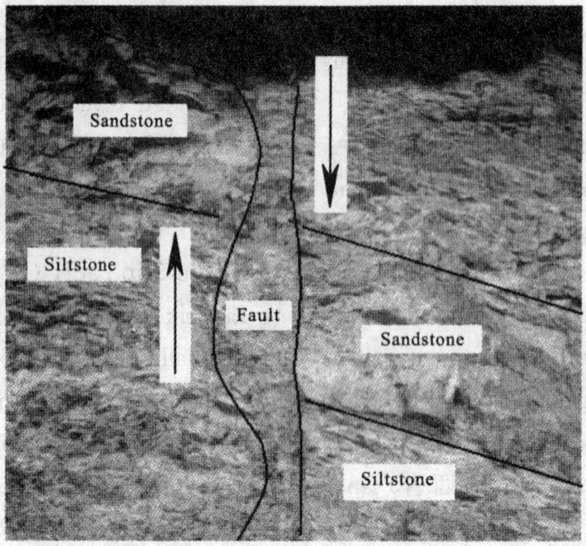

Figure 8. 7 A view of the nearly vertical reverse fault in a roadcut on the west side of Fenghai hydropower station, Yongan (福建永安,丰海),2005

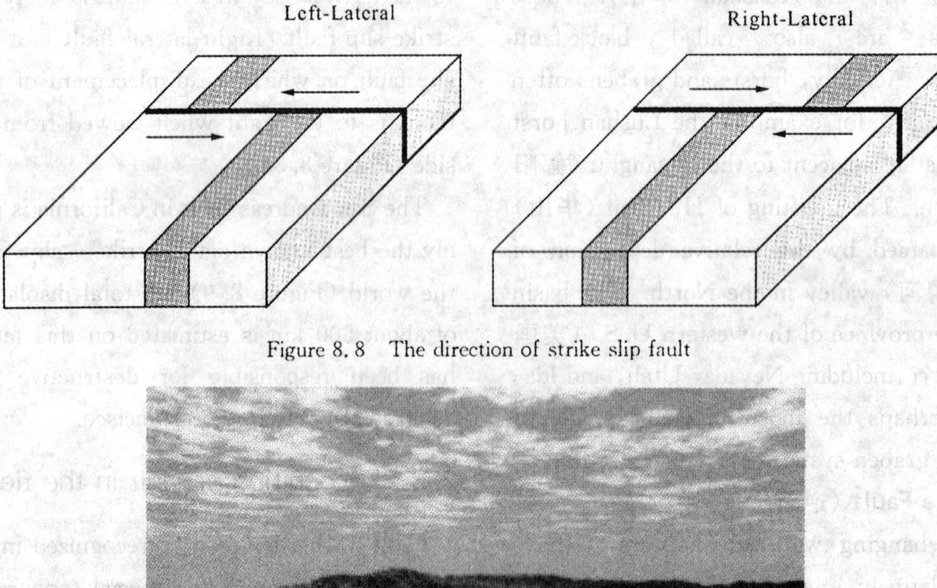

Figure 8. 8 The direction of strike slip fault

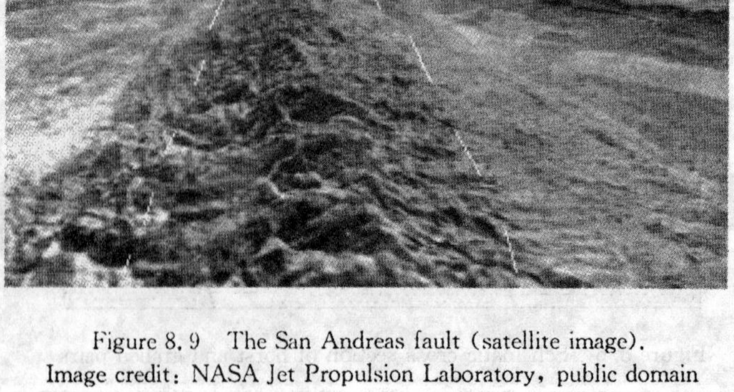

Figure 8. 9 The San Andreas fault (satellite image).
Image credit: NASA Jet Propulsion Laboratory, public domain

Figure 8. 10 Fault scarp in Jiuzhaigou(九寨沟), Photo by Chen Fuxing(谌福兴)

fault rocks and offset geological bodies are especially useful to recognize the fault.

Fault scarp (断层崖) (Figure 8. 10) is a steep slope or cliff formed when movement along a fault exposes the fault surface. A graben is often bounded by a distinct fault scarp on each side. If a ridge is cut through by a normal fault, the hanging wall block will show a triangular fault plane (断层三角面). Fault sag pond is small, enclosed depression along an active fault. Sag ponds and springs often distribute linearly along the fault lines. Surface expression of faults, such as fault scarp and fault sag ponds, is the result of neotectonic movement.

The fault rock consists of angular fragments of rocks. Linear distribution of such rocks may indicate the existence of fault. Be careful not to confuse fault rocks with sedimentary breccias and conglomerates.

If the continuity of sedimentary layers, intrusives and anyother geological bodies are broken, a fault may exsit. Acturally, the cross-cutting relashionship between diffenrent geolgoical bodies can be used to determine the age of a fault. In Figure 8. 11, the youngest rocks cut by the fault belong to Permian system and the oldest rock cutting through the fault belong to Triassic system, so the age of fault must be between Permian period and Triassic period.

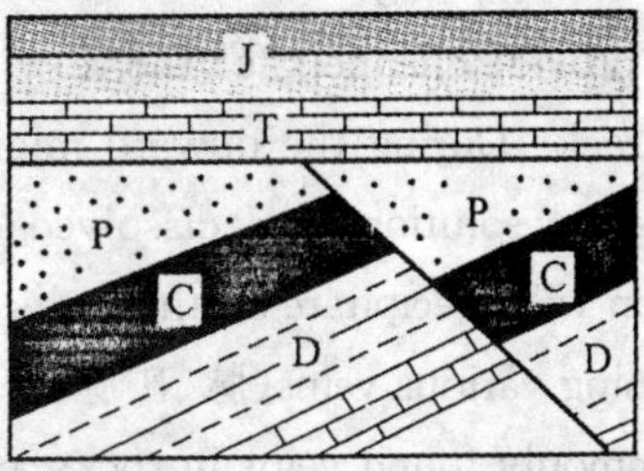

Figure 8. 11 The age of fault

3 Joints

Joints are regularly spaced fractures or cracks in bedrock along which no appreciable displacement occurs. Joint is perhaps the most commonly found geological structure. Nearly all the rocks exposed to the Earth's surfaces are fractured by joints.

In the field, joints usually occur in parallel with each other patterns. These adjacent joints with similar geometry and orientation are called a joint set(节理组). Two or more sets of joint formed at approximately the same time make up a joint system(节理系).

Figure 8.12 A: Vertical joints in nearly horizontal sedimentary rock in Zhangjiajie (张家界). The pinnacles are due to erosion along the vertical joints
B: Close-up view of the rocks and joints. Photo by Dr. Yu Shan (俞姗), 2005

The columnar structure found in basalt lava is sometimes described as columnar jointing. The hexagonal columns are surrounded by smooth joints as a result of contraction of a cooling, solidified lava flow.

Joints provide channels for the flow of water through rock. During the upward migration of hydrothermal solution, the dissolved elements in solution will precipitate on the walls of joint, thus forming various veins(脉、矿脉). The two most commonly found veins in rocks are quartz vein and calcite vein. Gold ores can also be formed by this way.

Joint plays a vital role in shaping some special landforms. The forest of sandstone in Zhangjiajie(张家界)(Figure 8.12) was the result of combination action of erosion and joint. The irregular blocks formed by intersection joint systems are more readily attacked by various weathering agents. With time, the weathered materials will be carried away by streams and remaining resistant sandstones will form towering columns.

4 Fold

Folds are produced by bending of various surfaces (Figure 8.13). When rocks deform in a ductile manner, instead of fracturing to form faults, they may bend or fold. Folds in rocks vary in size from microscopic crinkles to mountain-sized folds. They occur singly as isolated folds and in extensive fold trains of different sizes, on a variety of scales.

The study on the shape, distribution, and mechanism of folds is very important to show the deformation history of certain regions. Folds are also closely related to the formation of ores. At the same time, folds are a very important factor influencing the engineering geological environments.

4.1 Basic Types of Folds

There are lots of different types of folds. But according to the bending of single folded layer, folds can be classified into two basic types: anticline and syncline (Figure 8.14). An upward closing fold is called an anticline while a downward closing one is called a syncline.

Anticline is folded with old rock in the core (An easy way to remember this is "antique" = old = *anti* cline = old rock will be in center.), and syncline with younger rocks in the

Figure 8.13 Anticline in Fenghai, Yongan (福建永安，丰海), 2005

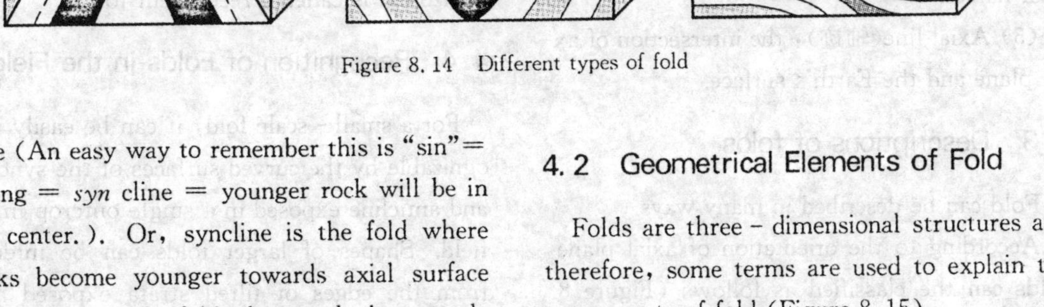

Figure 8.14 Different types of fold

core (An easy way to remember this is "sin"= young = *syn* cline = younger rock will be in the center.). Or, syncline is the fold where rocks become younger towards axial surface and anticline is the fold where rocks become older towards axial surface. Folds in which rock layers on both sides of the fold are horizontal but at different levels are called monoclines (挠曲、单斜). In the field, monoclines are often observed as rock layers which dip uniformly in one direction.

4.2 Geometrical Elements of Fold

Folds are three - dimensional structures and therefore, some terms are used to explain the components of fold (Figure 8.15).

(1) Core(褶皱核部): the central part of the fold .

(2) Limb(褶皱翼): The sides of a fold are called limbs.

(3) Hinge(枢纽): The line joining points of greatest curvature on a folded surface is called

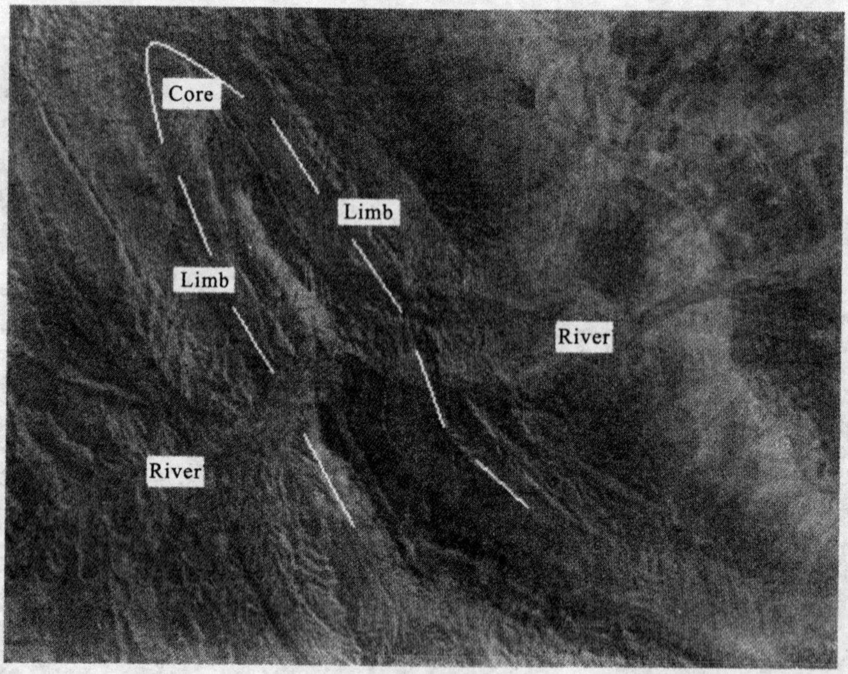

Figure 8.15 A large scale fold in Urumqi (乌鲁木齐)(satellite image)
Image credit: NASA - JSC - ES&IA, Public domain

the hinge of a fold. The hinge can be observed in naturally developed outcrop - sized folds.

(4) Axial Plane (轴面): The imaginary plane or surface that contains hinges or hinge lines on all surfaces involved in folding and which generally divides the fold into nearly equal halves.

(5) Axial line(轴迹): the intersection of axial plane and the Earth's surface.

4.3 Descriptions of folds

Fold can be described in many ways.

According to the orientation of axial plane, folds can the classified as follows (Figure 8.16):

(1) Upright fold(直立褶皱): often has a nearly vertical axial plane, with two limbs dipping in opposite direction.

(2) Inclined fold(斜歪褶皱): has an oblique axial plane. Two limbs dip in opposite direction but have different dip angles.

(3) Overturned fold(倒转褶皱): If the strata on one limb of the fold become nearly upside down, the fold is called an overturned fold.

(4) Recumbent fold(平卧褶皱): An overturned fold with an axial plane that is nearly horizontal is called a recumbent fold.

4.4 Recognition of Folds in the Field

For a small - scale fold, it can be easily recognizable by the curved surfaces of the syncline and anticline exposed in a single outcrop in the field. Shapes of larger folds can be inferred from the edges of tilted strata exposed at a number of different locations. But in most cases, due to the covering of vegetation or other reasons, it is not so easily to identify the fold.

Some of the landforms are closely related to folds, but landform itself can never be used as a reliable criterion to recognize folds. So, it is

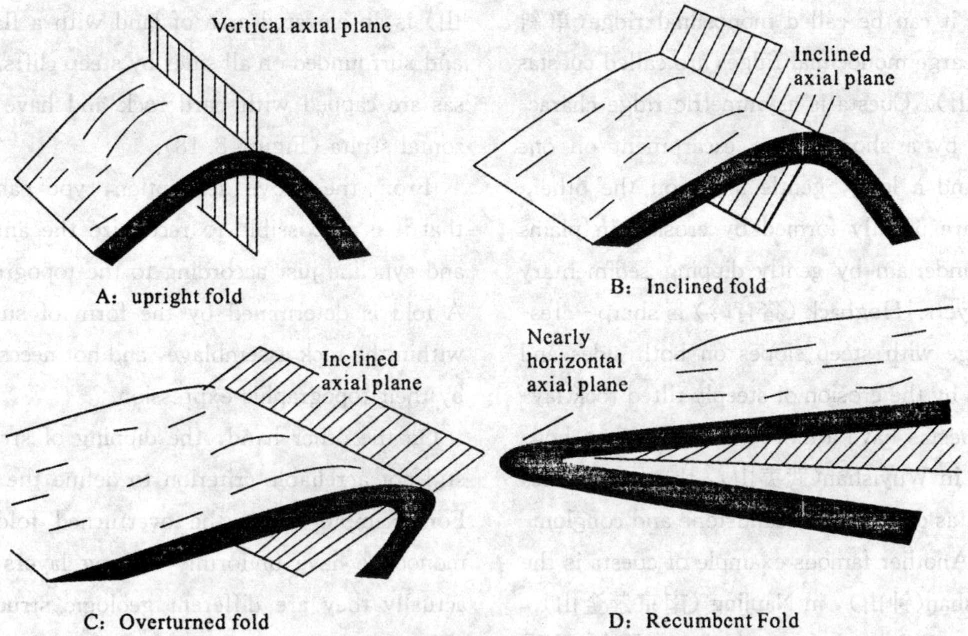

Figure 8.16 Classification of fold according to axial plane

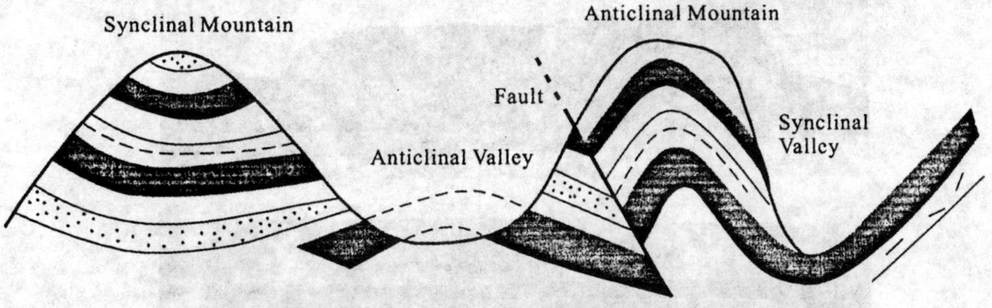

Figure 8.17 Relationship between fold and topography

important to realize that there is not necessarily a simple correspondence between the form of the landscape and the form of a geological structure. The axis of a syncline may lie along a topographic ridge in one place, and in a valley elsewhere. At the beginning of rock deformation, the anticlines will form ridges and synclines form valleys. They are called anticlinal mountains and synclinal valleys respectively. Due to fractures and joints developed during folding, however, the anticline is structurally weaker and easily to be eroded, while the synclines are hardened by compression. Given time, the synclines tend to form the ridges, with the streams occupying the anticlinal valleys. This situation is sometimes called "inversion of relief" or "inversion of topography" and the landforms are called " synclinal mountains" and "anticlinal valleys" (Figure 8.17).

The inversion of relief is very common near the Nanjing (南京) city. In this region, the anticline axes often form the flat bottom of valleys while syncline axes form the ridges.

Monocline is the geologic structure in which all the strata are inclined in the same direction. If the ridge is composed of uniformly inclined strata, it can be called monoclinal ridge(单斜山). Large monoclinal ridges are called cuestas (单面山). Cuesta is asymmetric ridge characterized by a short, steep escarpment on one side, and a long, gentle slope on the other. They are usually formed by erosion in plains areas underlain by gently dipping sedimentary rock layers. Hogback (猪背岭) is sharp-crested ridge with steep slopes on both sides and formed by the erosion of steeply tilted rock layers. Cuestas have a more gentle dip than hogbacks. In Wuyishan(武夷山), there is a series of cuestas composed of sandstone and conglomerate. Another famous example of cuesta is the Zhongshan(钟山), in Nanjing (南京紫金山).

A mesa (Spanish for "table", 桌子山, 平顶山) is an elevated area of land with a flat top and surrounded on all sides by steep cliffs. Mesas are capped with hard rock and have horizontal strata (Figure 8.18).

From the above description, you can find that it is impossible to recognize the anticline and syncline just according to the topography. A fold is determined by the form of surfaces within the rock assemblage, and not necessarily by their topographic expression.

On the other hand, the dipping of strata is still not a reliable criterion to define the fold. For example, both the overturned fold and monocline have uniformly dipping layers, but actually they are different geologic structures (Figure 8.19).

Figure 8.18 Mesa on the Colorado Plateau. Photo by Yu Shan(俞姗)

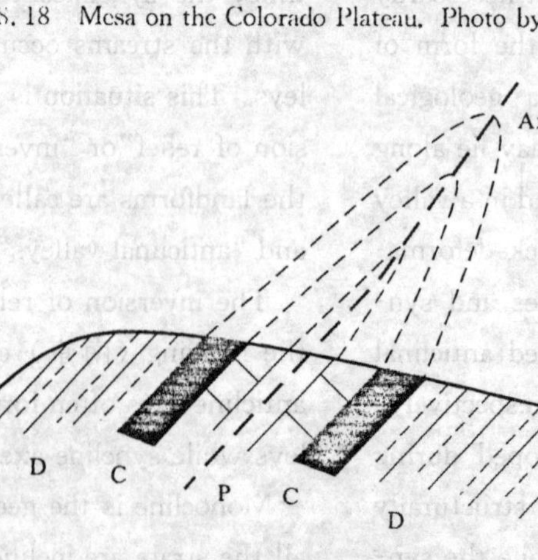

Figure 8.19 Overturned fold

Besides the shape of the fold, the correct way to distinguish the fold is to study the strata sequences (地层层序). The strata sequences of the same age will repeat symmetrically on two sides of fold along the dipping direction. Anticlines are folds in which the oldest rock lies in the center or core and synclines have the youngest rock in its center or core (Figure 8.20).

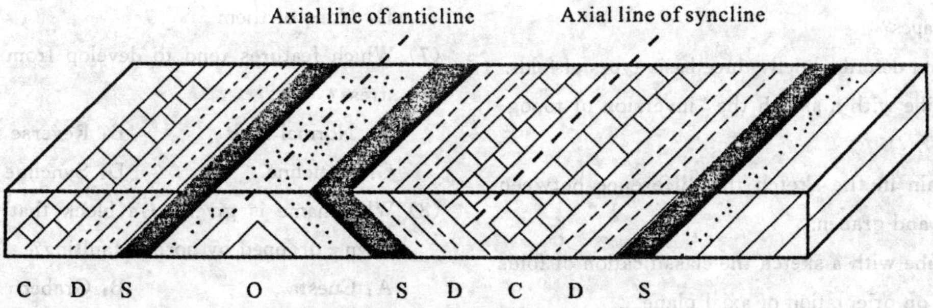

Figure 8.20 Strata sequences indicating folding

1. Answer the Following Questions

(1) Define the phrase "geological structure".

(2) Define the term neotectonic movement.

(3) Differentiate strike and trend.

(4) What role does joint play in sculpting some special landscapes?

(5) List and describe briefly three basic types of folds.

(6) Describe with a sketch the "inversion of topography".

(7) Explain in the sketch the difference between horst and graben.

(8) Describe with a sketch the classification of folds based on orientation of axial plane.

(9) Differentiate with a sketch the true dip and the apparent dip.

(10) Define active fault and relate it to the earthquake.

(11) Differentiate between strike and trend.

(12) How to distinguish between the hinge and axial line?

2. Multiple Choice

(1) Which term refers to line made by the intersection between a bedding plane and a horizontal plane?

A: Dip　　B: Trend

C: Strike　　D: Plunge

(2) Which type of structure is a fold with the oldest rock unit exposed in the center?

A: Anticline　　B: Syncline

C: Horst　　D: Graben

(3) What geologic structure is the Lushan(庐山) mountain an example of?

A: Cuesta　　B: Graben

C: Horst　　D: Hogback

(4) What geologic structure are the Wuyishan (武夷山) mountains an example of?

A: Cuesta　　B: Mesa

C: Horst　　D: Hogback

(5) The unique landscape of Zhangjiajie(张家界) is formed by ______.

A: Jointing and erosion　　B: Faulting

C: Folding　　D: Weathering

(6) The uplifting of Taishan (泰山) is due to the ______.

A: Extensional force

B: Compressional force

C: Shear force

D: None of them

(7) Which features tend to develop from tensional stress?

A: Normal fault　　B: Reverse fault

C: Anticline　　D: Syncline

(8) What name is given to a block that has been down-dropped by normal faulting?

A: Cuesta　　B: Graben

C: Horst　　D: Hogback

(9) A fracture that shows no appreciable displacement is called ______.

A: Tertiary　　B: Joint

C: Fold　　D: All of them

(10) Neotectonic movement is the tectonic movement taking place during which period?

A: Tertiary　　B: Cambrian

C: Jurassic　　D: Quaternary

(11) Orientation is the ______ of any of the geologic structure.

A: Shape　　B: Scale

C: Spatial position　　D: Length

(12) If the orientation of a plane is recorded as 133° ∠53°, the 133° and 53° refer to ______.

A: Strike and dip　　B: Dip and strike

C: Dip direction and dip　　D: Trend and plunge

(13) Which kind of folds has an upside down limb?

A: Upright fold　　B: Overturned fold

C: Inclined fold　　D: None of above

(14) Sites where rock is exposed at the surface are called ______.

A: Strikes　　B: Dips

C: Structures　　D: Outcrops

(15) Which type of structure is a fold with the oldest rock unit exposed in the center?

A: Anticline　　B: Syncline

C: Horst　　D: Graben

(16) Which statement is not right?

A: An active fault is a fault that has had displacement in the last 10 000 years.

B: The San Andreas fault in California is an active strike – slip fault.

C: An active fault has had displacement since the beginning of Quaternary period.

D: An active fault is the source of earthquake.

(17) Which statement is not right?

A: The limbs of an anticline dip away from axial plane.

B: The limbs of a syncline dip toward to axial plane.

C: Anticlines usually form mountains.

D: Synclines usually form mountains.

(18) True dip is ______ apparent dip.

A: Less than B: Great than

C: Equal to D: None of above

(19) Fault scarp is usually formed ______.

A: Normal fault

B: Reverse fault

C: Strike – slip Fault

D: All of above

(20) Which statement is not right ______.

A: Quartz vein is due to cooling and solidification of magma.

B: Quartz vein is formed by direct precipitation from hydrothermal solution.

C: Quartz vein always occur along pre-existing ruptures.

D: Gold can be found in quartz vein.

Chapter 9 Plate Tectonics

Key words

Continental drift theory	Sea-floor spreading theory	Asthenosphere	Lithosphere	Plate tectonics
大陆漂移说	海底扩张说	软流圈	岩石圈	板块构造
Divergent boundary	Convergent boundary	Transform plate boundaries		Continental shelf
离散性板块边界	聚敛性板块边界	转换边界		大陆架
Continental slope	Continental rise	Active continental margin	Passive continental margin	Rift
大陆坡	陆隆	主动大陆边缘	被动大陆边缘	裂谷
Subduction zone	Wilson Cycle	Mid—ocean ridges	Transform fault	Trench
俯冲带	威尔逊旋回	大洋中脊	转换断层	海沟
Benioff Zone	Island arc	Continental crust	Oceanic crust	
毕乌夫带、贝尼奥夫带	岛弧	陆壳	洋壳	

Plate tectonics is a theory of geology developed to explain the phenomenon of continental drift. Plate tectonic theory arose out of two separate geological observations: seafloor spreading and continental drift.

In 1912 the theory of continental drift was proposed by Alfred Wegener. After many years of heated debate, the theory was rejected by the vast majority of Earth scientists at that time, because Wegener's hypothesis lacked a geological mechanism to explain how the continents could drift across the Earth's surface as he proposed. However, during 1950s and 1960s, new information obtained about the nature of the ocean floor resurrected Wegener's theory. These discoveries and more led Harry Hess (1962) and R. Deitz (1961) to put forward similar hypotheses based on mantle convection currents, now known as "sea floor spreading". By 1968, the combination of "continental drift" and "an sea floor spreading" eventually led to the unfolding of a far more encompassing theory — a theory known as plate tectonics.

Plate tectonics is now almost universally accepted. Plate tectonics represents a scientific revolution as significant to geology as relativity(相对论) was to physics.

1 The Continental Drift Theory

The theory of continental drift has been put forward around for a long time. Sir Francis Bacon (培根) noted as early as 1620 that the continents seemed to fit together. In 1915 the German geologist and meteorologist Alfred Wegener (1880 — 1930) first proposed the theory of continental drift. He noticed that the shapes of continents on either side of the Atlantic Ocean seem to fit together (for example Africa and South America). He suggested that a very long time

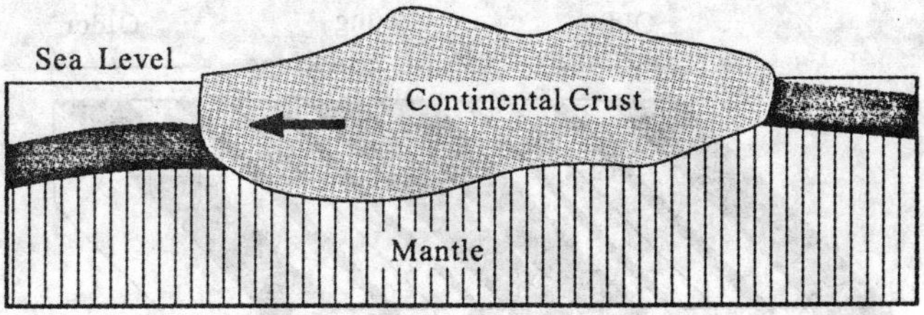

Figure 9.1 Continental curst plows through the oceanic

ago all the land that covered the Earth had been joined together into one huge continent which he called Pangea (meaning "all lands", 泛大陆、联合古陆). Wegener also used the distribution of fossils, a similar sequence of rocks at numerous locations, and ancient climates to support his idea. He speculated that the Earth took this form about 245 million years ago, during the Triassic period of the Mesozoic era. And then this supercontinent(超大陆) broke up and had drifted into the present day positions.

The greatest shortcoming was the lack of an adequate mechanism for moving the continents. Wegener proposed that the Earth's rotation and gravitational forces of the sun and the moon caused the continents to move, plowing through the oceanic crust and producing mountains on their leading edges (Figure 9.1). Geologists at that time understood enough about the strength of rocks to know that this was highly unlikely. Wegener's inability to provide an adequate explanation of the forces responsible for continental drift resulted in the scientific dismissal of his theories.

2 Sea-floor Spreading Theory

In 1929, about the time Wegener's ideas began to be dismissed, Arthur Holmes proposed the idea that the mantle undergoes thermal convection. As in heating a pot of water on a stove, convection currents are established within the Earth, which bring warm material to the surface and send cooler material to the interior. The mantle convection may be enough to carry the continents along in much the same way as a conveyor belt. This idea received very little attention at the time.

During the 1940s and 1950s, as a result of the development of echo sounders(回声测深仪) during World War Ⅰ, great advances were made in our knowledge of the sea floor and in the magnetic properties of rocks. Especially the discoveries of mid-oceanic ridges and magnetic stripes(磁条带) on the sea floor eventually led to the sea-floor spreading theory (Figure 9.2).

The mapping of the sea floor revealed that the mid-oceanic ridges rise 3 000 m from the ocean floor and are more than 2 000 km wide surpassing the Himalayas in size. The Iceland is part of the mid-Atlantic ridge and made of recently erupted basalt. A huge trench bisects these huge underwater mountain ranges. Seismic studies and researches on the heat flow indicate intense geological activity at the mid-oceanic ridges.

The volcanic rocks which make up the sea floor can record the orientation of Earth's magnetic field at the time the magma cools.

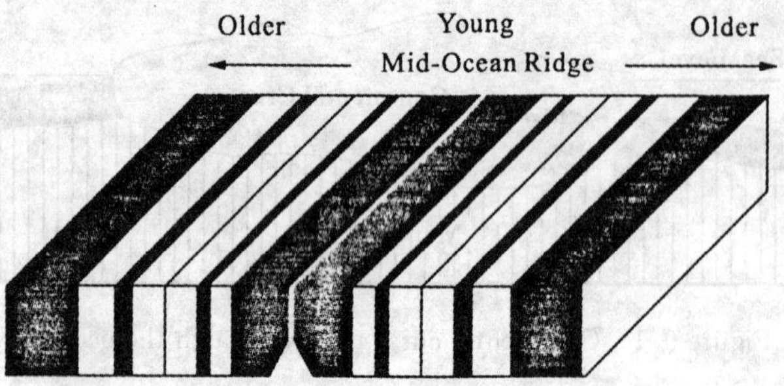

Figure 9. 2 Schematic diagram of magnetic anomalies

And periodically, the Earth's magnetic field reverses (The north magnetic pole becomes the south magnetic pole). In the late 1950s, scientists found the "stripes" of alternating magnetic bands parallel to the mid-oceanic ridges. Magnetic stripes, expressed as anomalies(异常), were bands of magnetism with one band in one direction then changed to other band in reversed direction, and then reversed again. Notice that the patterns on either side of mid-oceanic ridge are mirror images of one another.

Based on the above discoveries and the association of island arcs and oceanic trenches occurring together, Harry Hess (1962) and R. Deitz (1961) proposed the theory of sea-floor spreading. This theory was basically based on mantle convection proposed by Holmes over 30 years earlier but with more evidences. Sea-floor spreading theory stated that magma from the mantle rises to create new ocean floor at mid-ocean ridges. On each side of the ridge, sea floor moves from the ridge towards the deep-sea trenches, where it is destroyed and recycled as it is pushed back into the mantle. The entire sea floor is recycled every 200 million years. As the sea floor spread, it took the continents with them.

3 Plate Tectonics

The combination of continental drift with sea-floor spreading led to the theory of plate tectonics.

The synthesis began in 1965 when Tuzo Wilson introduced the term plate for the broken pieces of the Earth's lithosphere. Then, in 1967, two British geophysicists, Dan MacKenzie and R. L. Parker, and an American geophysicist, independently proposed the theory of plate tectonics, meaning plate "construction." In the same year, Jason Morgan proposed that the Earth's surface consists of 12 rigid plates. Two months after Morgan, Xavier Le Pichon published a map indicating the probable locations of the plates.

In geologic terms, a plate is a large, rigid slab of solid rock. The word tectonics comes from the Greek root "to build" (A character in Homer's Iliad, the carpenter Tekton). Plate tectonics means plate structure, or how the Earth's surface is built of plates. The theory of Plate Tectonics suggested that the Earth's lithosphere is split up into seven large plates and a few smaller ones that are moving relative to one another as they ride atop hotter, more mobile asthenosphere. The movement of plates is driven by mantle convection. Plate interactions produce earthquakes, volcanoes, and mountains.

Plate tectonics is now almost universally accepted. The theory has unified the study of the Earth by drawing together many branches of the Earth sciences and provides unified simple explanations for so many widely disparate phenomena: earthquakes, volcanoes, oceanic trenches, mountain building, and many other geologic problems.

According to chemical composition, the outermost layers of the Earth can be divided into three chemical layers: the core, the mantle and the crust.

According to physical properties, the outermost layers of the Earth can be divided into lithosphere and asthenosphere (Figure 9.3).

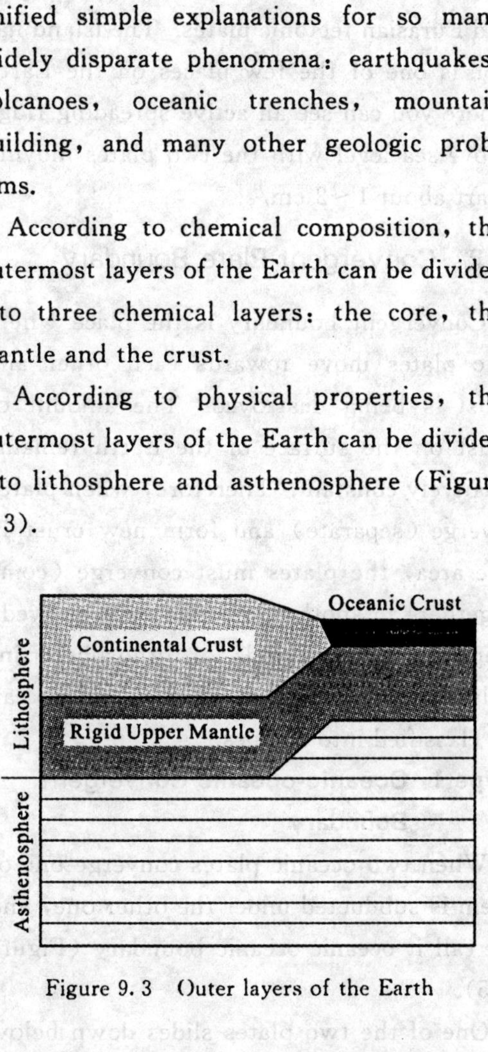

Figure 9.3 Outer layers of the Earth

The lithosphere (from the Greek, lithos, stone) is the rigid outermost layer made of rigid crust and the solid portion of uppermost mantle. The lithosphere, the "plate" of the plate tectonic theory, is typically 50 ~ 100 km thick and is broken into large plates. Plates are rigid bodies of rock that essentially float atop a region of partial melt called the asthenosphere.

The asthenosphere (from the Greek, asthenos, devoid of force, weak) is a part of the upper mantle that exhibits plastic properties and can flow slowly, at a rate of a few centimeters per year. It is located below the lithosphere, between about 100 ~ 250 km deep. The flow of the asthenosphere is part of mantle convection, which plays an important role in moving lithospheric plates.

4 Types of Plate Boundaries

Plate boundaries are found at the edge of the lithospheric plates and are of three types, convergent, divergent and transform. These different margin types are associated with characteristic geological processes. These boundaries contain almost all the Earth's earthquakes, volcanic activities and wide zones of deformation.

4.1 Divergent Plate boundary

A divergent plate boundary is a spreading boundary where the tectonic plates are separating and new crust is being created (Figure 9.4). The great majority of the Earth's volcanism occurs along divergent plate boundaries, most of which are under the ocean.

Due to the mantle convection, the lithosphere is pulled apart along parallel normal faults that tilt slightly outward from each other. When the plate on either side of fault zone moves away, the block between the faults drops down and cracks. The sinking of the block forms a central valley, typical-

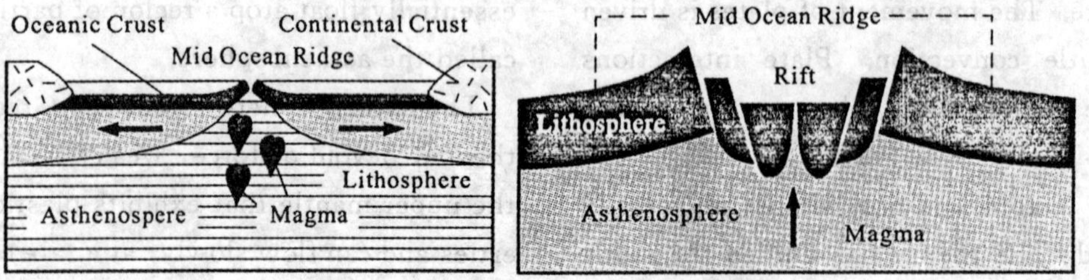

Figure 9.4 Divergent plate boundary

ly 30 to 50 km wide, called a rift (裂谷) or graben. As the magma forces its way upward along the cracks, a new oceanic crust is created. The newly formed oceanic lithosphere is pushed aside in opposite directions.

The Mid-Atlantic Ridge (大西洋中脊) and East Pacific Rise (东太平洋洋隆) provide good examples of this type of plate boundary. The Mid-Atlantic Ridge neatly divides the Atlantic Ocean into two equal halves, and the geography of the ridge is almost exactly parallel to the coastlines of Africa and South America as well as Europe and North America. A mid ocean ridge is an uplifting of the ocean floor formed by the upwelling of magma beneath the ocean bed. The topography and age distribution of the ocean floor are symmetrical along the axis of the ridge. The youngest rocks are at the mid ocean ridge. It is the longest mountain range on the Earth. The 84 000 km long Mid Ocean Ridge, described as the wound of the Earth that never heals, winds around the globe from the Arctic Ocean to the Atlantic Ocean, around Africa, Asia and Australia, under the Pacific Ocean and to the west coast of North America.

Iceland lies over the Mid-Atlantic Ridge, the boundary between the North American and Eurasian tectonic plates. The island nation is one of the few places on the Earth where you can see an active spreading ridge above sea level with the two plates moving apart about 1～2 cm/a.

4.2 Convergent Plate Boundary

Convergent boundary is the place where two plates move towards each other and crust is being destroyed. The amount of crust on the surface of the Earth remains relatively constant. Therefore, when plates diverge (separate) and form new crust in one area, the plates must converge (come together) in another area and be destroyed. Depends on the kinds of lithosphere involved, the convergent plate boundary can be classified into 3 types.

Type 1: Oceanic-oceanic Convergent Boundary

When two oceanic plates converge one of them is subducted under the other one, and we call it oceanic-oceanic boundary (Figure 9.5).

One of the two plates slides down below the other at an angle of around 45° and is incorporated into the Earth's mantle along a subduction zone. One oceanic plate tips down and slides beneath the other plate forming a deep ocean trench. Trenches are

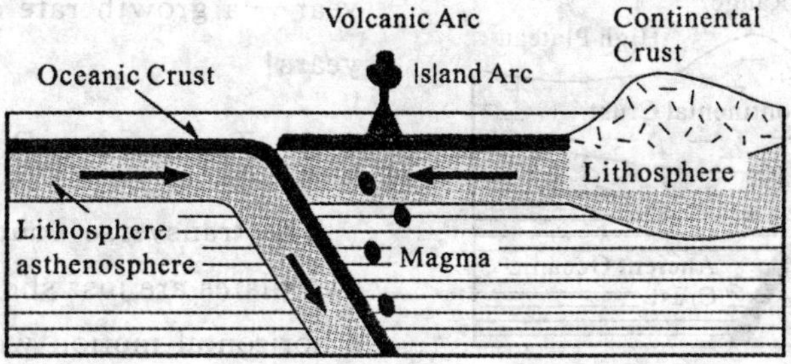

Figure 9.5 Oceanic-oceanic convergent boundary

the deepest parts of the ocean floor and are created by subduction. The Marianas Trench in the western Pacific plunges to a depth of over 11 000 m.

The Benioff Zone is a sheet of earthquake foci that dips downward at about 50° to depths of about 300 km, starting from an oceanic trench. In fact, it is the active seismic zone in a subduction. These earthquakes can be produced by slip along the subduction thrust fault or by slip on faults within the downgoing plate as a result of bending and extension as the plate is pulled into the mantle. Benioff zone can be found along the oceanic - oceanic convergent boundary as well as oceanic - continental boundary.

Subduction processes in oceanic-oceanic plate convergence can result in the formation of volcanoes. As the subducting plate descends into the mantle, it gets hot and starts to melt partially generating magma. This magma is less dense and rises to the surface to form a chain of volcanic islands called volcanic island arc. Volcanic island arcs are closely parallel with the trenches. Chains of islands are found throughout the oceans and especially in the western Pacific margins. Japan, Aleutians, Philippines and Indonesia are some examples.

Type 2: Oceanic-continental Convergent Boundary

This is the boundary where the oceanic plate subduct underneath the denser continental (Figure 9.6).

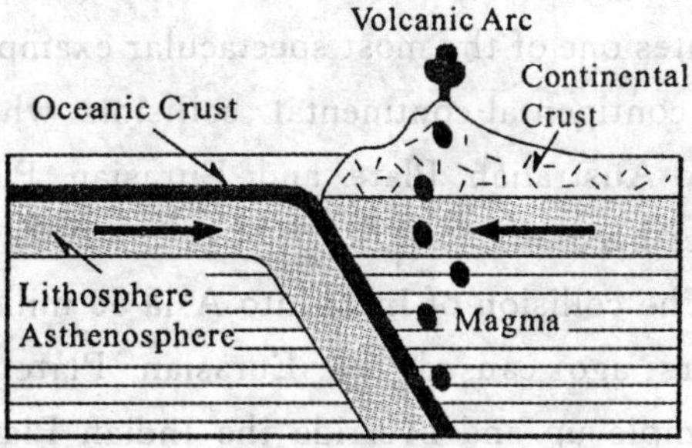

Figure 9.6 Oceanic-continental convergent boundary

An excellent example of this kind of boundary is the west coast of South America along the Peru-Chile trench where the oceanic Pacific Plate is pushing into and being subducted under the South American Plate. In turn, the overriding South American Plate is being lifted up, creating the towering Andes Mountains. The magma chain forming here is also called continental volcanic arc (山弧).

Type 3: Continental-continental or Collisional Plate Boundaries

Collisional boundary is a special type of convergent boundary where two continental plates collide (Figure 9.7). As the two

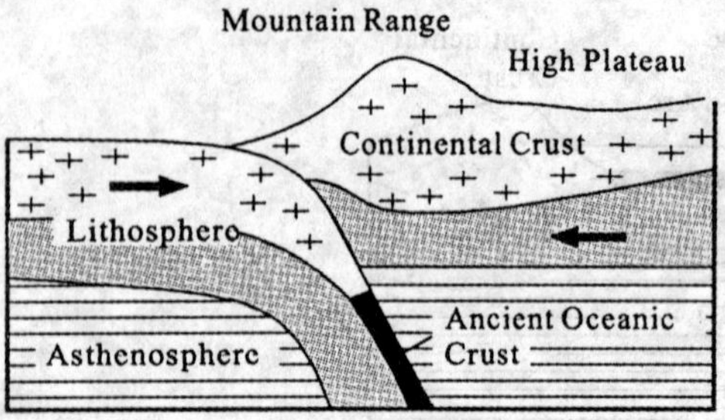

Figure 9. 7 Collisional boundary

continental plates plow into one another, a very high mountain range is created. Actually, nearly all of the mountains are the suture zones(缝合线)of ancient plate boundaries!

The Himalayan mountain chain demonstrates one of the most spectacular examples of continental-continental collision where Indo-Australian Plate and Eurasian Plate meet head-on.

The collision of India into Asia 50 million years ago caused the Eurasian Plate to crumple up and override the Indian Plate. The collision marked the beginning of the rapid uplift of the Himalayas and the Tibetan Plateau. Most of this growth occurred during the past 10 million years. The Himalayas continue to rise more than 1 cm a year — a growth rate of 10 km in a million years!

4. 3 Transverse Boundaries

The transverse boundaries are where the two plates are just sliding past each other in a horizontal motion without creating or destroying the oceanic crust.

This boundary is the third type of plate boundary which is also called the transform fault. Transform fault is a particular type of strike-slip fault. The actual movement of a transform fault is opposite to its apparent displacement because of the interplay of spreading and faulting between tectonic plates (Figure 9. 8).

The San Andreas Fault zone is an example of this type of boundary where the Pacific Plate on which Los Angeles sits is moving slowly northwestward relative to the North American Plate on which San Francisco sits. It is famous for generating many of the larger quakes in California, including the world-renowned San Francisco earthquake of 1906. Earthquakes along such transform faults are shallow focus earthquakes.

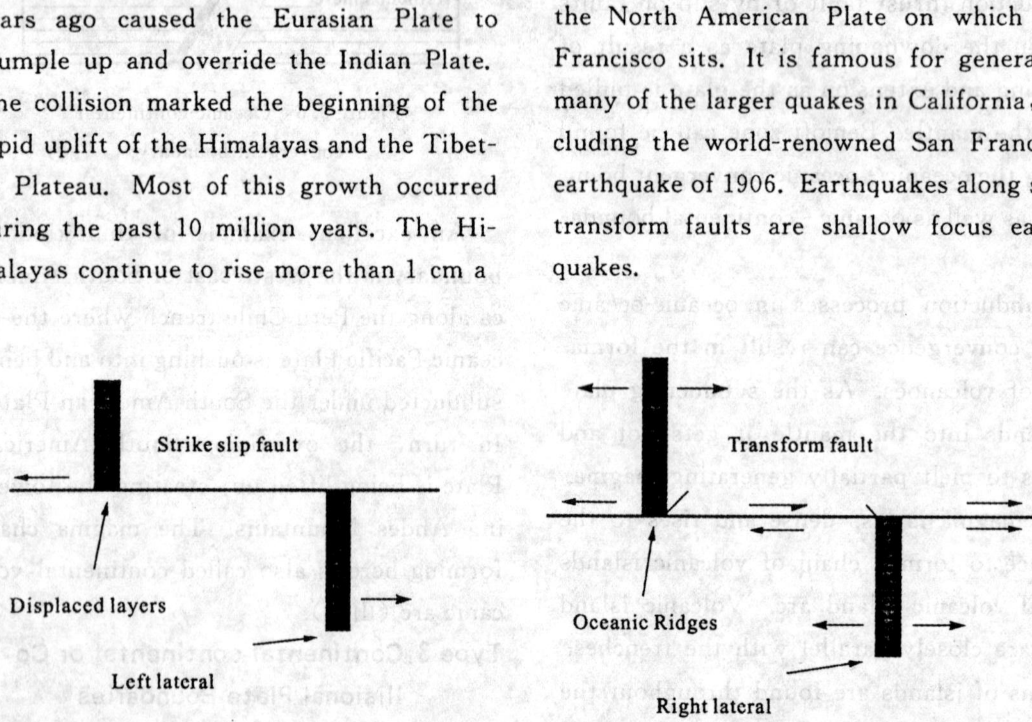

Figure 9. 8 Differences between strike-slip fault and transform fault

5 Earth's Major Plates

The surface of the Earth is divided into approximately seven large "plates", plus a number of smaller ones (Figure 9.9). The individual plate consists of both oceanic and continental portions.

- **North American Plate** North America, western North Atlantic and Greenland
- **South American Plate** South America and western South Atlantic
- **Antarctic Plate** Antarctica and the "Southern Ocean"
- **Eurasian Plate** eastern North Atlantic, Europe and Asia except for India
- **African Plate** Africa, eastern South Atlantic and western Indian Ocean
- **Indian-Australian Plate** India, Australia, New Zealand and most of Indian Ocean
- **Pacific Plate** most of the Pacific Ocean (and the southern coast of California)

6 Continental Margins

The top layer of the Earth's surface is called the crust (it lies on top of the plates). Continental crust is composed of granitic rocks which are made up of relatively lightweight minerals such as quartz and feldspar. By contrast, oceanic crust is composed of basaltic rocks, which are much thinner and denser than continental crust. Continental crust is lighter and never destroyed and is permanent. Oceanic crust is heavier so it can sink below continental crust. It is constantly being formed and destroyed at ocean ridges and trenches.

Continental margins are those areas which separate the continental crust and oceanic crust and defined as the steep slopes of the margins of the land masses below the ocean surfaces, forming 7%~8% of total ocean area. Typical continental margins should include continental shelf, slope and rise.

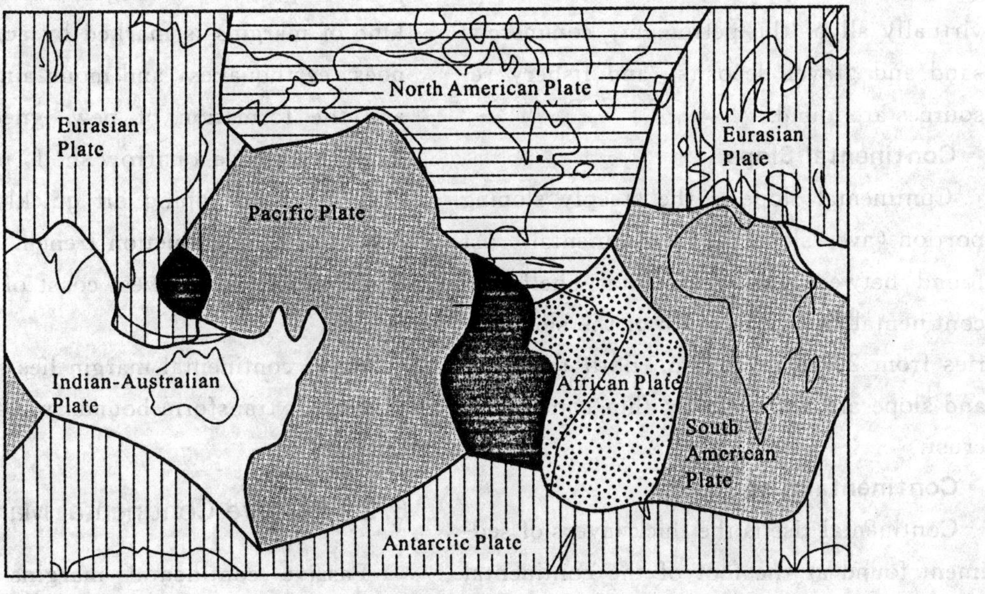

Figure 9.9 Major tectonic plates of the world

Table 9.1 The differences between continental crust and oceanic crust

Type of Crust	Average Thickness (km)	Average Age	Major Component
Continental Crust	33	3 billion years	Granite
Oceanic Crust	10	200 million years	Basalt

• **Continental Shelf**

Continental shelf is the gently dipping shallow submerged margin of the continents that extends from the shore to an abrupt break in slope known as the shelf edge. This nearly level area (ave. 1°~2° slope) of the continental crust has surface layers composed of sediment or sedimentary rock. Continental shelf is structurally part of the continental landmasses despite the fact that it is under water.

The water depth over the shelf is generally less than 200m. Continental shelves vary in width from almost zero up to the 1 500 km wide.

The continental shelves are the regions of the oceans best known and the most exploited commercially. It is this region where virtually all of the petroleum, commercial sand and gravel deposits, and fishery resources are found.

• **Continental Slope**

Continental slope is the steeply sloping portion (ave. slope 4°) of continental crust found between the continental shelf and continental rise. The width of the slope varies from 20km to 100km. Both the shelf and slope are underlain by thin continental crust.

• **Continental Rise**

Continental rise is the thick layers of sediment found at the foot of the continental slope. The depth of the rise ranges from 2 000m to 5 000m deep. Its breadth is up 300km wide. Although the rise is considered a component of the continental margin it actually lies on and is part of the ocean floor.

The continental shelf, slope, and rise together are called the continental margin. The shape of the continental margin reflects its tectonic evolution and subsequent geologic history. Continental margins can be of two forms.

6.1 Active Continental Margins

Active continental margins are also known as Pacific-type margin. It is a continental margin characterized by active tectonic movement at the leading edge of a continental plate where oceanic lithosphere is subducted. Because overlying plate is pushed up by subducting oceanic plate, this kind of margins is marked by active volcanoes, earthquakes, and mountain building, and the formation of new igneous rock. They often have a narrow shelf, narrow and deep slope, dropping off quickly into the depths of the subduction trench. An excellent example is the west coast of South America.

Active continental margin lies along convergent or transform boundary plate boundaries.

6.2 Passive Continental Margins

Passive continental margins are also known as stable continental margins or Atlantic - type margins (Figure 9.10). They

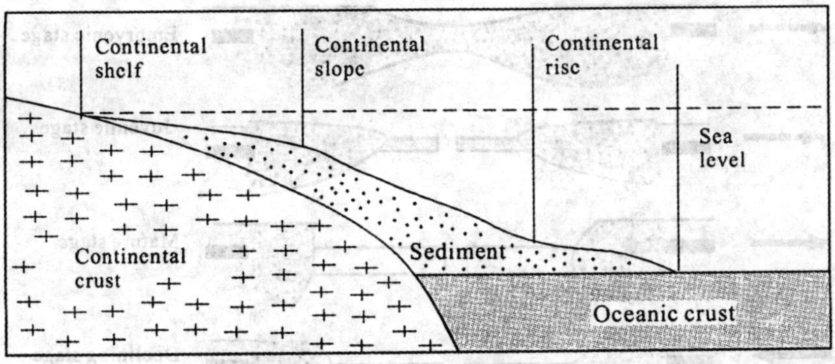

Figure 9. 10 Passive continental margin profile

are the trailing edge of a continental plate consisting of a broad continental shelf and a continental slope and rise. Passive margins are margins where the continent and the adjoining ocean floor are part of the same plate. Because it does not lie along active plate boundaries, tectonic activities, such as earthquake, active volcano and mountain building are minimal. These margins represent sites of former continental rifting, where the crust had been stretched and faulted. For example, the divergence of the African and South American plates along the mid-oceanic ridge resulted in the development of passive continental margins on either side.

Due to the well developed continental shelf, the passive continental margins often characterized by accumulation of sediments with great thickness.

The east coast of the North American is the good example of passive margin because it does not lie at a tectonic plate boundary.

7 Wilson Cycle

The Wilson Cycle is the cycle of events in the ocean from seafloor spreading (opening) to subduction (closing). It was proposed by famous Canadian geologist J. Tuzo Wilson, one of the fathers of plate tectonics. He developed the idea that throughout the history of the Earth the continents have combined and separated and then combined again and the process of opening and closing ocean basins is cyclic with a full cycle lasting something like one half or two billion years. This cycle was postulated to explain the numerous parallel mountain chains on the different continents.

The following Wilson Cycle model follows the series of cross sections constituting the Wilson Cycle (Figure 9. 11). Be aware that the divisions are arbitrary and it is much more important to understand how one process changes naturally into the next one because the Earth is an ongoing series of processes.

7. 1 Embryonic Stage (胚胎期)

Stage of formation of continental rifting. Due to the movement of mantle, the continental crust will be stretched and finally rifted. Break-up of the continent produces rift valleys. Current example is the East African Rift (东非大裂谷).

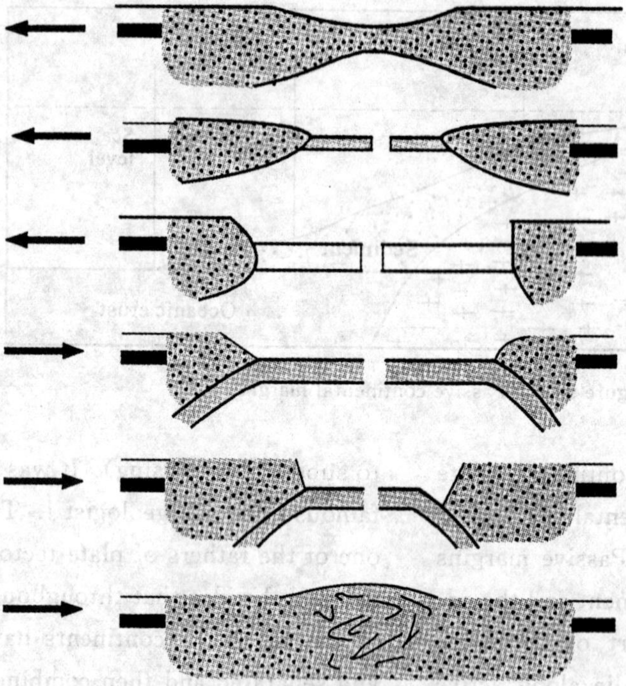

Figure 9. 11 Wilson Cycle

7. 2 Juvenile Stage (幼年期)

Stage of young ocean basin. At this stage, one continent has been separated into two continents, and a new ocean basin is generated between them. Modern examples are the Red Sea.

7. 3 Mature Stage(成熟期)

Stage of well-grown ocean basin. The ocean basin widens, sometimes, to thousands of miles. As long as the ocean basin is opening we are still in the opening phase of the Wilson Cycle. The typical example is the Atlantic Ocean today.

7. 4 Declining Stage(消亡期)

Stage of shrinking ocean basin and the closing phase of the Wilson Cycle. At this stage divergence stops and the two continents begin to move toward each other. A subduction zone (new convergent plate boundary) forms where the oceanic crust is oldest. Once the subduction zone is active the ocean basin will eventually disappear. The typical example is the Pacific Ocean today.

7. 5 Remnant Stage (残留海期)

Stage of shrunken ocean. Ocean basin dies and becomes a sea. Typical example is Mediterranean Ocean.

7. 6 Suturing Stage (缝合线期)

Stage of vanished ocean basin. At this stage, sea has gone and two continents collide and mountains form. The European and Asian land masses collided as part of the assembly of Pangea, thereby creating the Ural Mountains.

8 Cause of Plate Tectonics

Over the years, the plate tectonic theory has been used to explain mountain building, volcanic activity, earthquake and a host of other geological phenomena. But, some basic features of this theory have remained unresolved even today. One of the important but still unresolved problems is what forces are responsible for driving the plates?

Plate tectonic theory requires some system of horizontal forces that can cause plates to collide and combine or break up. The hypothesis of mantle convection proposed by Arthur Holmes nearly 90 years ago is still be employed to explain the mechanism of plate tectonic theory.

Mantle convection is essentially motion induced by buoyancy, with hot lighter material rising and cold denser material sinking. The convection currents within the mantle can drag and move the lithospheric plates above the asthenosphere, provided that asthenosphere could flow under stress like a liquid even though it is solid.

The scale of convection in the mantle is still unclear. Presently, two modes of mantle convection are competing for recognition. One of them, the two-layer mantle convection mode, regards the Earth's mantle to be split into two convecting layers in which convection in the lower mantle drives convection in the upper mantle, while the second or the "whole mantle convection" considers that the mantle convects as a whole.

Some other two important hypotheses were put forward to complement the mantle convection theory. "Ridge-push" hypothesis suggests that the ridge is topographically high, thus gravity pushes the oceanic lithosphere "downhill" towards the trench. And the "Slab-pull" hypothesis regards that subducting cold slab of oceanic lithosphere is denser than the surrounding warmer asthenosphere, thus it pull the rest of the lithosphere downward with it as it subducts.

The actual situation may be that: the combination of dragging the lithosphere along the top of asthenosphere, ridge push, and slab pull all appear to be contributing factors to the cause of plate tectonics (Figure 9.12).

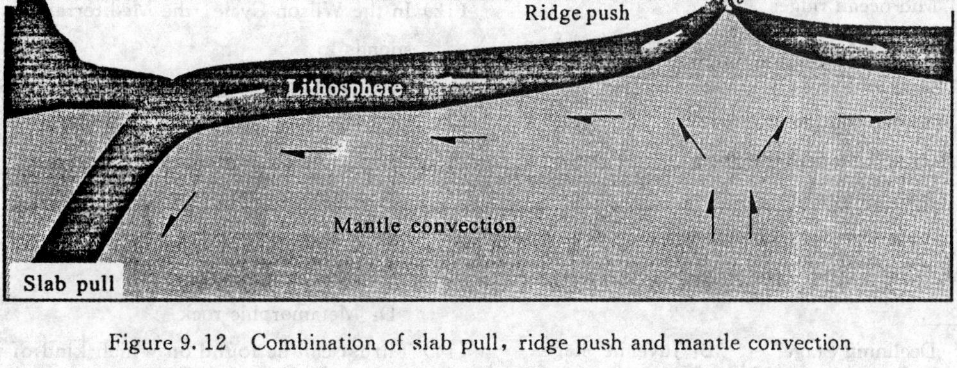

Figure 9.12 Combination of slab pull, ridge push and mantle convection

1. Answer the Following Questions

(1) Describe briefly the main ideas of "continental drift" proposed by Alfred Wegener.

(2) Define the Benioff Zone and relate it to the earthquake.

(3) Define the plate tectonic.

(4) Describe with a sketch the main characteristics of passive continental margin.

(5) Describe with a sketch the main characteristics of active continental margin.

(6) Explain the Wilson Cycle and list its different stages.

(7) What is continental shelf?

(8) Describe the divergent plate boundary.

(9) Describe in detail the convergent plate boundary.

(10) Name the major Earth's plates.

(11) Differentiate the asthenoshpere and lithosphere.

(12) What models have been proposed to explain the driving mechanism for plate motion?

2. Multiple Choice

(1) Active continental margins constitute most of the rim of what ocean?

A: Atlantic B: Pacific

C: Indian D: Arctic

(2) Which geologic feature is associated with divergent plate boundaries?

A: Deep-focus earthquakes

B: Deep-sea trenches

C: Island arcs

D: Mid-ocean ridges

(3) What is the east coast of the Taiwan (台湾) an example of?

A: Passive continental margin

B: Active continental margin

C: Convergent plate boundary

D: Transform plate boundary

(4) In the Wilson Cycle, the Pacific corresponds to ______.

A: Declining Stage B: Juvenile Stage

C: Remnant Stage D: Mature Stage

(5) Active continental margins constitute most of the rim of ______ ocean,

A: Atlantic B: Indian

C: Pacific D: Arctic

(6) What is the west coast of the Europe an example of?

A: Passive continental margin

B: Active continental margin

C: Convergent plate boundary

D: Divergent plate boundary

(7) Oceanic crust is composed mainly of ______.

A: Granite

B: Basalt

C: Sedimentary rock

D: Metamorphic rock

(8) Which geologic feature is not associated with convergent plate boundaries?

A: Deep-focus earthquakes

B: Deep-sea trenches

C: Island arcs

D: Mid-ocean ridges

(9) The single supercontinent proposed by Wegener is called ______.

A: Pangea B: Eurasia

C: Gondwanaland D: Laurasia

(10) Almost all deep-focus earthquakes occur along or near ______ plate boundary.

A: Divergent B: Convergent

C: Transform D: All of above

(11) Passive continental margins constitute most of the rim of ______ ocean.

A: Atlantic B: Indian

C: Pacific D: Arctic

(12) In the Wilson Cycle, the Mediterranean corresponds to ______ .

A: Declining Stage B: Juvenile Stage

C: Remnant Stage D: Mature Stage

(13) Continental crust is composed mainly of ______

A: Granite

B: Basalt

C: Sedimentary rock

D: Metamorphic rock

(14) Thrust can be found on which kind of plate boundary?

A: Divergent B: Convergent

C: Transform D: All of above

(15) The Benioff Zone is the zone of ______.

A: Volcano B: Earthquake

C: Rift D: Ridge

(16) Which one is not island arc?

A: Japan B: Iceland

C: Indonesia D: Aleutians

(17) What mechanism did Wegener use to explain the break up of Pangea?

A: Mantle convection

B: The rotation of the Earth

C: Gravitational force

D: Sea floor spreading

(18) Which of these phrases is not true for a passive continental margin?

A: These margins represent sites of former continental rifting.

B: Earthquake, active volcano and mountain building are minimal.

C: They generally have a broad continental shelf.

D: They are boundaries between different plates.

(19) Iceland is an (a) ______.

A: Island arc

B: Volcanic arc

C: Exposed ocean ridge

D: Submerged continent

(20) Graben can be found on which kind of plate boundary?

A: Divergent B: Convergent

C: Transform D: All of above

(21) The original mantle convection theory was devised by ______.

A: Harry Hess B: Alfred Wegener

C: J. Tuzo Wilson D: Arthur Holmes

(22) What type(s) of evidence did Wegener use to support his theory?

A: Matching rocks and mountain rages on different continents

B: Geographic fit of the continents on either side of the Atlantic

C: Matching fossils on different continents

D: All of the above

(23) The average thickness of continental crust is about ______.

A: 33 km B: 76 km

C: 10 km D: 5 km

(24) Which of these is continental volcanic arc?

A: Hawaii B: Philippines

C: Iceland D: Andes

(25) The Himalayas rise about ______ a year.

A: 1 cm B: 1 mm

C: 1 dm D: 1 m

(26) The continents drift about ______ a year.

A: 2 cm B: 2 mm

C: 2 dm D: 2 m

(27) Which one doesn't belong to Eurasian Plate?

A: Eastern North Atlantic

B: Europe

C: Asia

D: Western North Atlantic

(28) The term "plate" in the plate tectonic theory refers to ______.

A: Lithosphere

B: Crust

C: Asthenosphere

D: Rigid upper part of mantle

Chapter 10 Weathering and Soil

Key words

Weathering	Physical Weathering	Ice wedging	Crystallization	Thermal expansion and contraction	
风化	物理风化	冰劈作用	结晶	热胀冷缩	
Unloading	Chemical Weathering	Dissolution	Oxidation	Hydrolysis	Carbonation
卸载作用	化学风化	溶解作用	氧化作用	水解作用	碳酸化作用
Hydration	Differential weathering	Spheroidal weathering		Regolith	Soil
水合作用	差异风化	球形风化		风化壳(层)	土壤
Humus	Root wedging	Exfoliation		A-Horizon	B-Horizon
腐殖质	根劈作用	叶(页)状剥落,层裂		A层	B层

Weathering encompasses all the processes that cause rocks to fragment, crack, crumble, or decay. It is the breakup of rocks by physical or chemical ways due to exposure to the atmosphere. Because temperatures and pressures of the Earth surface are totally different from the condition under which most of the rocks and minerals are formed, the exposed materials are easily attacked by various chemical and physical surface processes. Thus, the weathering process takes place.

The hard minerals and rocks will be reduced to sediments eventually due to weathering. Those sediments are a major source for erosion and deposition. Some distinctive landforms, such as karst topography, are produced mainly by weathering. The rock fragments caused by weathering can be loosened and taken away by erosional agents, such as running water, wind and glacier. Weathering products, such as the clay, salt, iron ores and bauxite can be of considerable economic significance.

There are three broad categories of mechanisms for weathering: chemical, physical and biological.

1 Physical Weathering

Physical weathering is the breakdown of minerals or rocks by physical methods without any significant change in the chemical composition of the rock. Igneous rocks are very susceptible to mechanical weathering.

1.1 Thermal Expansion and Contraction

This kind of weathering occurs where rocks and minerals break up due to diurnal temperature fluctuations. The repeated heating by day and cooling by night cause the outer layers of the rock to expand and contract alternately. Since rock is a poor conductor of heat, the outer parts of a rock expand more than its interior. This leads to stresses being set up, which in turn causes the outer layers of rock peel off (Figure

10. 1).

Differential expansion and contraction may also be due to the variance in mineral composition in rock. Different minerals have different absorptive properties and expand at different rates. Therefore, in a rock consisting of many different minerals, rupturing can occur at the various mineral boundaries.

1. 2 Ice Wedging

Ice wedging is the most widespread form of weathering in cold temperate regions and mountain tops where temperatures fluctuate above and below freezing point. It is the top-down growth into soil of vertical wedges of ice. It occurs as the result of repeated freezing and thawing of ice in pore space or fractures. When water freezes, for example in a rock crack, it expands by 9%. The pressure produced by the ice wedge commonly rises to 8～18 kg/cm^2 and will cause the rocks to crack and open (Figure 10. 2). The initial size of the fracture is not important — over time the fracture will expand as the freeze/thaw cycle continues. The rocks will eventually be shattered along the cracks or joints into angular pieces.

In mountainous regions these particles may accumulate at the bottom of steep slopes forming talus slope (Figure 10. 3). Ice wedging is prominent in high-elevation, mountainous regions as well as in canyons during the wintertime. In engineering, ice

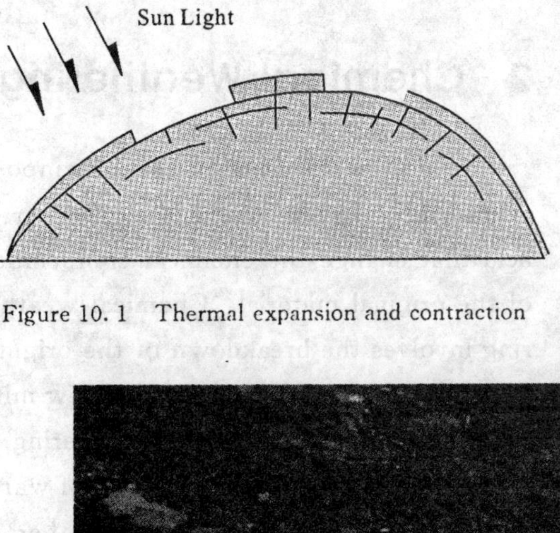

Figure 10. 1 Thermal expansion and contraction

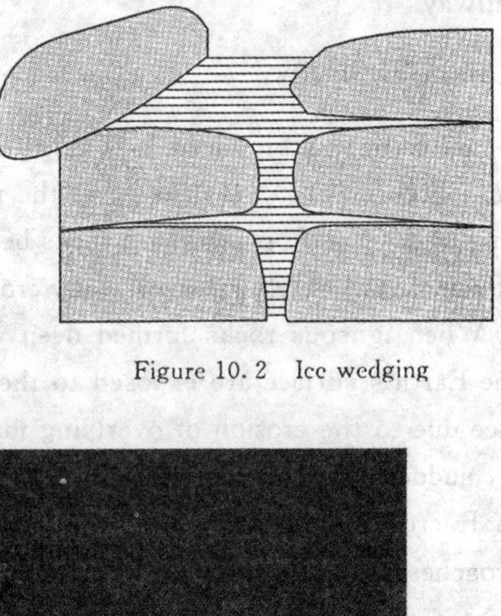

Figure 10. 2 Ice wedging

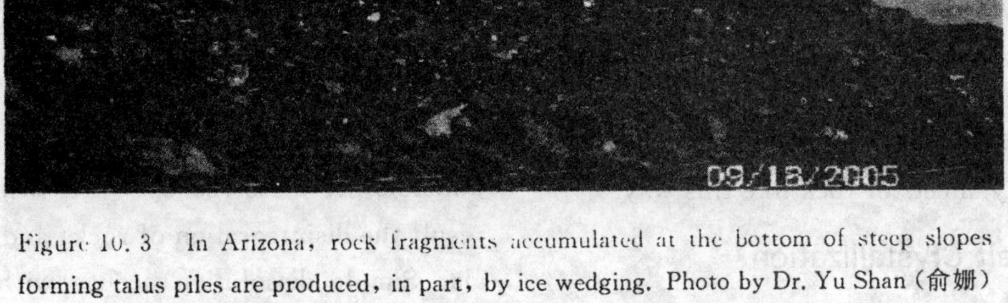

Figure 10. 3 In Arizona, rock fragments accumulated at the bottom of steep slopes forming talus piles are produced, in part, by ice wedging. Photo by Dr. Yu Shan (俞姗)

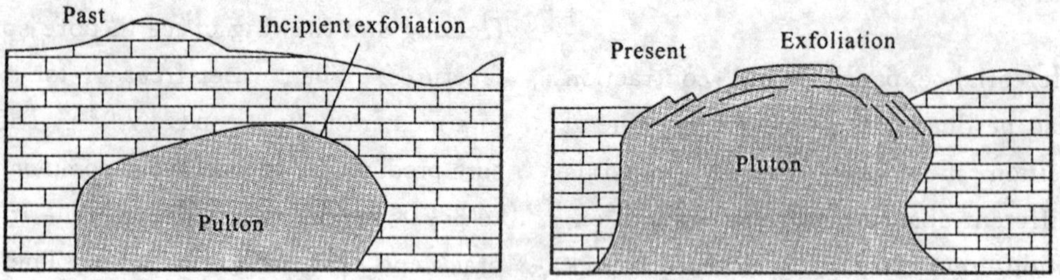

Figure 10.4 Exfoliation

wedging may be a serious problem sometimes. For example, the engineers are always pested by the ice when they want to construct road in cold and high-altitude areas, such as Tibet plateau(青藏高原). That is because the water accumulated in the roadbed will expand as it freezes, thus lifting and cracking the road surface, and sometimes destroy the bases of bridge and railway.

1.3 Exfoliation

Exfoliation is produced by unloading(卸载) (Figure 10.4). Unloading is the releasing of downward pressure on rocks because of removal of overlying material by erosion.

When igneous rocks formed deep under the Earth's surface are exposed to the surface due to the erosion of overlying material, sudden pressure changes cause the rocks to fracture horizontally. As the rock approaches the Earth's surface, the number of fractures will increase. In this case, huge sheets of rocks peel off like onion layers. Eventually large sheets of rock may"slide"off the rock creating a rounded upper surface, called exfoliation domes(叶状剥落穹隆、页状剥落穹隆). Granite tends to open and peel away sheets of rock like this way.

1.4 Salt Crystallization

In desert environments, rocks can be weathered by growth of salt crystals. Because the evaporation is very high in deserts, the groundwater containing dissolved salts will be drawn into the cracks of the rock. As the water evaporates, the dissolved salts will crystallize. Crystallization of the salts with a dramatic increase in volume will exerts an excessive pressure against the rock, forcing cracks open and breaking up the rock.

2 Chemical Weathering

Chemical weathering is the decomposition of rocks by the action of air, water or acid that change the chemical composition of the original material. Chemical weathering involves the breakdown of the original minerals within a rock to produce new minerals. The rate of chemical weathering is greatly accelerated by the presence of warm temperatures and moisture. A number of different processes can result in chemical weathering.

2.1 Dissolution

Dissolution occurs when rocks and/or minerals are dissolved by water which can result the disintegration of rocks and minerals. Simple dissolution occurs with easily dissolved substances like salt and gypsum.

2.2 Oxidation

Oxidation is the process by which an atom or ion loses an electron, and thus attains a positive charge. Oxidation reaction occurs upon contact of the rock with oxygen in air and water. One of the excellent examples is the reaction of oxygen with iron in minerals to form iron oxide (氧化物) minerals, such as hematite (赤铁矿). It is not unusual to see red-colored rocks in tropical or subtropical regions which have high temperatures and precipitation. For example, the red cliff in Wuyishan(武夷山,碧水丹山) is the result of oxidation of iron contained in rocks.

$4FeO + 3H_2O + O_2 \longrightarrow 2\,Fe_2O_3 3H_2O$

Gray-green iron Ⅱ oxide ⟶ rusty iron Ⅲ oxide

2.3 Carbonation

Carbonation is the reactions of minerals in rocks with carbonic acid (H_2CO_3) which naturally forms when carbon dioxide dissolves in moisture. Calcium carbonate (limestone) is particularly vulnerable to the carbonation. The carbonation of limestone results in karst topography. Carbon dioxide (CO_2) from the air is dissolved in rainwater to create a weak acid, carbonic acid (H_2CO_3).

$H_2O + CO_2 \longrightarrow H_2CO_3 \longrightarrow H^+ + HCO_3^-$

Rain + Carbon dioxide ⟶ Carbonic acid ⟶ Hydrogen ion bicarbonate ion that dissolves limestone:

$CaCO_3 + H^+ + HCO_3^- \longrightarrow Ca(HCO_3)_2$

Limestone ⟶ Calcium bicarbonate(soluble and leached away)

2.4 Hydrolysis

Hydrolysis is the reaction of minerals in rocks with acid water to form other products. It occurs in the form of replacement of mineral cations (e.g., Ca^{2+}, Fe^{2+}, Na^+, K^+, Al^{3+}) by hydrogen ions (H^+) from acidic water. It is the most significant chemical weathering process.

For example, the hydrolysis of orthoclase feldspar ($KAlSi_3O_8$) produces clay minerals such as kaolinite ($Si_4Al_4O_{10}\{OH\}_8$)(高岭石).

$4KAlSi_3O_8 + 4H^+ + 2H_2O \longrightarrow 4K^+ + Si_4Al_4O_{10}\{OH\}_8 + 8SiO_2$

Orthoclase + Hydrogen ion + Water ⟶ Potassium + Kaolinite + Quartz

In tropical climates, the clay mineral may then be broken down further forming bauxite(铝土矿), the ore from which aluminum is commercially extracted.

2.5 Hydration

Hydration is the disintegration of the rock or mineral due to the volume change caused by absorption of water into a mineral's crystal structure.

An example is the hydration of anhydrite (硬石膏) turning into gypsum(石膏) with a volume expansion of 35%.

$CaSO_4 + 2H_2O \longrightarrow CaSO_4\,2H_2O$

Anhydrite + Water ⟶ Gypsum

3 Biological Weathering

Biological weathering is the weathering process resulting from the action of organic materials. Biological weathering is not really a process of it's own, instead, they can assist in breaking down rock into sediment or soil by physical or chemical ways. Biological weathering agents involved in the weathering of rocks range from large mammals through to microfauna, from large plants through to bacteria.

3.1 Root Wedging

Root wedging occurs when plant roots grow into rock fractures causing further widening and possibly breaking the material apart. Trees growing in cliffs and hard bedrock exposed areas often sink their root into existing cracks and joints. As the root grows it forces the fracture to expand and finally breaks the rock into smaller pieces.

Root wedging is very important for soil development.

3.2 Animal Activity

Earthworms and termites make tunnels in the ground. Hazards caused by mice digging the holes contribute much to the deterioration of grasslands in some regions of Northern China. Human activities, including road construction, expanding of city during urbanization and agricultural production, cause the rock disintegration and shape the landforms.

3.3 Microorganisms

The biological weathering of microorganisms is mainly caused by corrosive acids and other materials produced during metabolism (新陈代谢) and humification (腐殖质化作用). Microorganisms modify rates and mechanisms of chemical and physical weathering and clay growth, thus playing fundamental roles in soil and sediment formation.

Lichens (地衣) are ubiquitous on exposed rock surfaces and help to chemically weather the underlying stone and to physically damage the stone by plucking out grains as they dry out. They often attack silicates and phosphates with organic acids.

Bacteria contained in soils and rocks play an important role in the decomposition of minerals. Bacteria occur even several kilometers deep in the lithosphere. They can convert atmospheric nitrogen to ammonia (氨), or oxidize it to nitrous acid (亚硝酸盐) or nitric acid (硝酸盐).

Biological weathering results in the formation of humus. Humification is the process by which organic matter decomposes to form humus. Humus is the organic matter that remains after most of the added plant and animal residues have decomposed.

4 Factors that Influence Weathering

The nature and rate of weathering at any one location is determined by the interaction between geological and environmental characteristics. Factors that play key roles are:

4.1 Climate(气候)

The most important factor affecting weathering processes is climate. Climate has three major components for weathering. The first is the temperature. As temperature increases, the chemical weathering of the rocks and minerals in the soil will be faster. Especially in tropical areas, the chemical weathering is extremely fast because the chemical reactions will be taking place faster and the reactions will occur throughout the year. In cold areas physical weathering will be more active. Along with temperature is the factor of precipitation or rainfall. Water is needed for nearly all of the chemical reactions that take place. Heavy rainfall promotes chemical weathering. The last factor is organic activity.

The influence of organism is very complex. Animal burrowing and root growing of plant can break up the rocks and minerals. Organisms can influence the local moisture and therefore enhance weathering. The carbon dioxide produced by plants can enhance the chemical processes like solution.

In general, weathering rates are lowest in cold dry climates and highest in warm, humid climates. For example, limestone in a dry climate is very resistant to weathering, in a tropical climate weather, it is very rapidly to be weathered.

4.2 Topography (地形)

Topography, or the relief of the land, can have a significant impact on weathering. Irregular, steep topography exposes more rock to weathering than horizontal surfaces. In fact, some landforms, such as karst topography, are created by the physical or chemical decomposition of rock through weathering.

4.3 Rock Type and Structure (岩石类型和构造)

Different rocks are composed of different minerals, and each mineral has a different susceptibility to weathering. For example sandstone consisting only of quartz is very stable on the Earth's surface, and will not weather at all in comparison to limestone, composed entirely of calcite, which will eventually dissolve completely in a wet climate.

Bedding planes, joints, and fractures, all provide pathways for the entry of water. A rock with lots of these features will weather more rapidly than a massive rock containing no bedding planes, joints or fractures.

If there are large contrasts in the susceptibility to weathering within a large body of rock, the more susceptible parts of the rock will weather faster than the more resistant portions of the rock. This will result in differential weathering. Excellent examples of differential weathering occur in red cliff in Wuyishan(武夷山) and the Grand Canyon in U. S. A (Figure 10. 5).

Another excellent example of joints influencing weathering is the spheroidal weathering. Spheroidal weathering is a form of

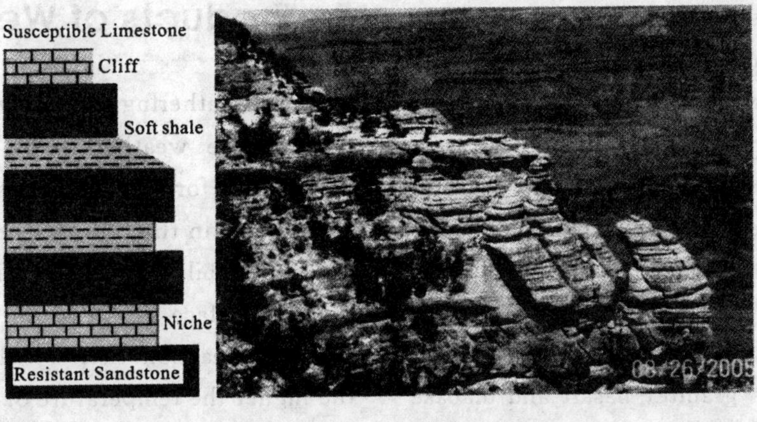

Figure 10. 5 The differential weathering in Grand Canyon. Photo by Dr. Yu Shan(俞姗)
The varied nature of the slopes of the Grand Canyon are expressions of the resistance of the different kinds of rock to erosion. The steeper cliffs are formed in sandstones and limestones. And the gentler rubbly slopes formed on shale.

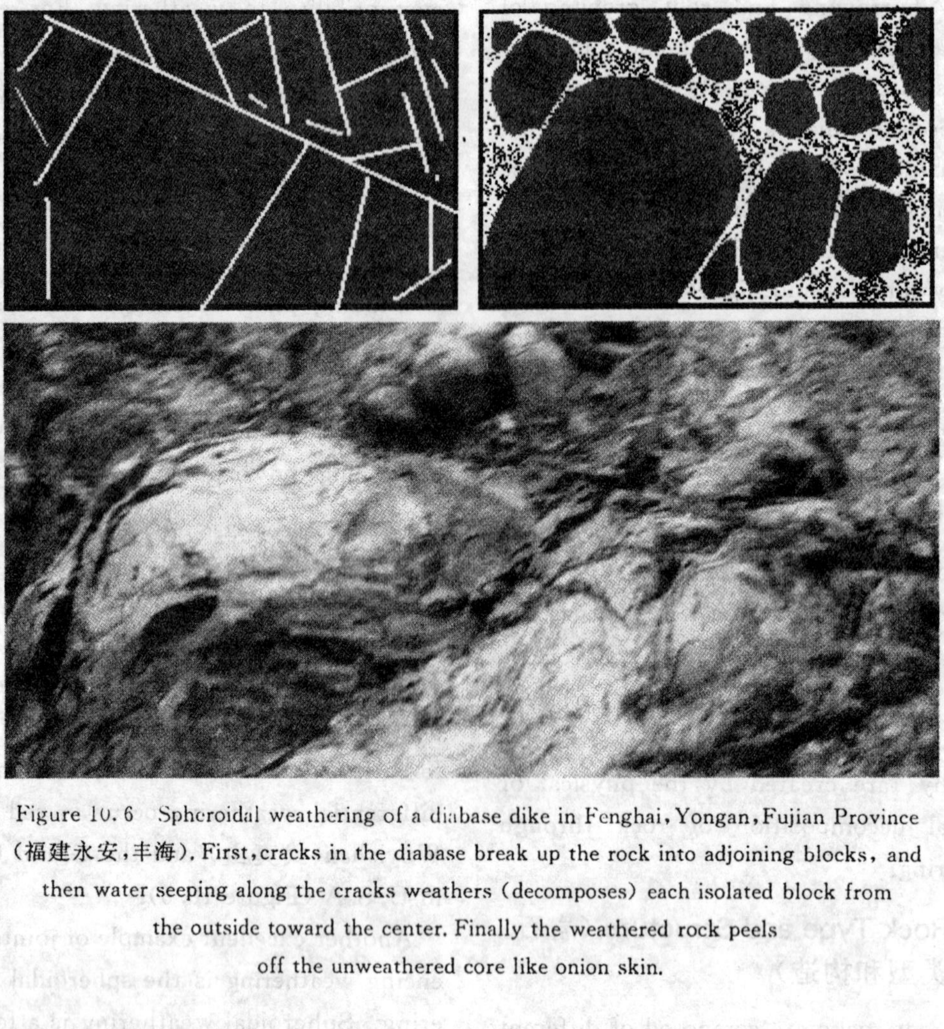

Figure 10.6 Spheroidal weathering of a diabase dike in Fenghai, Yongan, Fujian Province (福建永安,丰海). First, cracks in the diabase break up the rock into adjoining blocks, and then water seeping along the cracks weathers (decomposes) each isolated block from the outside toward the center. Finally the weathered rock peels off the unweathered core like onion skin.

chemical weathering enhanced by joints, which produces rounded surfaces on weathered rock (Figure 10.6). If joints and fractures in rock form a 3-D network, the water can penetrate more easily along these fractures. Thus the sharp corners and edges of rock tend to weather more rapidly and eventually the rock will weather into a spherical shape.

Those rocks with massive structure (块状构造), such as granite, basalt and diabase are readily to undergo spheroidal weathering, especially when they are fractured by joints.

5 Products of Weathering

Rock weathering is mainly accomplished through the weathering of mineral. The major rock-forming minerals actually are those listed in the Bowen's Reaction Series, but the stability of these minerals on the Earth's surface is found to be inverse of Bowen's Reaction Series. That is because the higher the temperature of formation of a mineral, the more unstable it is on the Earth's surface. Hence olivine weathers the most rapidly and the quartz is the most resistant to weathering. And that is why the quartz is one of the most abundant minerals in sedimentary rocks (Figure 10.7).

The final products of weathering are classified into three different groups.

5.1 Clastic Particle(碎屑物质)

It is also known as sediment, the loose material released by physical weathering. They are solid particles from mechanical weathering. Clastic particles include both the rock fragments and mineral particles. They are the main sources of clastic sedimentary rocks and main components of soil.

5.2 Soluble Materials(溶解物质)

Soluble material is the product of chemical weathering. They include Ca^{2+}, K^+, Na^+, Mg^{2+} and silica (SiO_2), which are usually dissolved in water. Dissolved ions will be transported by rivers to the sea, and will become part of the salts in the sea. They are the main sources of chemical sedimentary rocks.

5.3 Insoluble Residue(不溶残余物)

Insoluble residue (Table 10.1) are those aluminum-rich or iron-rich minerals, such as the clay mineral, bauxite and limonite(褐铁矿). They are the end products of weathering, i. e. the most stable form of silicate mineral (硅酸盐矿物) in the low temperature, high moisture sedimentary environment.

Let's take the granite as an example to show the procedure and products of weathering.

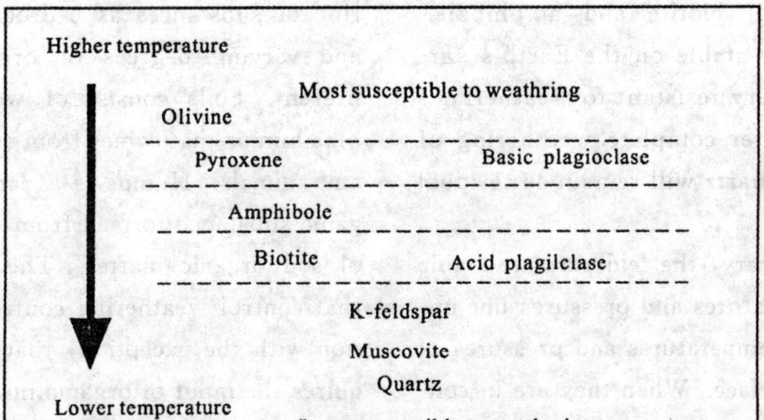

Figure 10.7 The weathering of minerals

Table 10.1 Products of weathering

Clastic material	Insoluble residue	Soluble material
Quartz	Clay minerals, quartz	Silica
Feldspars	Hematite	K^+, Na^+
Amphibole. Rarely olivine	Limonite	Ca^{2+}, Mg^{2+}
Various-sized rock fragments		

• **Physical Weathering**

When it is exposed the Earth's surface, the granite will be weathered firstly by various physical weathering processes. The rock is fractured and sometimes break up into different sized fragments. The rock fragments will be reduced in size furthermore by ongoing weathering. Once the rock is fractured by joints, the chemical and organic weathering will take place. In arid areas, the various-sized rock fragments and unweathered mineral particles will accumulate at the foot of hills or others low-lying region to form sediment. Physical weathering results in increased surface area for chemical reactions to occur on.

• **Chemical Weathering**

Unweathered granite consists mainly of plagioclase, K-feldspar, quartz, and lesser amounts of biotite and amphibole. Quartz is very stable on the Earth's surface, and is very resistant to weathering. Therefore, after complete weathering of granite, the quartz will remain as residual minerals.

On the contrary, the feldspars are stable at high temperatures and pressures but unstable at the temperatures and pressures of the Earth's surface. When they are in contact with acid water, the feldspars will undergo hydrolysis to form kaolinite (clay) and Na and K ions.

Feldspars and clay is similar in composition, but the latter are more stable under conditions at the Earth's surface. The clays form the insoluble residues and the Na and K ions will be dissolved in water. The biotite and amphibole are very unstable and easily weathered to form clays, dissolved ions and other materials.

6 Regolith and Soil

Regolith is the blanket of loose, unconsolidated rock particles (of various sizes) produced by weathering, that commonly lies on top of bedrock. The regolith may be a few centimeters to tens of meters thick.

The term soil is used in more than one sense. To an engineer it is a synonym for regolith, the aggregate of all loose rock material above bedrock. In geology, soil is usually defined as that portion of the regolith which is capable of supporting plant life. This definition excludes the lower part of the residual regolith.

Soil is an unconsolidated material formed from the chemical and/or physical weathering of substances as bedrock or outcrops and varying degrees of organic material present. Soils consist of weathered rock plus humus that comes from decaying plants and animals. Humus is a dark colored organic substance formed from decomposition of soil organic matter. The same factors that control weathering control soil formation with the exception, that soils also requires the input of organic material as some forms of carbon. Soil can be found in situ (in place, 原地) or transported.

Soils develop over long periods of time, perhaps ten thousands of years, as a response to the soil forming factors. With time, soils generally become deeper and develop distinct layers or horizons. The soil profile has three horizon types, from the surface downwards called A, B and C. Horizons can usually be distinguished by color differences, but closer study also shows

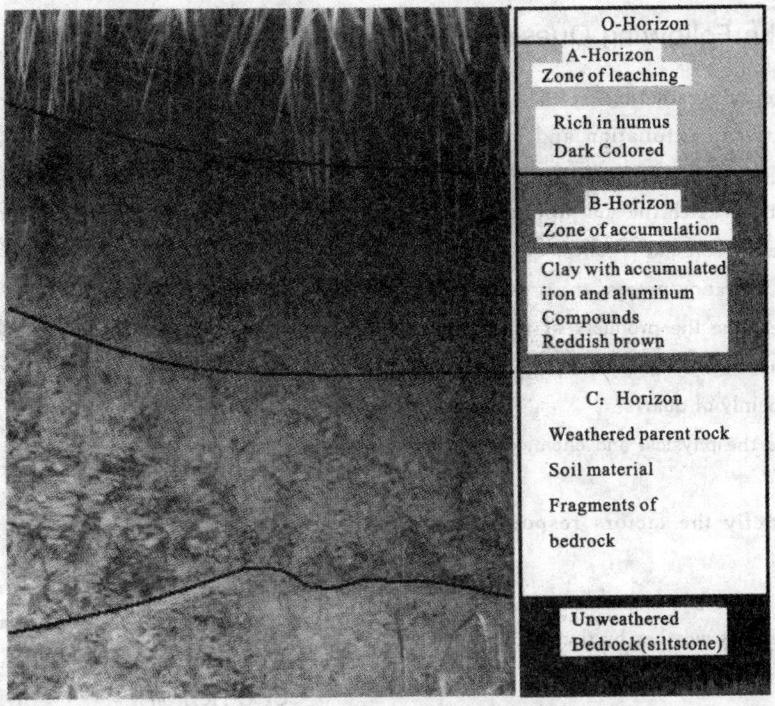

Figure 10.8 Soil profile

differences in chemical and physical properties. Individual horizons have defining characteristics. The arrangement of these horizons in a soil is known as a soil profile (土壤剖面 Figure 10.8).

6.1 O-Horizon(腐殖质层)

O-Horizon is the organic layer or humus layer. It is composed of organic debris, dead leaves and other plant and animal remains. In general, it is only a few cm thick.

An organic layer developed mainly from the humification process.

6.2 A-Horizon (A层、淋滤层、表土层、残积层、淋溶层)

A-Horizon is the zone of leaching. This is the dark colored top layer of soil (topsoil, 表土, usually less than 20 cm) that contains much humus.

A horizon is the zone of leaching where soluble ions and other particles, such as clay minerals, iron oxides, and dissolved calcite, are leached (淋滤)downward due to the downward movement of water or strongly acid solutions generated in the O-horizon.

6.3 B-Horizon(B层、淀积层、次土层、淋积层)

This is the white or red layer of subsoil (次土层) where leached minerals from horizon A precipitate. B-Horizon consists of clays, iron and aluminum oxides. This horizon represents a zone of accumulation of materials leached out of the A horizon.

6.4 C-Horizon(C层、母质层、母岩层、底土层)

This layer consists of slightly weathered parent material and rock fragments. It cannot support plant life.

1. Answer the Following Questions

(1) What is chemical weathering?

(2) Differentiate the exfoliation and spheroidal weathering.

(3) Describe with a sketch the soil profile.

(4) Differentiate the soil and regolith.

(5) Explain the hydration process.

(6) List and describe the products of weathering. What are the weathering products of sandstone consisting mainly of quartz?

(7) Differentiate the physical and chemical weathering.

(8) Discuss briefly the factors responsible for weathering.

(9) What is ice wedging?

(10) How is the rock weather by thermal expansion and contraction?

2. Multiple Choice

(1) A blanket of loose rock debris that covers large areas of the Earth's surface is referred to as ______.

A: Regolith　　B: Bedrock

C: Outcrop　　D: Soil

(2) Topsoil refer to ______.

A: A-Horizon　　B: B-Horizon

C: C-Horizon　　D: O-Horizon

(3) Which of the following is NOT a type of physical weathering?

A: Dissolution

B: Ice wedging

C: Root wedging

D: Thermal expansion.

(4) Hematite is the result of the chemical weathering of iron-rich minerals by the process of ______.

A: Dissolution　　B: Hydration

C: Oxidation　　D: Hydrolysis

(5) Which one is the most susceptible rock to chemical weathering?

A: Granite　　B: Sandstone

C: Mudstone　　D: Limestone

(6) Spheroidal weathering is a form of ______ weathering.

A: Physical　　B: Chemical

C: Biochemical　　D: All of them

(7) Which rock is readily attacked by acid water?

A: Granite　　B: Sandstone

C: Mudstone　　D: Limestone

(8) Which of these best describes weathering?

A: The removal of material by water or air.

B: the collapse of rock cliff by gravity.

C: The chemical breakdown of a rock.

D: the disintegration and decomposition of a rock on the surface of the Earth.

(9) Which of these minerals is most stable on the Earth's surface?

A: Potassium feldspar　　B: Quartz

C: Mica　　D: Plagioclase

(10) Which horizon is the zone of accumulation?

A: A-Horizon　　B: B-Horizon

C: C-Horizon　　D: O-Horizon

(11) To an engineer soil is a synonym for ______.

A: Rock　　B: Humus

C: Earth material　　D: Regolith

(12) Clay can be produced from plagioclase by a process of chemical weathering, called ______.

A: Oxidation　　B: Hydration

C: Hydrolysis　　D: Carbonation

(13) Which horizon has the highest content of humus?

A: A-Horizon　　B: B-Horizon

C: C-Horizon　　D: O-Horizon

(14) Which process can cause the outer layers of rock peel off?

A: Ice wedging

B: Differential weathering

C: Solution

D: Thermal expansion and contraction

(15) The caves in karst areas are formed by ______.

A: Oxidation　　B: Hydration

C: Hydrolysis　　D: Carbonation

(16) The red cliff in Wuyishan (武夷山) is the result of ______.

A: Oxidation　　B: Hydration

C: Hydrolysis　　D: Carbonation

(17) Some granite hills show a rounded shape with

nearly horizontal layered structure. It is the result of ______.

A: Exfoliation

B: Spheroidal weathering

C: Deposition

D: Differential weathering

(18) Unevenly cliff surface consisting of alternate layers of weak and strong rocks is the result of ______.

A: Exfoliation

B: Spheroidal weathering

C: Hydration

D: Differential weathering

(19) The most of clay minerals, iron oxides, and dissolved calcite are found in ______.

A: A-Horizon B: B-Horizon

C: C-Horizon D: O-Horizon

(20) Which one is the most stable mineral on the Earth's surface?

A: Salt B: Clay mineral

C: Plagioclase D: Mica

(21) The products of weathering are ______.

A: Sediments

B: Soil

C: Regolith

D: All of the above

(22) Hydrolysis is a process whereby water reacts with a mineral to produce new mineral by-products.

A: True B: False

(23) Which horizon is the zone of accumulation?

A: A-Horizon B: B-Horizon

C: C-Horizon D: O-Horizon

Chapter 11 Streams

Key words

Slopewash	Meandering streams	Braided streams	Discharge	Downcutting erosion	Rapid
坡积物	曲流	辫状曲流	流量	下蚀	急流
Waterfall	Lateral erosion	Channel bar	Point bar	Longitudinal profile	Flood plain
瀑布	侧蚀、旁蚀	心滩	点沙坝	河流纵剖面	河漫滩
River delta	Headward erosion	River capture	Degradation	Base level of stream	Runoff
河流三角洲	溯源侵蚀	河流袭夺	去均夷化作用	河流侵蚀基准面	(地表)径流
River terrace	Abandoned meander	Incised meander	River system	Coriolis force	Meander core
河流阶地	废弃河曲	深切曲流	水系	科里奥里力	离堆山
Drainage basin	Oxbow lake	Free meander	Alluvium	Rill flow, Gully flow	Headwater
流域	牛轭湖	自由河曲	冲积物	沟谷流水、洪流	河流源头
Aggradation	Slope wash	Alluvial fan	Diluvium	Graded river	Natural levee
加积	坡面冲刷	冲积扇	洪积物	均衡河流、均夷河流	自然堤
Capacity	Competence	Side bar	Saltation	Accumulation terrace	Abrasion
搬运量	搬运能力	边滩	跃运	堆积阶地	磨蚀作用
Terrace scarp	Terrace tread	The Davisian cycle		Planation surface	Peneplain
阶地坡	阶地面	戴维斯地貌循环说		夷平面	准平原

Streams are general terms applied to all bodies of water flowing in channels regardless of their size. Thus streams may vary in width from a few centimeters to several kilometers. Rivers are the largest streams and creeks are medium sized streams. Rivers always begin at their source, or headwaters, and flow downhill. Precipitation (rainfall and melted snow) moves downslope across the surface and eventually gathers into rivers. Rivers and other streams are major molders of modern landscapes, as they erode and transport the materials of the land gradually down to the sea. Stream is also a key factor influencing the human society. Most human population centers are located next to streams and some of the great rivers in the world are the cradles of ancient civilizations. Some of the earliest human construction projects were associated with the transportation and storage of water.

A river or stream is dynamic through time. Change is one of the most common features associated with river and stream channels. In general, most streams are continually changing position and shape as a consequence of hydraulic forces acting on their beds and banks. Basic natural factors that govern the stream development are geologic, hydrologic, hydraulic, and geometric factors.

1 Introduction

1.1 Stream Development

Precipitation is the start of land runoff. Most of precipitation evaporates back into the air and a small portion sinks into the ground and moves slowly downslope as groundwater(地下水). The water that does not seep into the ground, but flows atop the soil is called runoff. Runoff develops gradually from sheet flow, rill flow, gully flow and river.

• **Sheet Flow** (片流)

Sheet flow is the movement of water over slopes as a thin, broad layer of water. Sheet flow occurs over plane surfaces, when runoff water flows in a thin uniform sheet. The shallow flow of water on driveways and large sloping expanses of lawn are also the examples of sheet flow.

Erosion is the removal of the surface layer of the Earth's crust by the natural agents of wind, water and ice. The erosion of the Earth's surface starts from rain. The raindrop has an impact on the soil surface and detaches the particles. The detached soil particles are susceptible to transportation. Raindrop erosion is called splash erosion(溅射侵蚀). Sheet erosion is the erosion caused by the sheet flow when the shallow flow of water runs off the land. Sheet flow is laminar (layered,层流), so particles can only be displaced but not suspended. Laminar flow is a type of flow in which the surface, bed, and internal flow vectors are all parallel to one another(Figure 11.1).

The sheet flow erosion (slope wash) is sometimes powerful enough to overcome the soil's resistance and manages to carry a great deal of surface material down slope. Sheet flow and sheet erosion usually occur on surfaces that have a relatively flat slope. On steeper slopes, this generally occurs at the top of the slope prior to the water collecting into concentrated flow. Slopewash (坡积物)is soil and rock material that have been moved down a slope by running water.

• **Rill Flow**

Due to irregularities in the soil surface and uneven topography, sheet flow will usually begin to concentrate in the low spots of the irregular contours of the surface where the flow picks up velocity and erosive energy because of gravitational forces. Once the deepened sheet flow reach a critical depth where the laminar flow cannot be maintained and turbulent flow begins to develop, rill flow takes place. The rill flow is in the form of "turbulent flow". Turbulent

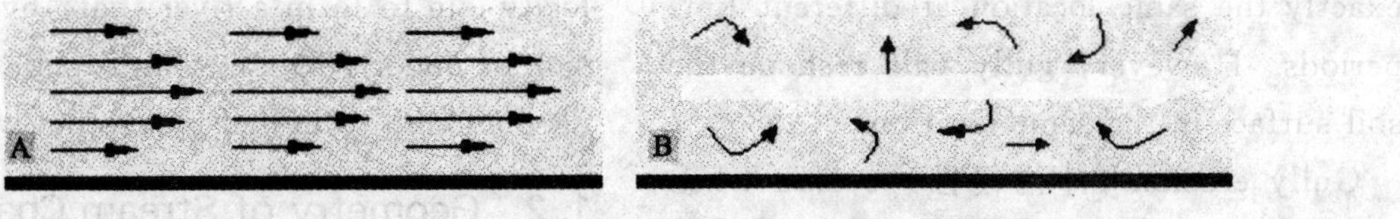

Figure 11.1 A. Laminar flow: Fluid layers move at different speed but still slide over one another without fluid exchange between layers.
B. Turbulent flow: Flow generally moves left to right but is highly disorganized

flow（紊流）is the motion of a fluid having local velocities and pressures that fluctuate complicatedly and randomly. In turbulent flow, unsteady vortices appear on many scales and interact with each other. Turbulent flow can keep sediment in suspension longer than laminar flow and aids in erosion of the channel bottom.

Rills（细沟）are small but well defined channels which are at most only tens of centimeters in depth. Rill erosion is the erosion that develops as sheet flow begins to concentrate into small grooves. Flow concentration occurs more rapidly on bare ground, sloping ground, and where long distances are involved. Rill erosion often occurs with sheet erosion and is commonly seen in paddocks of recently cultivated soils following high-intensity rainfall. It is easily identified as a series of little channels or rills up to 30 cm deep.

Sheet and rill erosion is still rampant in the some regions of China, especially in Northwestern China.

• **Gully Flow**

Gully flow occurs as the flow in rills comes together in larger and larger channels. Both the rill flow and gully flow carry the ephemeral flow and erode the soil. The difference between a rill and a gully is size and duration. Rill frequently changes its place, meaning that rill does not form in the exactly the same location at different time periods. However, gully will reshape the soil surface in the same location.

Gully erosion（冲沟侵蚀）occurs when water flows in narrow channels during or immediately after heavy rains or melting snow. Gully extends upstream as a result of waterfall undercutting and gravitational slumping of the gully head and widen through lateral erosion, where water undercutting causes subsequent slumping of the sides. Although gully flow has more energy to erode the soil and transport the sediments, in most areas more soil is lost through sheet erosion and rill erosion.

Diluvium (Lat. for deluge, from diluere, to wash away) is superficial deposits formed by flood. The term was formerly given to the boulder clay deposits, supposed to have been caused by the Noachian deluge. Alluvium is young sediment-freshly eroded rock particles that have come off the hillside and been carried by slow and steady stream water. Alluvial fans are fan-shaped deposits of water-transported material (alluvium+diluvium) and they typically form at the base of topographic features where there is a marked break in slope (Figure 11.2). This feature is formed when the speed of flow is suddenly greatly reduced when a stream rushing down the gully or valley reaches the gently sloping valley floor. When this happens, the stream is unable to carry much of its load and this sediment (usually sands and gravels) is laid down in a fan shape. Alluvial fans are quite common in arid regions where water is lost to evaporation and infiltration into coarse surface material when the stream exits the mountain front.

Runoff water of sufficient volume and velocity join to form a river that, by the erosion of underlying earth and rock, becomes deep enough to be fed by ground water.

1.2 Geometry of Stream Channels

The term river describes water moving through a well-defined channel. A river starts on hillsides as small channels, or rills.

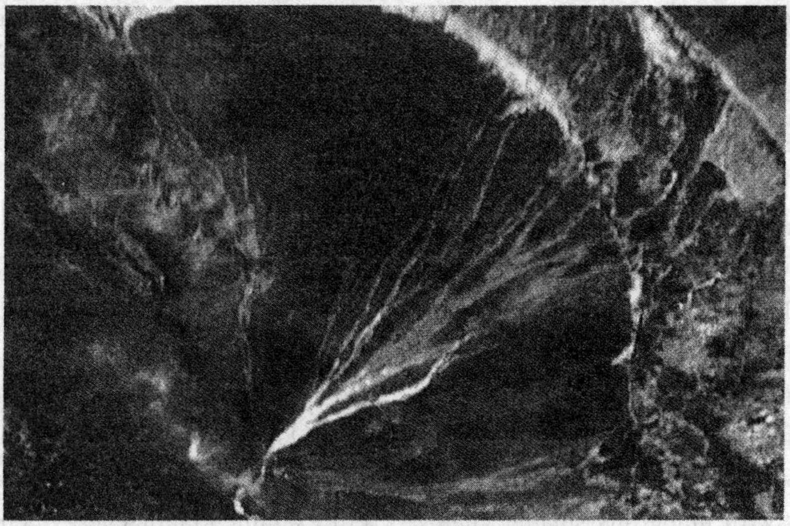

Figure 11. 2 Alluvial fan of Shule River (甘肃疏勒河)
Image credit: NASA-JSC-ES&IA, Public domain

The rills and gullies combine to make larger channels or tributaries that eventually come together, forming distinct streams. The beginning of a river is called its headwater or source. The place where the river empties into a larger body of water is called its mouth. One walks upriver when following a river towards its source, downriver when following it downhill. A tributary is a stream or river that flows into a larger stream or river.

Stream channel is the long trough-like depression that is normally occupied by the water in a stream. Stream channel (河道、河谷)is a kind of valley(峡谷). The bottom of a valley is called its floor (谷底). A valley's sides are called valley walls or valley slopes (谷坡). A valley wall is the side slope of a stream. The lateral confines (channel margins) during all but flood stage are known as the stream banks (河岸). As a general rule, the bed (河床)is that part of the channel below the "normal" water line, and the banks are that part above the water line.

The stream can be classified geometrically in two ways. One is channel cross-sectional shape(横切面), and another is stream pattern. There are three types of cross-sectional profile(横剖面) of valley, that is, V-shaped(V 形谷), U-shaped(U 形谷) and plate-shaped (碟形谷) valleys. The U-shaped and plate-shaped valleys are sometimes collectively referred to as flat-bottomed valleys(平底谷)(Figure 11. 3).

Stream patterns can be broadly classified as straight, meandering, braided or some combination of these classifications. The categories are not unique, for a stream can possess characteristics of straight, meandering, and braided in different reaches of its course.

• Straight Channel Stream

Straight stream channels are rare. Where they do occur, the channel is usually controlled by a linear zone of weakness in the underlying rock, like a fault or joint system.

• **Meandering Streams(曲流)**

Meander is a loop shaped bend in a river. It is characterized by a steep, laterally eroded outside bank, and a gently sloping slip off slope on its inside bank. Streams at lower elevations typically have a sinuous course characterized by numerous broad curves (meanders). The term meander comes from the word *rniandrus*, the name of a tortuous stream in Turkey which is known today as Menderes.

• **Braided Streams(辫状曲流)**

It is a river with numerous smaller channels which are continually dividing and reuniting. In between the individual channels, small islands tend to form, and these may be permanent enough for vegetation to colonize upon them (Figure 11. 4).

1. 3 Stream Kinetic Energy(动能)

Stream water possesses potential energy (势能), or the energy of position. As the water begins to flow down hill, this potential energy is converted into kinetic energy (动能), or the energy of motion. This kinetic energy can be used up in various ways. For example kinetic energy is used when the

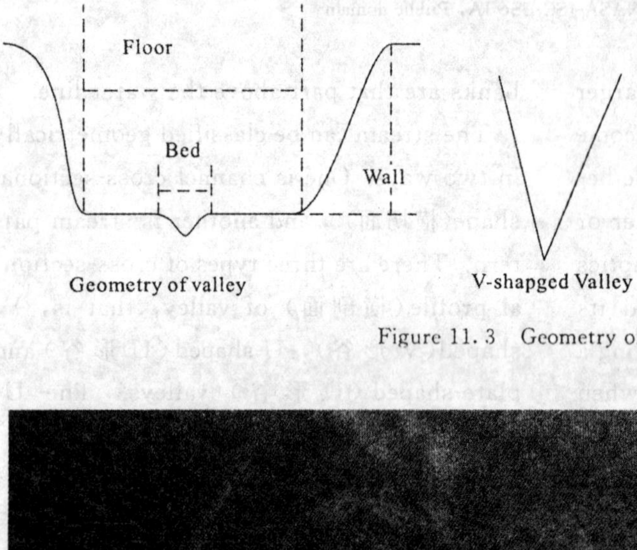

Figure 11. 3 Geometry of valley

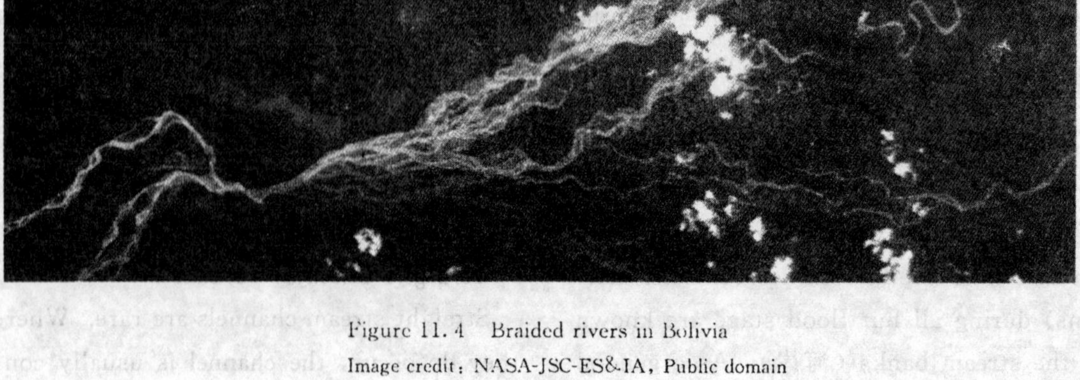

Figure 11. 4 Braided rivers in Bolivia

Image credit: NASA-JSC-ES&IA, Public domain

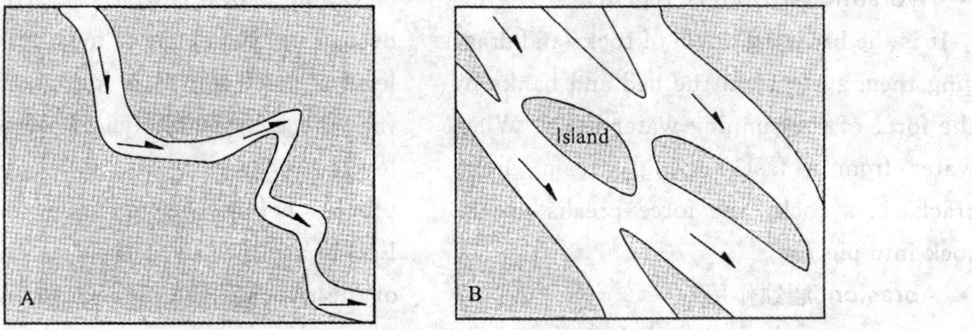

Figure 11.5 A: Meander B: Braided stream

water erodes away the surface of the stream channel it is passing through or moves sediments along the streambed. Hydroelectric power uses the kinetic energy of moving water to make electricity. Kinetic energy can be increased by having more water in the stream and or by having it move faster.

$$E = 1/2\ QV^2$$

Where Q is the discharge and V is the velocity.

Stream velocity is controlled by slope, channel shape and stream size, as well as the irregularities in the stream channel caused by resistant rock. Each of these factors is highly variable, both regionally and locally. These foctors are definitely interrelated, each working with the others to regulate stream flow. If all other factors were equal, stream velocity would be greatest in streams with the steepest gradients. For the most part of stream channel, the velocity increases downstream even as stream gradient decreases because stream size increases (due to the addition of water from tributary streams) and decreasing channel roughness.

Discharge is the amount of water flowing past a given point in a given amount of time.

$$Q\ (m^3/s) = A\ (m^2) \times V\ (m/s)$$

Where A is the cross-sectional area and V is the velocity. The units of discharge are m^3/s.

The volume of the discharge will be determined by factors such as climate, vegetation, soil type, drainage basin relief and the activities of man. Discharge increases downstream because more water is added through rainfall, tributary streams, or from groundwater seeping into the stream. As discharge increases, generally the velocity of the stream also increases.

Though the Nile River (尼罗河) is the world's longest river, South America's Amazon is wider and has the greatest discharge of any river.

2 Stream Erosion

Erosion is the mechanical process of gradual wearing something down or the condition in which the Earth's surface is worn away by the action of water and wind. Streams are a major force of erosion and river erosion is the gradual removal of rock material from the riverbanks and bed. A stream erodes rock/sediments in three ways.

• **Hydraulic Action**(水力作用)

It is the breaking down of rocks and dragging them away from the bed and banks by the force of the running water itself. When water from a fast moving stream enters cracks in a rock, the force breaks up the rock into pieces.

• **Abrasion**(磨蚀作用)

It is the grinding of rock fragments carried by the river against the streambed and banks of the river. The river uses its load to grind against the bed and sides and dislodge the materials and carry them away. This action both widens and deepens the channel. The power of the grinding is especially powerful in floods when large fragments of rock are carried along the riverbed eroding the bed and the banks.

• **Dissolution**(溶蚀作用)

It is the process by which river water reacts chemically with soluble minerals in the rocks and dissolves them. For example, when a river stream flows over an area of limestone (calcium carbonate), it erodes and dissolves the limestone by reacting chemically with it.

According to the erosion direction, the stream erosion can be summarized as three different ways.

2.1 Downcutting Erosion

Downcutting erosion, also known as downward erosion, can result in a deepening of the stream channel and valley. The downhill movement of stream water has potential energy as well as kinetic energy, thus it can erode the bottom of river.

The amount and rate of downcutting erosion are dependent on the base level of the stream. Base level is the theoretical limiting elevation beneath which a stream cannot deepen its channel. For most streams, base level is sea level. It is impractical to think that the continents could be eroded to sea level, because uplift will interrupt the process before the end point is achieved. Local base level is defined as the elevation of the river or lake into which the tributary flows. Hence, the base level for the Hanshui River (汉水) in Northwestern Hubei (湖北) is the level of the Changjiang River (长江). The land in Northwestern Hubei (湖北) cannot be reduced in elevation below the level of the Changjiang River (长江). Sometimes variations in bedrock and topography, such as the resistant body of rock and natural or artificial dams, can also result in a local base level above the ultimate sea level. For example, the building of Three Gorges Dam (三峡大坝) can change the base level of the upper reaches of Changjiang River. Due to the rising of local base level caused by construction of Sanmenxia (三门峡) Reservoir and the excessive deposition of sediments in the reservoir, the Guanzhong region (关中地区) has been frequently flooded by Weihe River (渭河), which empties itself into Yellow River at Tongguan (潼关), a city to the west of Sanmenxia (三门峡) (Figure 11.6).

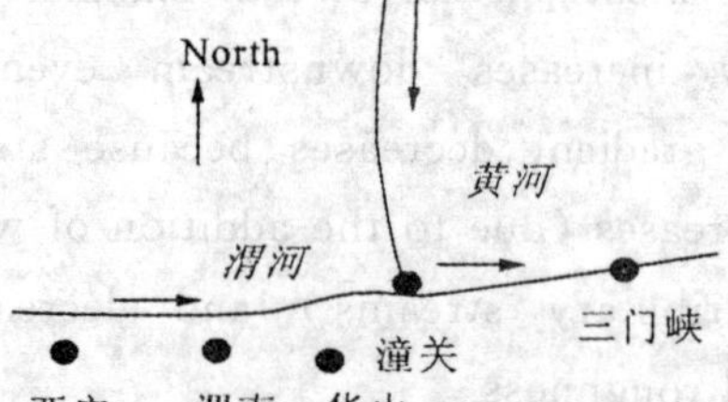

Figure 11.6 Map of Weihe River

The longitudinal profile of a stream is a concave upward curve plotted by elevation versus distance, showing a steep gradient near its headwaters and a gentle gradient near its mouth (Figure 11.7). With the variation of topography and rocks, the stream will adjust its base level dynamically by adjusting its long profile. In the example here, due to the construction of dam, the erosion will occur downstream from the dam and deposition will take place in the reservoir causing the gradient to become lower. In general, it reflects dynamic equilibrium between tectonic uplift, erosion, transport, and deposition. Base level is the lowest level of longitudinal profile.

In reality, of course, the smooth concave profile is rarely achieved: Irregularities such as resistant rock outcrops result in a "step" in the profile, marked by waterfalls or rapids.

Rapid is a part of a river where the current is very fast and forms when there is a resistant rock dips downstream and river flows on it. However, not only resistant rocks can make it happen. Many factors including bedrock condition, gradient of riverbed, water volume may lead to the formation of rapid.

A waterfall is usually a geological formation resulting from water flowing over an erosion-resistant rock formation that forms a sudden break in elevation (Figure 11.8). Waterfalls come in all shapes and sizes and have a variety of ways of forming.

The stream waterfall is usually formed by rocks that alternate in resistance to stream erosion (Figure 11.9). If a weaker bed of rock is downstream of a more resistant rock, the weaker bed will be eroded faster, leaving a steep scarp and a waterfall. Often, the rock strata just below the more resistant shelf will erode out to form a shallow cave-like formation known as a rock shelter (also known as a rock house) under and behind the waterfall (Figure 11.10). In time, the

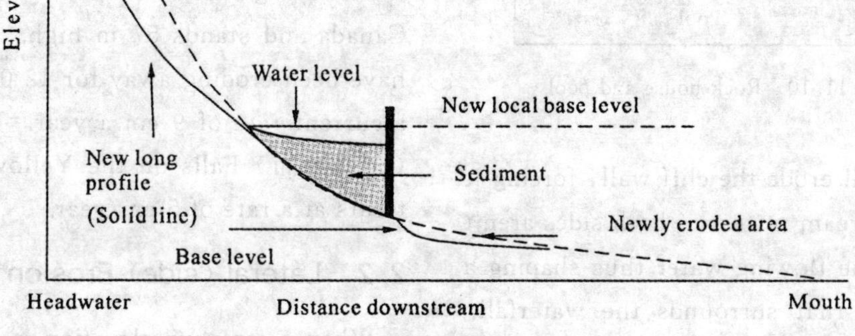

Figure 11.7 Typical long profile

Figure 11. 8 Shuzheng Falls in Jiuzhaigou (九寨沟树正瀑布). Photo by Chen Fuxing(谌福兴)

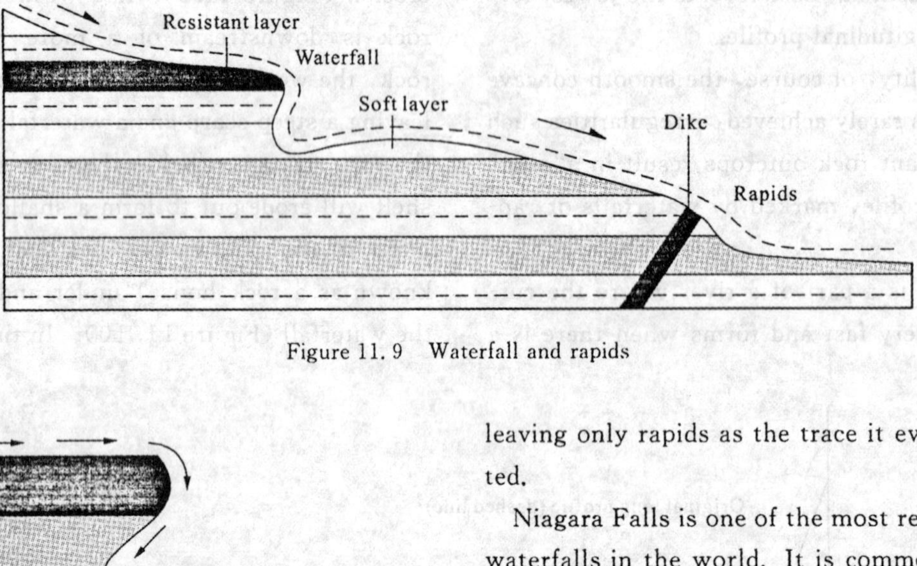

Figure 11. 9 Waterfall and rapids

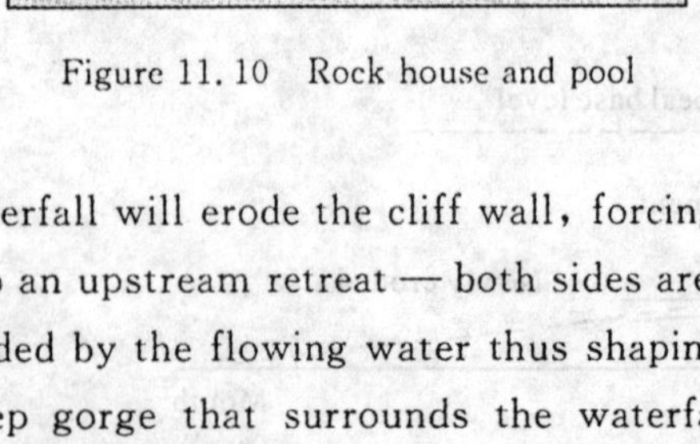

Figure 11. 10 Rock house and pool

waterfall will erode the cliff wall, forcing it into an upstream retreat — both sides aren't eroded by the flowing water thus shaping a steep gorge that surrounds the waterfall. As this action continues, eventually after many years, the waterfall will disappear leaving only rapids as the trace it ever existed.

Niagara Falls is one of the most renowned waterfalls in the world. It is commonly referred to the honeymoon capital of the world. Niagara Falls divides the U. S. and Canada and stands 57 m high. These falls have been eroding away for 12 000 years at a current rate of 9 cm a year. The Hukou (壶口瀑布) Falls on the Yellow River retreats at a rate of 5 cm/year.

2. 2 Lateral (side) Erosion

When a stream's elevation is close to its base level, downward erosion will give way to lateral erosion. Lateral erosion undermines part of the bank or valley wall, leading to lateral migration of the channel and widening of the valley.

Lateral erosion is caused by the combination of centrifugal force and Coriolis force.

Coriolis force is an apparent force due to the Earth's rotation. Coriolis isn't even a real force, since it doesn't make anything speed up or slow down — it only explains why things appear to speed or slow as the world spins out from under them. The effect of the Coriolis force is an apparent deflection of the path of an object that moves on the Earth's surface (Figure 11.11). On the Earth an object that moves along a north-south path, or longitudinal line, will undergo apparent deflection to the right in the Northern Hemisphere and to the left in the Southern Hemisphere. The object does not actually deviate from its path, but it appears to do so because of the motion of the coordinate system. Even the straight stream channel will be bended by the lateral erosion due to the action of Coriolis force.

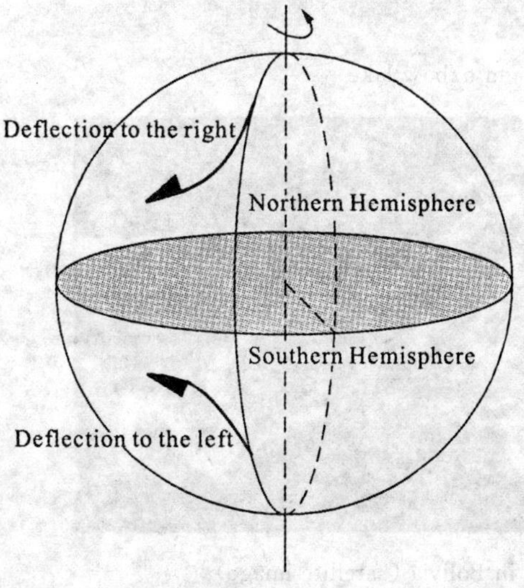

Figure 11.11 Coriolis force

Once the channel is slightly bended, the centrifugal force will take effects. Centrifugal force causes the water surface elevation to be highest at the outside of the bend and lowest at the innerside of the bend. As gravity pulls the additional mass of water at the outside of the bend downward, a rolling, helical spiral is created (Figure 11.12). This downward erosive force, coupled with the stream velocity, can undercut the toe of the bank. The most severe erosion will occur immediately downstream from the point of maximum curvature.

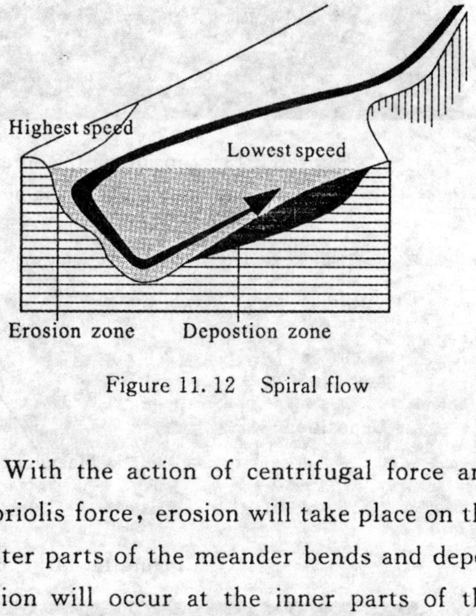

Figure 11.12 Spiral flow

With the action of centrifugal force and Coriolis force, erosion will take place on the outer parts of the meander bends and deposition will occur at the inner parts of the bends (Figure 11.13). During the lateral cutting, the river widens its bed and valley and finally V-shaped valley begins to give way to a flat-bottomed valley. At sometime, the stream channel develops a sinuous course that forms exaggerated loops and bends called meander. Once a stream channel begins to meander there is a strong tendency to maintain and increase the curve. A meander will become more pronounced as the streamline shifts between the two banks. In this way the stream channel tends to migrate laterally in the direction of the outside curve and the curve or meander grows. River cliff or bluff is a steep, near vertical cliff on the outside bend of a meander. It is created by the lateral erosion of the meander as the outside bend of a meander is the river's fastest area of velocity.

Free meanders refer to meanders occurring in unconsolidated alluvium, free to migrate and develop waveforms without constraints from valley walls (Figure 11. 14). Oxbow lake is a crescent shaped abandoned meander loop (Figure 11. 15). When a mea-

Figure 11. 13 Erosion on the outer parts of the meander bends and deposition at the inner parts of the bends. Photo by Dr. Yu Shan (俞姗)

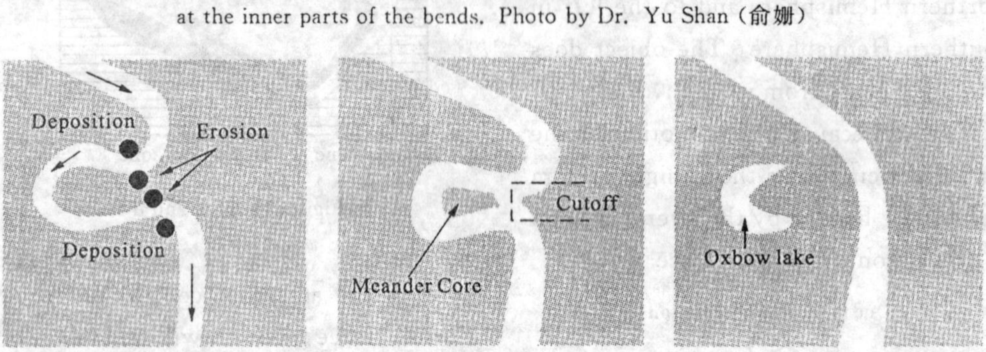

Figure 11. 14 Free meander and oxbow lake

Figure 11. 15 Meander and oxbow lakes in Bolivia (satellite image)
Image credit: Nasa-JSC-ES & IA, Public domain

nder has almost formed a complete loop, the narrow neck of land may be cutoff in flood conditions to form a new, shorter channel, allowing the river to cut the bend. Silt begins to build up at the entrances to cutoff section, eventually creating an ox-bow lake.

2.3 Headward Erosion

Headward erosion is the deepening and widening of the channel uphill, resulting in making the valley longer. This headward erosion occurs where the valley profiles are the steepest, and where the valleys are occupied only by intermittent streams. Waterfall retreat is a local headward erosion.

The streams also simultaneously erode headward extending their length as they gnaw away into the surrounding hillsides and plateaus. Continued headward erosion may connect different channels and result in the development of river system and drainage basin.

A river and all of its tributaries is called a river system (Figure 11.16). Drainage basin is the land drained by a particular stream or and its tributaries. It may be grouped with other drainage basins to form a larger drainage basin. The size of a drainage basin can vary from being as small as a few square miles or kilometers to as large as part of a continent. Basins can be drained by a single river or stream, or can consist of a huge network of rivers, including thousands of streams and creeks. There is no limit to how many tributaries a stream may have. The vast area drained by the Changjiang River(长江) covers about 1.8 million km^2 in area, which is about one fifth of the total area of the whole country. Divide is the boundary line or highland between adjacent drainage basins. The most important divide in China is the Qinling-Dabie(秦岭-大别山). Another example of divide is the continental divide of North America, which separates streams that ultimately empty into one ocean (the Pacific Ocean) from those that ultimately empty into another (the Gulf of Mexico).

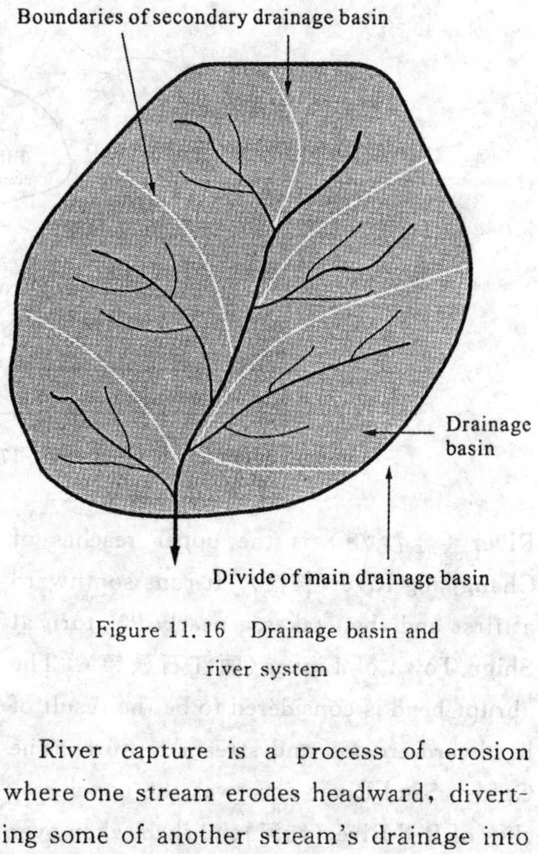

Figure 11.16 Drainage basin and river system

River capture is a process of erosion where one stream erodes headward, diverting some of another stream's drainage into its own channel (Figure 11.17). The stream with water being diverted is the captured stream(被夺河)and the stream diverting other stream's water is the capturing stream(袭夺河). Where the capture has occurred there is a sharp bend, called the elbow of capture(袭夺弯). The remaining captured stream is called be-headed. River capture creates a short streamless valley referred to as a wind gap(风口). Golden Sand

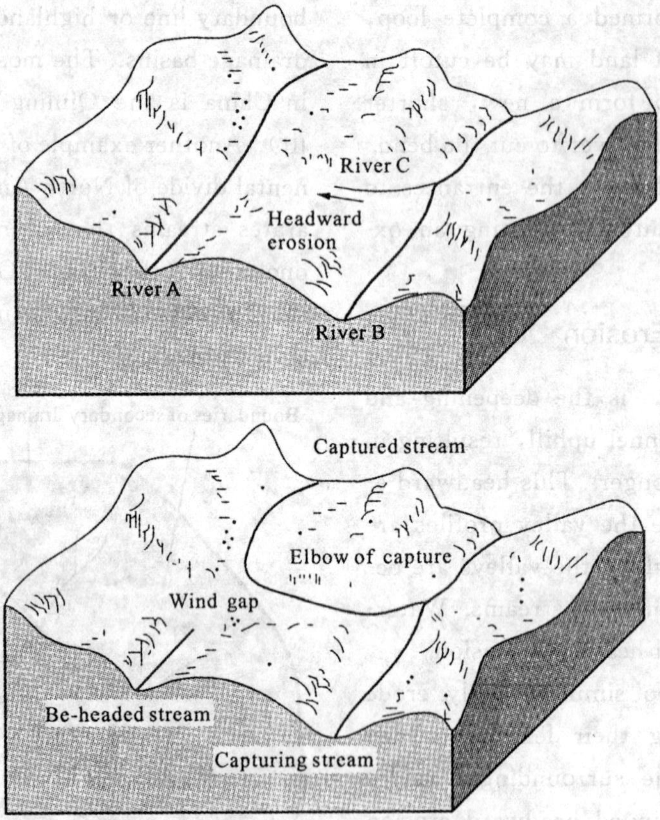

Figure 11.17 River capture

River (金沙江) is the upper reaches of Changjiang River(长江). It runs southward at first and then takes a nearly 90° turn at Shigu Town of Lijiang(丽江石鼓镇). The abrupt bend is considered to be the result of headward erosion and stream capture. The Golden Sand River once was the upper reaches of Red River(红河)and then taken over by Changjiang. However some geologists claimed that the Shigu bend is controlled by fractures, because there is no reliable evidence to support the hypothesis of headward erosion.

3 River Transportation

A river uses its energy to carry or transport eroded materials such as mud, sand, boulders and dissolved materials. These materials are called its load(搬运物). According to the transportation processes, loads can be classified as bed load(底运物, moving across the bed of the channel), suspended load(悬运物) and dissolved load(溶解物质). Stream loads are transported by the following ways (Figure 11.18).

3.1 Traction(拖运)

Traction is the movement of particles by rolling and sliding along the river bed.

3.2 Saltation(跃运、跳运)

Saltation is a"leap-frog"motion of material as it bounces downstream. Large particles such as gravel and coarse sand are lifted and dropped along the river, so they bounce

Figure 11.18 Stream transportation

along the river bed in a series of jumps. Bed loads are transported by traction and saltation.

3.3 Solution(溶运)

Solution is the means by which dissolved particles are carried in the stream.

3.4 Suspension(悬运)

Suspension is the means by which small particles of rock and minerals are carried up in the body of the stream. Smaller particles such as clay, silt and fine sand are carried along without contact with the riverbed. These materials are heavier than water, but they are stirred up and kept from sinking by the turbulence of the stream flow. Materials carried in suspension usually form the greatest part of the total river load.

In general, stream transportation of sand and gravel is accomplished as bed load (movement by traction or saltation). Silt and clay are transported by suspension in the water. Dissolved load is invisible — it is the ever-present soluble material which results from chemical weathering of the rocks along the channel. Suspension and solution comprise the bulk of a stream's load.

The carrying power of a stream is indicated by both the total amount of sediment in the stream and by the size of the particles being moved by the stream. The carrying power depends upon the speed of the stream and it's discharge. Competence is measured by the largest particle size the river can carry and will increases with velocity of river. And capacity is the total load of sediment that the stream can carry and increases with discharge. Capacity and competence are not necessarily related. Large slow rivers (Changjiang, 长江) have high capacity but low competence. Mountain streams have low capacity but high competence.

4 Stream Deposition

Deposition is the process when soil and regolith moved by erosional forces is set back down. Leaving the high elevations of the mountains and hills and entering the flat plains, the river slows down. Once the river slows down, the pieces of sediment have a chance to fall to the river bottom and be "deposited". This is river deposition.

River deposition is caused by variation of velocity at different parts of channel, decreasing in discharge and/or increasing in carried loads. Aggradation refers to the vertical build up of the riverbed due to deposition of solid materials. When sediment and gravel deposits in the lower river and essentially raises the riverbed, aggradation occurs.

Landforms produced by deposition include.

4.1 Channel Bar

Channel bar or mid-channel bar is the sand deposit in the course of stream. There are many causes of bar formation, such as the widening of the channel and tributary entrances. Here we will take the widening channel as an example to demonstrate the formation of channel bar (Figure 11.19).

When stream flows from a narrow channel to broad channel, the stream velocity will decrease rapidly, resulting in the deposition of coarse sediments at the center of channel. The deposition of coarser materials in the center of the channel symbolizes the initiation of bar growth. Once the bar is large enough to direct flow towards the adjacent banks, bank erosion occurs. This divergence shifts the maximum flow velocity to the edges of the channel. Due to the various reasons, flow on one side of the bar will become more dominant and eventually the bar becomes attached to the floodplain at other side. In this case, an asymmetrical bar referred to as a side bar will form. The Orange Bar in Changsha(长沙橘子洲) and Parrot Bar in Wuhan(武汉鹦鹉洲) may perhaps be the most famous channel bars in our country. Unforturenately, the Parrot Bar now has been submerged under the water. Hundreds of years later, another channel bar may accumulate adjacent to the former famous Parrot bar. Who knows?

4.2 Point Bar

Sand bar developing on the inside bank of the meander bend while the channel is migrating and the outside bank is eroded. It is covered by water only during a flood. Location of point bar is clearly influenced by the meandering path. As the water spirals toward the inside of the meander, it is slowed by frictional drag imposed by the bed of the channel. This causes deposition of alluvium on the inside bank to form a bar.

4.3 Flood Plain

Floodplain is the flat low-lying area located on either side of a river channel and subject to inundation under flood-stage conditions (Figure 11.20). Streams naturally meander and the meandering process eventually results in the evolution of a floodplain adjacent to the stream channel, except where valley characteristics prevent floodplain formation.

Floodplain is usually a constructional landform built of sediment deposited during repeated flooding and lateral migration of the stream. In a flood, the level of the water may rise and overflow its bank . At first the

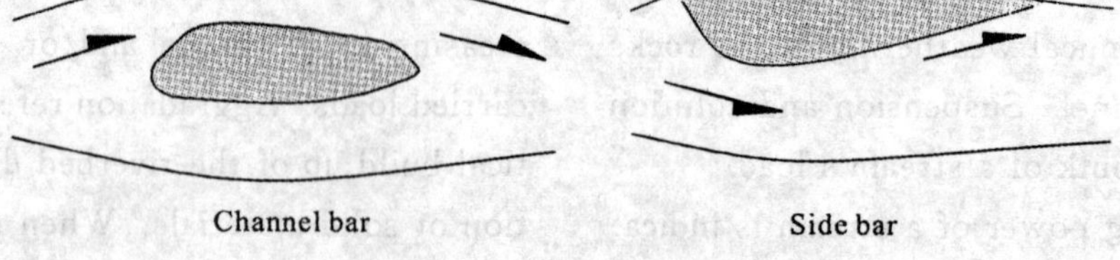

Figure 11.19 Stream deposition

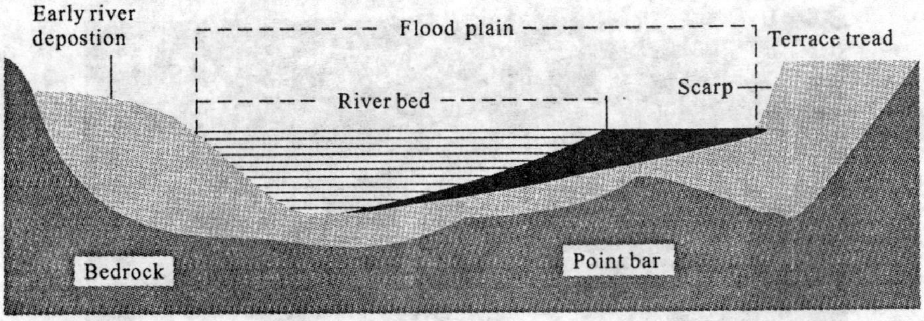

Figure 11.20 Floodplain

velocity of the flood will be very higher, but it will decrease gradually as the water flows out over the gentle gradient of the floodplain. This will result in fine-grained sheet deposition of suspended silts and sands on the floodplain. Repeated flood cycles over time can result in the deposition of many successive layers of alluvial material. The flood plain differs, however inasmuch as it is not altogether flat. It has a gentle slope downstream. Floodplains may be extremely broad, as in the case of the lower reaches of Changjiang(长江), or quite narrow, as in the case of the Nujiang(怒江).

At the onset of a flood the most abrupt decrease in velocity takes place immediately adjacent to the channel. Here the greatest thickness of flood-transported sediment tends to accumulate. These sediments build up a long broad low ridge or embankment of sand and coarse silt, forming natural levees. Note that the levees we see along the Yellow River in Henan Province (河南省) are not natural levees, but man-made levees, built to protect the floodplain from floods.

Alluvial plain is a broad flat area covered by river deposition. Because the stream is constantly migrating, the floodplains will extend its scope gruadully. Eventrually the floodplains will change into a broad flat alluvial plain.

4.4 River Deltas

When a stream or river enters into a large body of water, it slows down and drops its bed and suspended loads producing the delta, a body of land, mostly made of mud.

Named after the Greek alphabet"delta"for its commonly triangular shape, deltas come in a variety of shapes and sizes. No delta is formed if there is an ocean or tidal current strong enough to prevent sediment deposition. For example, due to the strong tidal current, there is no delta formed at the mouth of Qiantangjiang(钱塘江).

5 Degradation

River longitudinal profiles can be excellent indicators of tectonic processes. In tectonically stable areas river profiles are commonly concave-upward, with the slope of the river decreasing downstream. At this stage, it is neither incising nor aggrading its floodplain over a long period of time and a state of mass-balance equilibrium between erosion and deposition will be reached. Graded rivers are those rivers with a stable longitudinal profiles. However, changes in dis-

Figure 11.21 Incised meander

charge, velocity and channel shape, along with elevations due to tectonic movements can alter the long profile of a river until a river has had sufficient time to adjust itself to the new conditions. For example, if there is an increase in bed load, from deforestation or climate change, aggradation (deposition) may occur, elevating the stream bed and in turn increasing velocity and erosional capacity of the stream.

The fragile equilibrium achieved by graded river can also be disturbed by tectonic uplifting. Land uplifting lowers the stream profile level, causing streams to trench the landmass and carve steep-walled valleys. This is a process called degradation. Degradation can result in the forming of incised meanders and terraces.

A river flowing over a flat plain naturally forms sinuous loops called meanders. If the flat plain is subsequently elevated at an appropriate rate by tectonic forces and the resistance of the bedrock is higher enough to prevent the meanders from migrating, the river may cut down into its own bed as it follows the course of the original meander. The resulting steep-walled inner gorge is an incised meander (Figure 11.21). Abandoned meander is a former stream channel that was cut off from the rest of the river due to lateral erosion. The isolated hill surrounded by abandoned meander is called the core of an abandoned meander. Those hills not being cut off from main incised meander is meander core (Figure 11.22). However, some geologist uses the core of meander as a synonym of core of an abandoned meander.

The upper reaches of Changjiang River in

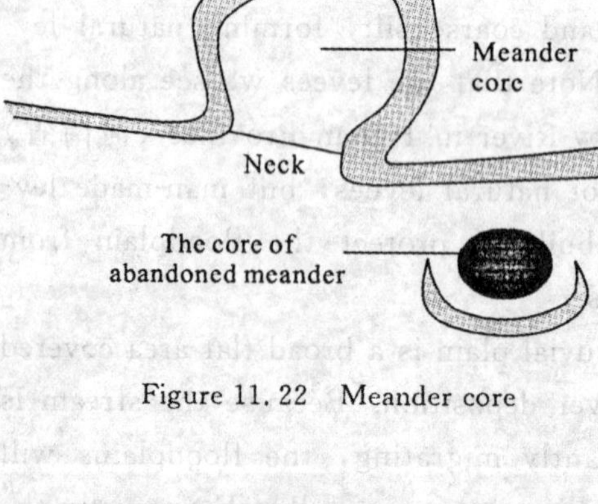

Figure 11.22 Meander core

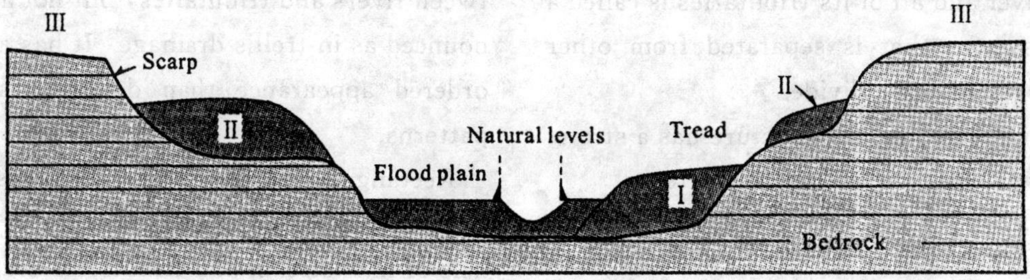

Figure 11. 23 River terraces

Sichuan Province(四川省) are featured by numerous abandoned meanders and abandoned meander cores.

River terrace (from. L. terra the Earth) is a remnant of an exposed former flood plain isolated by down-cutting of the river due to tectonic uplifting. Terraces that are above the current level of the river are the locations of river at an earlier time. It is formed by alternating periods of erosion and deposition.

River terrace generally has a flat terrace tread (阶地面) made up of sedimentary deposits, and a steep fore edge, referred to as terrace scarp(阶地壁、阶地坡). River terraces can extend along the river channels symmetrically or asymmetrically. A terrace may be made of river deposits such as gravel or sand, or it could be cut by the river on bedrock. Many rivers are bordered by a series of terraces at different levels, indicating the flood plains at successive periods in their history. From the lower to the higher, the different terraces are labeled terrace Ⅰ, terrace Ⅱ, terrace Ⅲ and so on.

According to the comprising material, the terrace can be classified as follows (Figure 11. 23):

- **Accumulation terrace** (Ⅰ in figure 11. 23) (堆积阶地). It consists of alluvial deposits. There is no exposure of bedrock.
- **Basement terrace** (Ⅱ in figure 11. 23) (基坐阶地) The basement is covered by the alluvial deposits. Bedrock exposes at the foot of terrace.
- **Erosional terraces** (Ⅲ in figure 11. 23) (侵蚀阶地) The terraces tread is bedrock. Only small amount of alluvial deposits is scattered on the tread.

6 River System Pattern (水系型式)

River evolution has a close relationship with geological development history. The course of a stream channel is guided and shaped by the nature and arrangement of the underlying bedrock and local geological structure.

The channel location and trending are controlled by local geological structure, such as fault, joint and fold axis. Thanks to its lower resistance to the water erosion, structurally weak geologic zone (地质构造薄弱带) is usually followed by the channel of rivers. That is why some of the channels are straight and turn abruptly. For example, Mulanxi(木兰溪) in Xianyou, Fujian(福建仙游) is structurally controlled by the east-west striking fault, "Zhangping-Xianyou Fault"("漳平—仙游东西向断裂").

A river and all of its tributaries is called a river system that is separated from other river systems by a divide.

Regional geological structure has a strong influence on the river system patterns.

6.1 Dendritic Pattern(树状水系)

Dendritic river system pattern is a tree like drainage pattern, which is common for most free flowing water drainage systems in the plains region. This basin also displays a pattern reminiscent of a tree leaf and its veins. Dendritic patterns form in sediment or soft uniform bedrock where rivers flow over a large area of the same rock type. That is, the subsurface geology has a similar resistance to weathering so there is no apparent control over the direction the tributaries take. Long tributary streams join the main stream at acute angle.

Dendritic systems are the most common form of drainage systems found on the Earth.

6.2 Rectangular Pattern(角状水系)

A drainage pattern in which the streams have many right-angle turns is called rectangular drainage. Or a stream drainage pattern dominated by right-angle contacts between rivers and tributaries, but not as pronounced as in trellis drainage. It has a more ordered appearance than dendritic stream patterns.

Rectangular pattern occurs in areas with regularly spaced faults or joints. The channels of mainstream and tributaries are made numerous cracks that form a grid. This pattern is common over certain types of rock, such as granite, in which cracks called joints develop to form a grid (Figure 11.24).

Figure 11.24 Joints control the rectangular pattern

6.3 Trellis Pattern(格子状水系)

It is a stream pattern that resembles a garden trellis; flows only in two orientations, more or less at right angles to each other. It often develops on parallel-folded topography where rivers may follow a straighter course along the softer bedrock, with hard rock on either side. The main channel of the stream usually resides in the valleys formed by synclines. Short tributary streams join the main stream at nearly right angles (Figure 11.25).

6.4 Radial Pattern(放射状水系)

It is the pattern where the rivers flow away from a central area of high ground. A radial pattern is evident when tributaries ra-

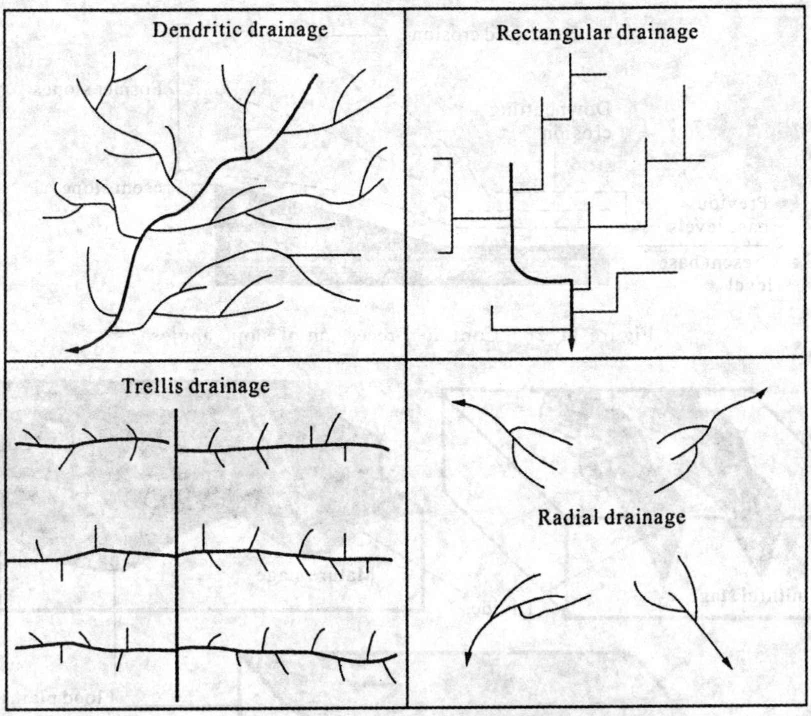

Figure 11. 25 River system pattern

diate outwards from a mountain or volcano in all directions. The tributary streams extend the headward reaches upslope toward the top of the volcano or mountain. This stream pattern is similar to the spokes of a bicycle tire.

6.5 Annular Pattern（环状水系）

Rivers that course down domes or basins generally flow in a ring-like arrangement known as an annular pattern. (Think of the annular tree rings that scientists count.)

7 Landscape Evolution（地貌演化）

Topography or landscape is the surface configuration of the Earth. The landscape can be shaped by any of the geological processes operating at or near the Earth's surface. However the most important landscape is that formed by stream erosion. With time as stream erode a landscape there is a definite progression followed by both the landscape itself and the streams that drain (and erode) it.

In 1890, the well-known geomorphologist W. M. Davis proposed a theory of landscape evolution, stating that a landscape passes through three general stages, being youth, maturity, and old age. This is called "The Davisian Cycle"(戴维斯循环说). These stages are defined by the degree of relief in the landscape, with youthful possessing the maximum amount, which diminishes through old age. Relief is the difference in elevation between the highest and lowest points in an area. The key point of the Davisian cycle is that landforms evolve through a continual recession of slope angles due to downcutting and headward erosion (Figure 11. 26).

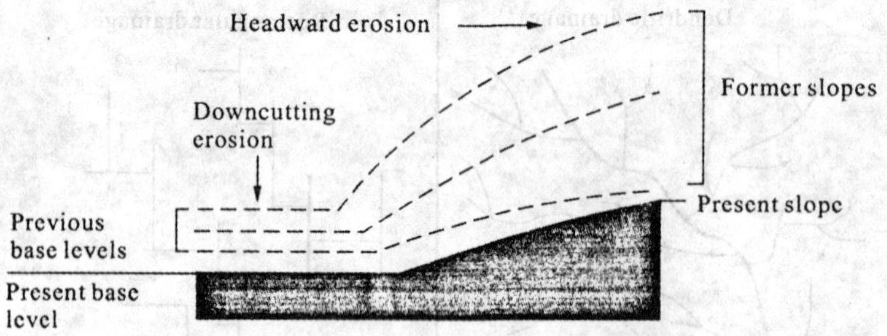

Figure 11.26 Continual recession of slope angles

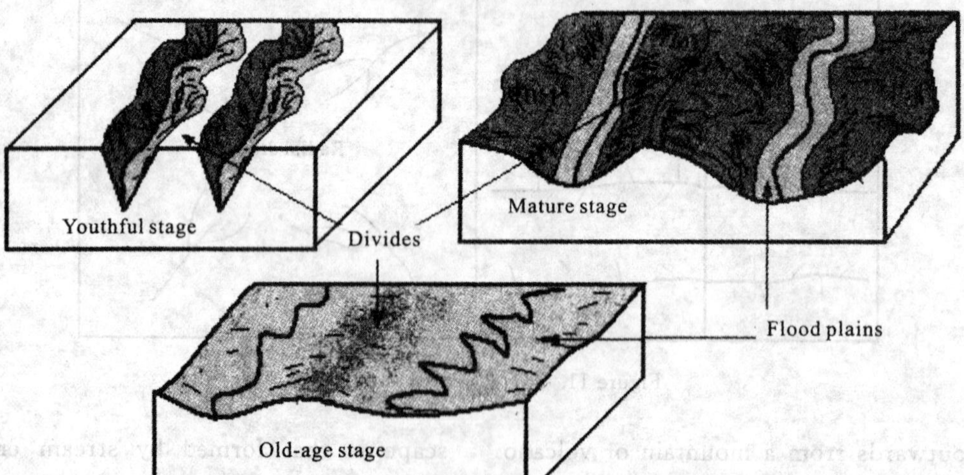

Figure 11. 27 The Davisian cycle

Though an over simplification of the true process, and one that is difficult to apply in many situations, it is useful for beginners to understand the evolution of landform (Figure 11.27).

7.1 Youthful Stage

Once a region is uplifted, the downcutting and headward erosions will work. This marks the beginning of youthful stage.

This stage is featured by V-shaped valleys, waterfalls, rapids and narrow channels. In addition to the steep gradient, there are no flood plains. Typical areas that are in the youthful stage of landscape are Wuyi Mountains (武夷山), Zhangjiajie (张家界) and the famous Three Gorges (长江三峡).

Landscapes in youthful stage are created by downcutting and headward erosions.

7.2 Mature Stage

Once the stream reaches base level, vertical-cutting ends and lateral erosion becomes predominate. Therefore, the youthful stage gives way to mature stage.

Stream in mature stage is featured by relatively wider flood plain, flat slope and broad-bended meander. Landscapes in mature stage are created by lateral erosion.

7.3 Old Stage

The old stage is characterized by extensive flood plains, wide shallow channels,

Figure 11. 28 The landscapes in Zhangjiajie (张家界，湖南) is a classic example of rejuvenation(回春地貌). Photo by Dr. Yu Shan (俞姗)

and flattened stream gradient.

The flat, featureless land surface at old stageis called a peneplain (the word meaning "almost a plain"). At some places, some isolated small round hills consisting of resistant rocks reside on the peneplain. Typical peneplain in our country is Huaibei peneplain (淮北准平原) in Xuzhou area (徐州地区).

Above theory is an oversimplified and idealized model for landscape evolution. Modern theories of landscape development suggest that landscape is a dynamic equilibrium between various factors. For example due to the tectonic movements, the landscape evolution can be interrupted at any stage. The abrupt uplifting can let the old age landscape to be eroded again. In such a case, the peneplain will be uplifted higher elevations. The elevated peneplain is called planation surface (Figure 11. 28).

1. Answer the Following Questions

(1) Differentiate the erosion and weathering.

(2) What is the Davisian cycle? Describe the different stages of Davisian cycle.

(3) What does a trellis drainage pattern tell about the rocks underneath it?

(4) Name and describe the four main ways in which a stream transports sediment.

(5) What is base level? What is longitudinal profile?

(6) What is alluvial fan? Describe the way by which the alluvial fan form.

(7) Name and describe five main patterns of river system.

(8) Name and describe three main patterns of river channel.

(9) Differentiate drainage basin and river system.

(10) What is headward erosion?

(11) Describe with sketches the river terrace and its classification.

(12) Under which geological condition can a waterfall form?

(13) What is river capture? What is windgap? What is be-headed river?

(14) Differentiate the flood plain and terrace.

(15) Differentiate the peneplain and plantation surface.

(16) Describe with a sketch the formation of an oxbow lake.

(17) Describe the formation of channel bar.

(18) Differentiate the competence and capacity.

(19) Differentiate the turbulent flow and laminar flow.

(20) What is river delta?

2. Multiple Choice

(1) Which type of drainage pattern is most likely to develop in uniform flat rocks, flat-lying sedimentary rocks?

A: Rectilinear　　B: Dendritic

C: Radial　　D: Trellis

(2) Which one is not a feature of young age river?

A: Headward erosion　　B: High relief

C: Canyon　　D: Flood plain

(3) Incised meander bend is the result of ______.

A: River capture　　B: Rejuvenation

C: Deposition　　D: Headward erosion

(4) What drainage pattern is the most common?

A: Rectilinear　　B: Dendritic

C: Radial　　D: Trellis

(5) The maximum water velocitys occur on the ______ of a meander bend.

A: Inside

B: Outside

C: None of the above

D: All of the above

(6) A moving object on the Earth will deflect to the ______ in the Northern Hemisphere due to Coriolis force.

A: Right　　B: Left

C: North　　D: West

(7) Qiantangjiang (钱塘江) lacks delta because ______.

A: No sediment contained in the river.

B: The wave and tide are so strong that carry all the sediments away.

C: The rocks of the coast are too strong to be weathered.

D: All of above.

(8) The erosion caused by the concentrated flow in a small channel on the slope is called ______.

A: Slope wash

B: Headwater erosion

C: Gully erosion

D: Rill erosion

(9) The most important divide in China is ______.

A: 秦岭-大别山　　B: 青藏高原

C: 黄河　　D: 长江

(10) A river and its tributaries are collectively called ______.

A: Drainage basin　　B: Divide

C: River system　　D: Water system

(11) The longitudinal profile of a stream is a ______ curve.

A: Concave upward

B: Convex upward

C: Concave downward

D: Horizontal

(12) The capacity of a stream is a measure of its ______.

A: Load B: Discharge

C: Velocity D: Volume

(13) The downward limit of a stream erosion is ______.

A: Longitudinal profile

B: Base level

C: River bed

D: Gravity

(14) Which of the following statements about flood plain is true.

A: Flood plain is the flat low-lying area located on only one side of a river channel.

B: All the rivers develop their flood plain.

C: Flood plain is inundated during flood time.

D: Flood plain is formed by the uplifting of river.

(15) Which of the following statements about river terrace is true.

A: Terraces are symmetrically distributed on either side of river channel.

B: River terrace is formed due to the depression.

C: Terrace at lower level is old.

D: Terrace is the exposure of former flood plain.

(16) Which one is not a feature of landscape in the old age stage?

A: Peneplain

B: Featureless surface

C: Extensive flood plains

D: Downward erosion

(17) As discharge increases, generally the velocity of the stream also ______.

A: Increases B: Decreases

C: Not changes D: None of above

(18) Diluvium is superficial deposits formed by ______.

A: River B: Flood

C: Alluvial fan D: Delta

(19) Graded river are those rivers with a stable longitudinal profiles.

A: Ture B: False

(20) Which following statements about river is not true.

A: Alluvial plain is a broad flat area covered by river deposition.

B: Drainage basin is the land drained by a particular stream or and its tributaries.

C: Degradation can result in the forming of incised meanders and terraces.

D: Graded rivers are those rivers with a concave downward longitudinal profiles.

(21) Which of the following statements is not true.

A: Orange bar (橘子洲) is a channel bar.

B: Landscape in Zhangjiajie(张家界) is in its old stage.

C: The base level for the Hanshui (汉水) is the level of the Changjiang River(长江).

D: Hukou Falls (壶口瀑布) will disappears in the future.

(22) Side bar developing ______.

A: On the inside bank of the meander

B: On the outside bank of the meander

C: In the center of channel

D: None of the above

(23) Ridges of sediment build up naturally on either side of a stream are called ______.

A: Levee B: Point bar

C: Channel bar D: Side bar

(24) The removal of weathered materials is ______.

A: Deposition B: Transportation

C: Discharge D: Erosion

(25) The abandoned river caused by stream capture are called ______.

A: Be-headed river B: Meander core

C: Oxbow lake D: Wind gap

(26) Which kind of river systems often develops on parallel-folded topography?

A: Rectilinear B: Trellis

C: Dendritic D: Radial

(27) The highland separating drainage basins is called ______.

A: Meander core B: Terrace

C: Divide D: Levee

(28) Competence is measured by the ______.

A: Grain size　B: Discharge
C: Load　D: Slope

(29) Which of the following statements about landscape evolution is true?
A: Stream in mature stage is dominated by lateral erosion.
B: The Davisian cycle is an oversimplified theory.
C: Landscapes in youthful stage are created by downcutting and headward erosions.
D: Landforms evolve through a continual recession of slope angles.
E: All of the above

(30) The primary process by which bed load is transported is ______.
A: Saltation　B: Precipitation
C: Suspension　D: None of above

(31) The local base level of Golden Sand River(金沙江)is ______.
A: 三峡水库　B: 东海
C: 洞庭湖　D: 汉水

(32) Sinuous course of a meander is caused by ______.
A: Head erosion　B: Lateral erosion
C: Stream capture　D: Downcutting

(33) Waterfall retreat is caused by ______.
A: Head erosion　B: Lateral erosion
C: Stream capture　D: Downcutting

(34) Alluvial plain is the combination of ______.
A: Alluvial fan　B: Flood plain
C: Point bar　D: Delta

(35) Incised meanders and terraces are the result of ______.
A: Aggradation　B: Degradation
C: Lateral erosion　D: None of them

(36) Terrace consisting of alluvial deposits is called ______.
A: Accumulation terrace
B: Basement terrace
C: Erosional terrace
D: None of them

Chapter 12 Groundwater

Key words

Groundwater	Porosity	Permeability	Water table	Saturated zone	Unsaturated zone
地下水	孔隙度	渗透性(率)	潜水面	饱和带	不饱和带
Aquiclude	Aquifer	Vadose zone	Discharge (area)	Recharge (area)	Vadose water
隔水层	含水层	包气带	排泄(区)	补给(区)	包气带水
Unconfined groundwater	Confined groundwater		Karren	Sinkhole	Tower karst
潜水、非承压水	承压水		溶沟	落水洞	塔状喀斯特
Perched groundwater	Hot spring	Stalactite	Stalagmite	Dripstone	Geothermal water
上层滞水	温泉	石钟乳	石笋	钟乳石，滴石	地下热水

The distribution of water on the land is dependent upon the complex interaction between atmosphere and oceans that we call climate. The circular path of the hydrologic cycle links evaporation of water vapor from oceans to condensation in the atmosphere and precipitation falling on land. Nearly one-third of all water falling as precipitation completes the circuit to the oceans by surface runoff in streams. Most of the rest returns to the atmosphere by evaporation or through the transpiration of plants. A slim fraction percolates below the surface to join the slow flow of groundwater that will inevitably return it to the ocean.

Groundwater is all water located under the Earth's surface or groundwater is water that comes from the ground. Many people think of groundwater as underground lakes or streams. These primarily exist where an area is underlain by cavernous limestone which will be discussed later. Most groundwater is simply water below the land surface that is filling all the spaces between rock gains or in the cracks in the rock. The water in lakes, rivers, or oceans is called surface water. Groundwater comes from rain and snow that soaks into the ground, and then moves through the soil into the groundwater system, where it eventually makes its way back to surface streams, lakes, or oceans. Water in a lake and river can also soak down into the ground and become groundwater.

Groundwater is one of the Earth's most valuable natural resources. Groundwater makes up only 0.6% of the total water of the Earth, but 30% of the world's freshwater available for human use.

1 Instruction

Groundwater distribution depends on the porosity and permeability of rocks or sediments. Rock and sediment contain free spaces between grains (pore spaces), in fractures, or in dissolved cavities (limestone), that may become filled with water.

1.1 Porosity

The spaces between soil or sediment particles and cracks in solid rock are called voids or pores. Infiltration, groundwater movement, and storage occur in these void spaces, so porosity is the capacity of rock or sediment to store water. The porosity of soil or geologic materials is the ratio of the volume of pore space to the total volume of material per unit volume.

A mathematical equation of porosity looks like this:

$$\text{Porosity or } n=(V_{\text{void}}/V_{\text{total}})\times 100\%$$

Porosity is often expressed as a percentage of rock or soil void of material, so multiply the answer by 100. For example, sand may have 40% pore space to 60% solid material, while granite may have 10% pore space to 90% solid rock (Figure 12.1). The sand is therefore more porous than the fractured granite.

Factors responsible for porosity include grain size, the degree of sorting, shapes of grain, and degree of cementation. Coarse-grained sediments have a higher porosity than fined-grained sediments because the grains do not fit together. That is why sandstone usually has a higher porosity than siltstone and claystone. Well-sorted sediments have a higher porosity than poorly sorted sediments because the spaces between coarser grains are filled by fine-grained fragments, that is matrix.

Igneous and metamorphic rocks usually have lower porosity due to their crystalline texture, in which minerals tend to be intergrown, leaving little free space. Only those highly fractured igneous and metamorphic rocks could have relatively higher porosity.

1.2 Permeability

Permeability is a measure of a soil's or rock's ability or capacity to transmit water. Degree of permeability depends upon the size and shape of the pores, the size and shape of their interconnections, and the extent of the latter.

A substance with high porosity does not necessarily possess good permeability. Only those porous rocks with well-interconnected pores have a high permeability. Formations that transmit fluids readily, such as sandstones, are described as permeable (可透水的) and tend to have many large, well-connected pores. Impermeable formations(不透

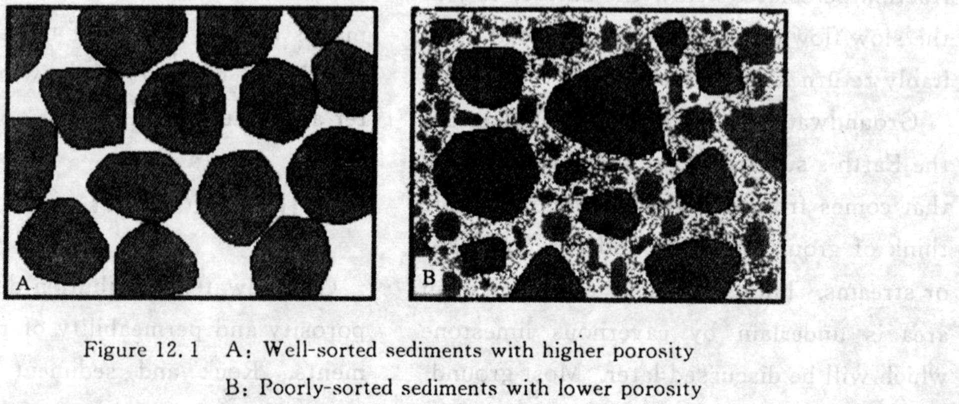

Figure 12.1 A: Well-sorted sediments with higher porosity
B: Poorly-sorted sediments with lower porosity

水建造), such as shales and siltstones, tend to be finer grained, with smaller, fewer, or less interconnected pores. Vesicular volcanic rock usually has a high porosity and low permeability because the vesicles created by bubbles are not connected to one another (Figure 12.2). In general, igneous and metamorphic rocks usually have low permeability as well as low porosity.

1.3 Water Table

Rainfall seeps underground through a process called percolation(下渗,渗透). Percolation is the physical process by which rainwater falling within a given area slowly travels downwards through the tiny spaces between rocks and soil particles. When water soaks into the soil, some stays near the surface absorbed onto the soil particles, forming soil moisture. The rest soaks down further due to gravity until it reaches the zone of saturation where all of the pore space is filled with water. The top of the zone of saturation is called the water table or groundwater table (Figure 12.3).

The water table is never flat (horizontal), due to the slow rate of motion of water through the saturated zone (on average about 4cm per day), so that the water table is always higher at higher elevations and near the source of the groundwater and lower near the outlet (river for instance or well). Generally, the water table is deeper beneath hills and shallower beneath valleys.

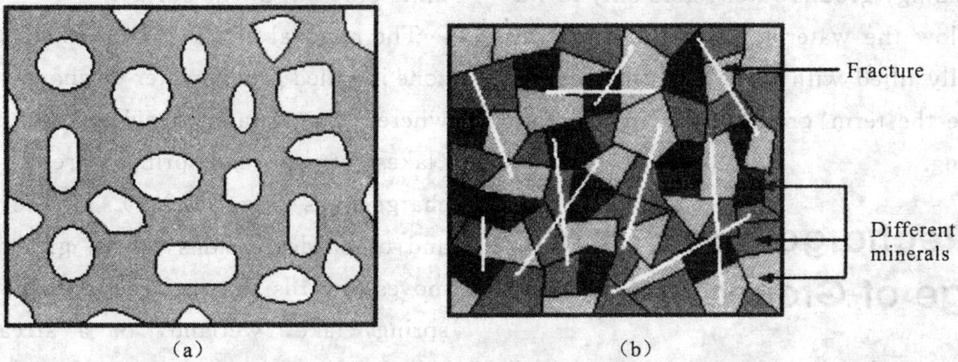

Figure 12.2 A: Vesicular volcanic rock with high porosity but low permeability
B: Igneous rocks with intergrown minerals have low permeability

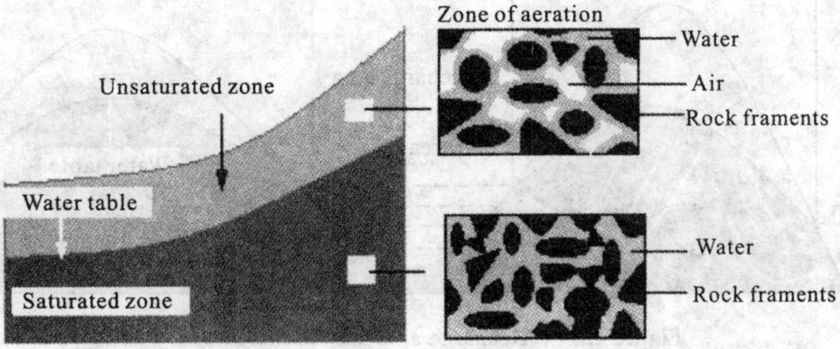

Figure 12.3 Water table

Water table also fluctuates according to season and rainfall. Ordinarily the water lies within a few meters of the surface, but it can be at the surface, as for instance, the bank of a river, or in some mountainous districts, it can lie at a depth of 100m or more.

The water table is the saturated level of the groundwater. The saturated zone is overlain by an unsaturated zone where pore spaces are only partially filled with water (Unfilled pore space contains air). Unsaturated zone can also be called as vadose zone (包气带).

The term "groundwater" is broadly defined as being any water contained both in saturated zone and unsaturated zone. Strictly speaking, groundwater refers only to water below the water table where the pores are fully filled with water. In this course, we use the term "groundwater" in its broad meaning.

2 Recharge and Discharge of Groundwater

As part of the water cycle, the groundwater is in constant motion, possibly very slowly. First the groundwater flows from the highlands to the lowlands due to the pull of gravity. But it can also move upward because it will flow from higher-pressure areas to lower pressure areas.

The groundwater recharge is the replenishment of groundwater into the saturated zone by infiltration or seepage from precipitation. Other sources of groundwater, besides precipitation, include lakes and rivers, artificial recharge (surface water purposely injected into groundwater system), agricultural irrigation and wastewater treatment systems.

Recharge area is a geographic area where water seeps into the saturated zone. Recharge areas usually coincide with topographically elevated regions where aquifer units crop out at the surface.

The removal of water from the saturated zone is called groundwater discharge. Areas where groundwater reaches the surface (lakes, streams and springs) are called discharge areas (Figure 12.4). Or the valleys and other depressions where groundwater moves towards the surface to escape into a spring, lake, wetland, or a stream are called discharge areas. Groundwater moves from the recharge areas to the discharge areas . In other words , groundwater moves

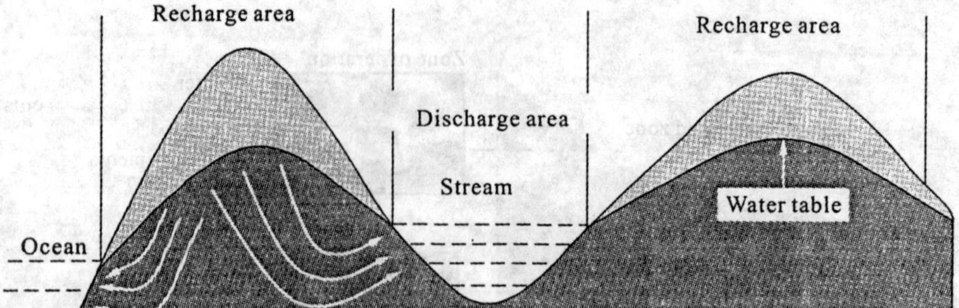

Figure 12.4 Recharge area and discharge area

from areas of higher elevation or higher pressure/hydraulic head (水头) to areas of lower elevation or pressure. Generally, the recharge areas are greater than discharge areas.

Groundwater can be discharged through evaporation, spring and pumping well.

An aquifer ("Aqua" means water, and "fer"is to carry), nature's underground water storage system, is an underground body of saturated rock or sediment through which water can move easily (Figure 12.5). It is typically characterized by high porosity and permeability values. An aquifer is composed of permeable rocks, such as sandstones or any of the brittle rocks that may acquire large interstices through fracturing. Poorly cemented sands and gravels are also good aquifers. To be an aquifer, it must store and transmit water at rates fast enough to supply reasonable amounts to wells and springs. Therefore, not all groundwater is stored in an aquifer.

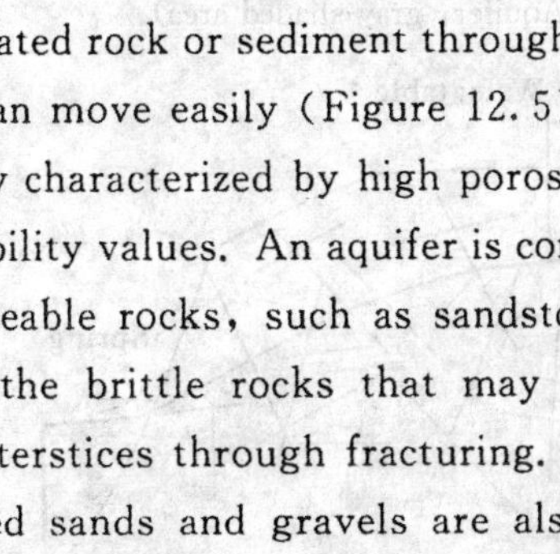

Figure 12.5 The saturated part of the permeable layer (beneath water table) is aquifer

Aquifers come in all shapes and sizes. Some may cover hundreds of miles while others may only cover a few square miles. Some areas may contain several aquifers located at different depths. Aquifers in the China are usually made from gravel, sandstone, limestone, or basalt (volcanic rock).

An aquiclude, or confining layer, is a layer of geologic material that impedes the movement of water into and out of an aquifer. Aquiclude is made up of low porosity and low permeability rock/sediment such as unfractured igneous and metamorphic rocks, shale, and unconsolidated sediments (clays).

A well is a human-made hole in the Earth from which fluid is withdrawn, or any other excavation that is dug or drilled deep enough to intersect the water table or hit an aquifer. An ordinary well is simply the one that if it is dug beneath the water table, water will fill the open space to the level of the water table, and then can be drawn out by pumping (Figure 12.6). Anyway, in some special cases, a well must be drilled into an aquifer to reach an adequate supply of water, or drilled to intersect a fracture beneath water table in igneous rocks.

A spring is an area on the surface of the Earth where the water table intersects the surface and water flows out of the ground (Figure 12.7). The overall land surface area where water seeps underground and contributes rainwater to a specific spring is called a spring's recharge basin.

Spring comes in many ways. However, the simplest spring is an gravity spring(重力泉), whose flow results directly from the force of gravity.

3 Storage Condition of Groundwater

According to the free spaces in which it is

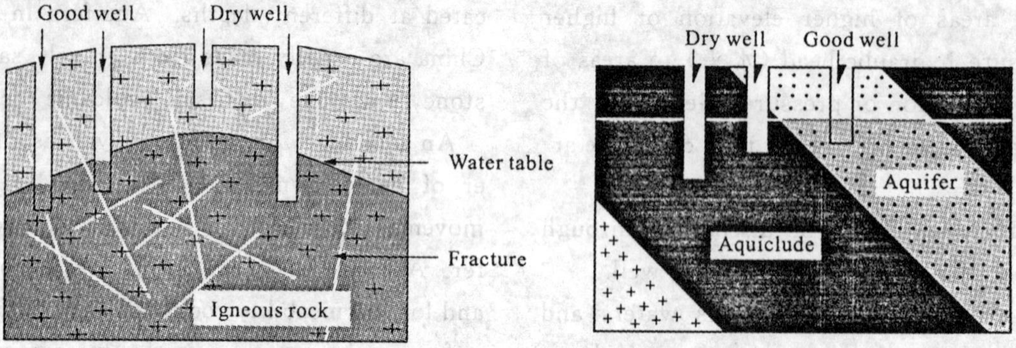

Figure 12.6 Left: A well must be drilled to intersect the fractures beneath water table in igneous rocks. Right: A well must hit an aquifer to obtain water (Aquifer: gray-shaded area)

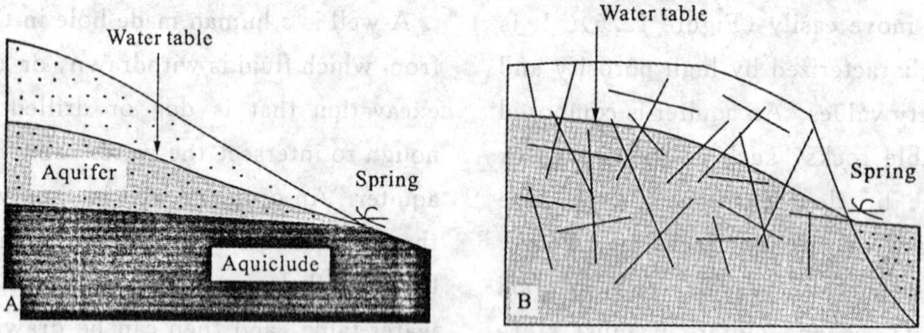

Figure 12.7 Two locations of gravity spring

A: Eroded aquifer; B: Fractured igneous rock

stored, the groundwater can be classified into pore water(孔隙水), fissure water (裂隙水)and karst water(岩溶水).

Pore water is the water filling the spaces between grains of sediment and rock. In three-dimensions all of the pore spaces are connected. Pore water distributes evenly and continuously with a layer and usually occurs in Quaternary unconsolidated sediments and some bedrock.

Fissure water occurs in fissure and fracture of rocks, especially in crystalline rock (igneous and metamorphic rocks). It is perhaps the only way for groundwater to be stored in igneous and metamorphic rocks. The crosscut fissures and fractures provide a channel for fissure water to travel in the saturated zone. The fissure water is strictly controlled by the distribution pattern, extent and density of fissures developed in rocks (Figure 12.8).

Karst water refers to that water existing in caves, fissures and underground rivers in soluble limestone. Karst water is a major geological factor shaping the karst landform. We will discuss it later in this chapter.

Based on the spatial relationship between water table and aquifer, the groundwater can be described as follows.

3.1 Vadose Water

Vadose water is the water occurring in the unsaturated zone between the ground surface and the water table, within the vadose zone. No matter how tightly (densely) the soil particles are packed, there will always be void spaces between them. The void spaces of soil in vadose zone are filled with varied amounts of air and water. The thin water films attracted to the grains by the forces of molecular attraction is soil moisture (土壤水,土壤湿度). All the soil particles are surrounded by water. Even in the desert when the soil is baked by the sun, soil particles continue to be coated by water. Water is just limited, it is not non-existent (Figure 12.9).

The zone immediately above the water table is known as the capillary fringe. This is the zone in which underground water is lifted against gravity by surface tension in passages of capillary size. The size distribution of the pores determines the extent and degree of the capillary fringe. In capillary fringe (毛细管带) the soil is actually saturated but at pressures less than atmospheric (under tension).

When the water from a rain shower infiltrates the soil, part of water will seep down into the saturated zone, and part of water will be held in place by the forces of molecular attraction. Discharge of water directly from the vadose-water includes evaporation, plants growing and transpiration of crops. In semiarid regions the consumption of vadose water by crops may be large and is of vital importance to agriculture . During

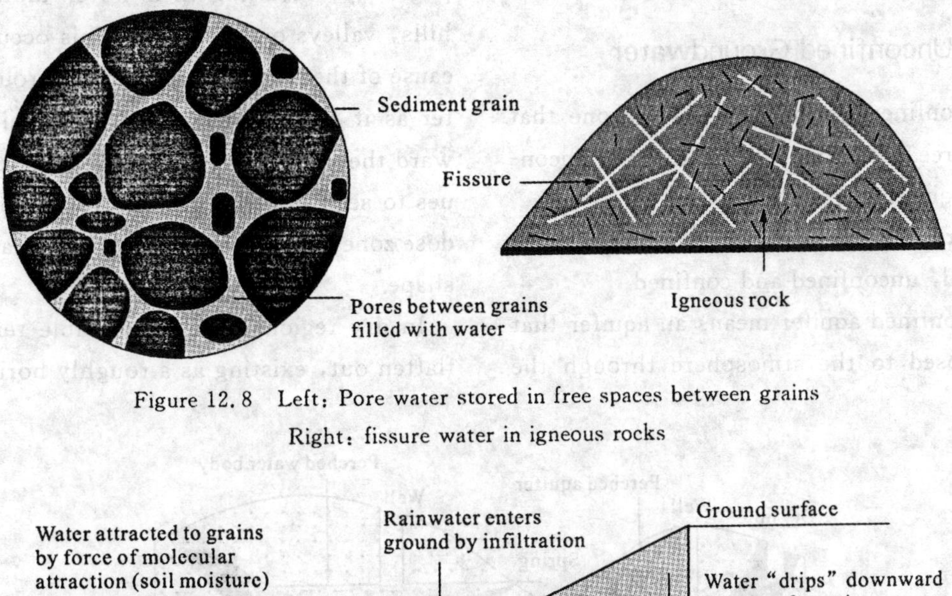

Figure 12.8 Left: Pore water stored in free spaces between grains
Right: fissure water in igneous rocks

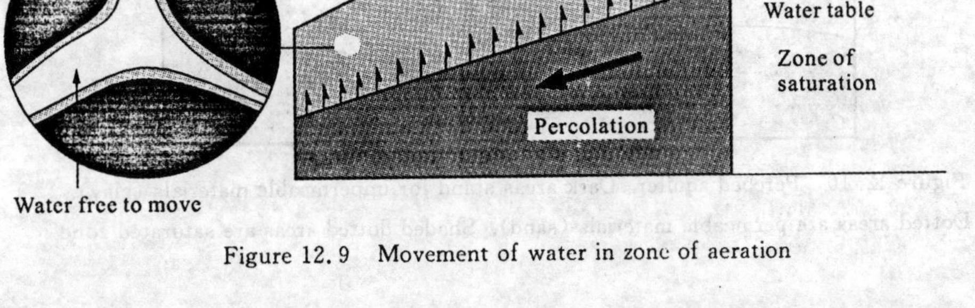

Figure 12.9 Movement of water in zone of aeration

long periods without rain the vadose zone may remain dry.

The discontinuous bodies of permeable and impermeable materials create perched water bodies. A perched aquifer is a saturated zone within the zone of aeration that overlies a confining layer. This aquifer sits above the main water table and"perched"atop a mostly impermeable layer within the zone of aeration. Ground water body in an aquifer is separated from the main groundwater below it by an impermeable layer.

Perched aquifer changes abruptly with seasonal fluctuations. Many aquifers are small and localized. A well may be drilled into a perched aquifer, but it usually only yields limited amount of groundwater(Figure 12. 10).

3. 2 Unconfined Groundwater

Unconfined groundwater is the one that has a free water table. It is stored in unconfined aquifer.

There are three main types of aquifers: perched, unconfined and confined.

Unconfined aquifer means an aquifer that is exposed to the atmosphere through the vadose zone which extends to the land surface. There is no continuous confining layers (隔水层,封闭层) between the zone of saturation and the land surface. Because the top of these aquifers is the water table, they also are called water table aquifers.

The water table marks the top of an unconfined aquifer. The hydrostatic pressure on the surface of the body of unconfined groundwater is equal to atmospheric pressure. Therefore, the height of the water table is the same as the water level in a well constructed in the unconfined.

In humid regions the water table, or the top boundary of unconfined groundwater tends to roughly parallel the overlying topography, resulting in a water table with hills, valleys or flat areas. This occurs because of the slow progress of the groundwater as it flows down through the hills toward the valleys. As long as water continues to seep into the ground from above vadose zone, the water table will maintain this shape.

In dry regions, the water table tends to flatten out, existing as a roughly horizontal

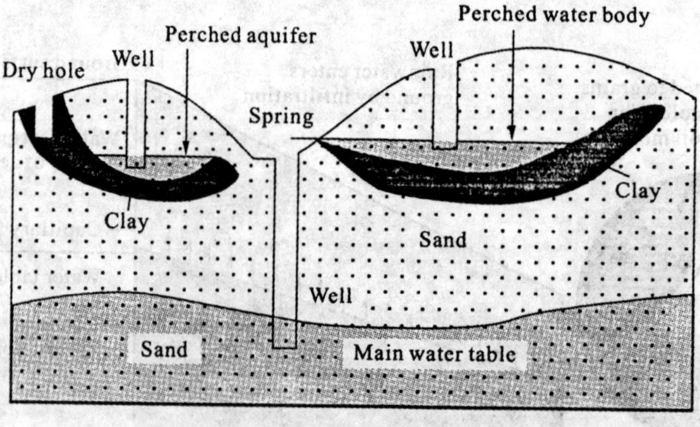

Figure 12. 10 Perched aquifer. Dark areas stand for impermeable materials (clay). Dotted areas are permeable materials (sand). Shaded dotted areas are saturated zone

plane beneath the level of most valleys.

Water table also changes seasonally in response to seasonally dry weather. During times of prolonged drought, water table levels may drop and wells may run dry(Figure 12.11).

3.3 Confined Groundwater

In some cases, groundwater can become confined between two stable impermeable layers. This type of enclosed water is sometimes called confined groundwater. The aquifer storing confined groundwater is called confined aquifer. In contrast to an unconfined aquifer, a confined aquifer is separated from other aquifers and the land surface by a confining layer.

The water body confined or sandwiched between two impermeable layers is under hydrostatic pressure which is greater than atmospheric pressure, because its roof is impermeable layers instead of the water table. Because of the high pressure, the water level in a well will rise to a level higher than the water level at the top of the aquifer. If a well is drilled into a confined aquifer, sometimes the water will flow out to surface of the Earth due to higher hydrostatic pressure (Figure 12.12).

Artesian well(承压井) is a well whose source of water is a confined (artesian) aquifer. The water level in artesian wells stands at some height above the water table. If the pressure surface (water table) is above the land surface, the well is a flowing

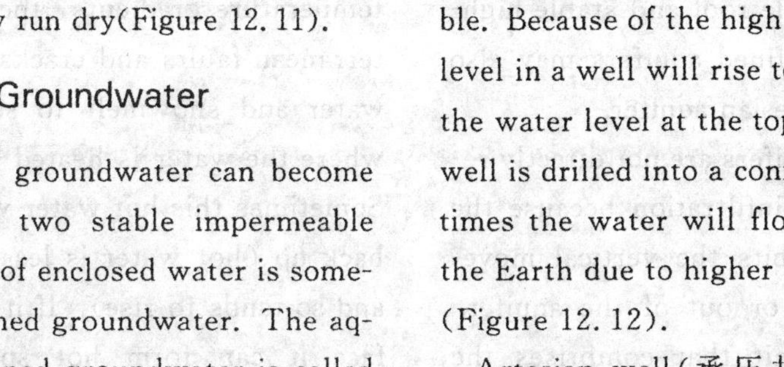

Figure 12.11 Seasonally change of water table

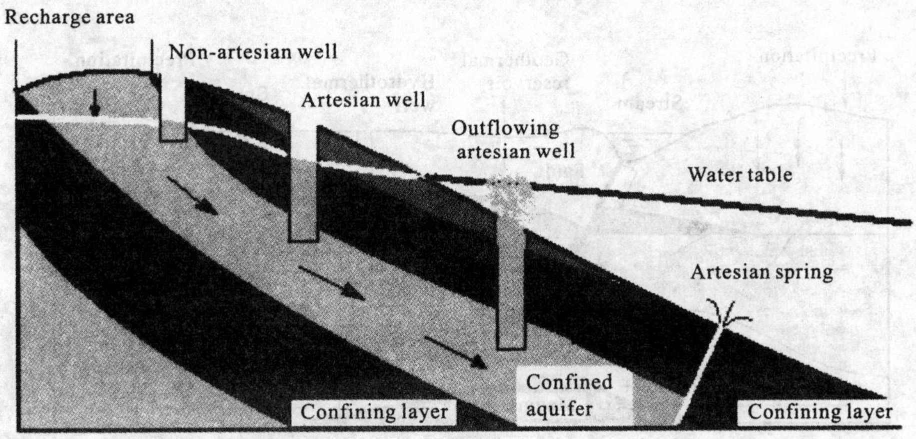

Figure 12.12 Confined aquifer and artesian well

artesian well. A water table higher than the well ensures water pressure will consistently force water into an artesian well. So the three conditions for artesian well are: an aquifer, an impermeable roof and stable higher water table. Confined aquifers may also be referred to as artesian aquifer.

Most confined aquifers are not directly recharged by vertical infiltration because the confining layer inhibits the vertical movement of water into or out of the aquifer. Only when rock unit that comprises the confined aquifer is exposed to the Earth's surface, the aquifer will be recharged. In such case, the "confined" aquifer is actually unconfined at the recharge zone. Otherwise confined aquifers need to be connected to an unconfined area through which recharge can occur.

4 Geothermal Water

Geo means Earth and thermal means heat. So geothermal water is the water naturally heated within the Earth's crust that is available at or near the surface, either through wells or in natural springs.

From the surface down through the crust the normal temperature gradient (地温梯度) in the Earth's crust is 17～30℃ per kilometer of depth. In some regions with high temperature gradients, there are deep subterranean faults and cracks that allow rainwater and snowmelt to seep underground where the water is heated by the hot rock. Sometimes this hot water will work its way back up (hot water is less dense than cold and so tends to rise). If it reaches the surface it can form hot springs (or warm springs) (Figure 12.13). If the ascending hot water meets an impermeable rock layer, however, the water is trapped underground, forming a geothermal reservoir. Most hot springs result from the interaction of groundwater with magma or with recently formed, hot, igneous rocks at shallow depth.

Geyser (Icelandic *Geysir*, hot spring in Iceland) is a periodic or sporadic jet of hot water and steam, driven by the boiling of groundwater in conduits in contact with hot subterranean rock. When the groundwater percolates downward and is heated, if the water becomes so heated that it builds ste-

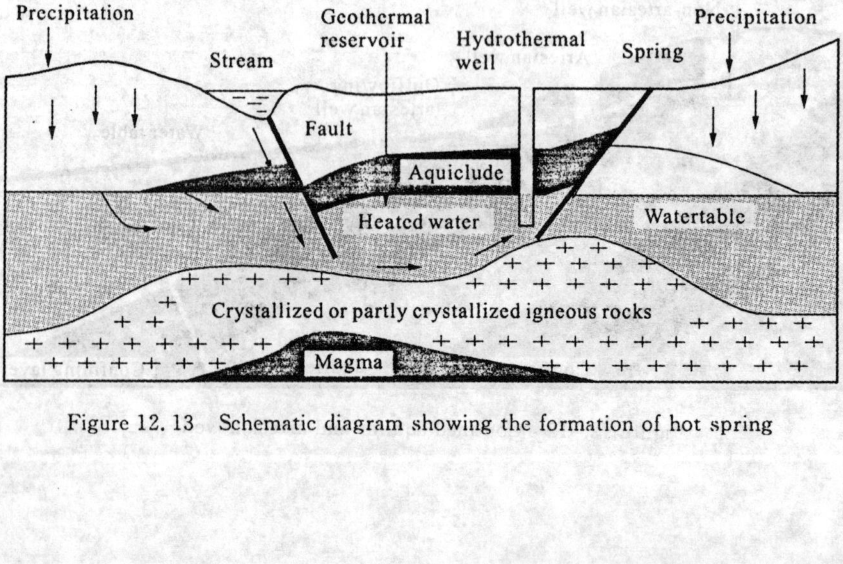

Figure 12.13 Schematic diagram showing the formation of hot spring

am pressure and erupts in a jet above the surface of the Earth, then a geyser comes into being. Geysers occur in many places where the surface is faulted above hot intrusions, such as in Iceland, New Zealand and Yellowstone Park.

It is important to note that the water ejected by hot springs is not directly released from molten magmas far below the surface. In fact nearly all of the water is recirculated groundwater.

Most of the hot springs and geothermal reservoirs occur in regions with active or geologically young volcanoes. These "hot spots" occur at plate boundaries or at places where the crust is thin enough to let the heat through. For example, Iceland, New Zealand and Yangbajing(羊八井) in Tibet, being noted for their hot springs, are located on or adjacent to plate boundaries.

The water and steam from hot springs and superheated geothermal reservoirs are the geothermal energy resources we use in many ways. Shallower reservoirs of lower temperature (21～149℃) are used directly in health, greenhouses, fish farms, and industry and in space heating systems for homes, schools and offices. High temperature reservoirs (120～370℃) are used to generate electricity in geothermal power plants. In recent years the depletion of fossil fuels has resulted in a renewed interest in utilizing the energy contained in geothermal water.

5 Karst Topography

The main difference between running water and groundwater is the fact that groundwater has no appreciable flow associated with it. The other difference is that more amount of groundwater remains in contact with rocks for longer time as compared to running water. These differences make dissolution as the main mode of erosion by groundwater instead of hydraulic action. Therefore two main geological actions of groundwater are erosion by dissolution and deposition by precipitation.

The dissolution by groundwater results in karst topography. Karst topography is a kind of highly irregular landforms developed mainly by groundwater dissolving soluble rock such as salt, gypsum and limestone. Salt karst and gypsum karst are seen only in some arid places. Pure limestones ($CaCO_3$) are the principal karst rocks, hosting the greatest extent and variety of features. About 10% of the Earth's land surface consists of soluble limestone, which can be easily dissolved by the weak solution of carbonic acid found in underground water. Karst topography is named for the Kras plateau region of eastern Italy and western Slovenia (Kras is Karst in German for "bleak, waterless place").

Erosion through the carbonation process is the basis of karst development. Carbon dioxide (CO_2) from the air is dissolved in rainwater to create a weak acid (pH 5), carbonic acid (H_2CO_3). The erosive power of the water may also be enhanced by combination of water with carbon dioxide from organisms and plants within the soil. Limestone is composed of calcium carbonate, which is only slightly soluble in pure water, yet easily dissolved by carbonic acid (H_2CO_3).

$$CaCO_3 + H^+ + HCO_3^- \longrightarrow Ca(HCO_3)_2$$

The solution percolates down through the

limestone bedrock through pores or cracks in the solid rock and moves through the aquifer as groundwater. Moving groundwater chemically dissolves joints and fractures in the limestone creating a topography dominated by caves, sinkholes, and underground rivers.

According to it's developing stages, the features of karst topography can be summarized as following six types:

5.1 Karren

As the surface water flow along the surface of rock, it dissolves and enlarges any pore or fracture in the rock and results in the formation of karren. Karren (German) is the subparallel grooves or channels formed on the surface of soluble bedrock by dissolution of a portion of the rock. Karren often takes the form of sharp-ridged grooves. The ridges between the grooves are stone buds(石芽). When in large scale and developed in groups, the stone buds can be called stone forest. Stone forest often develop in tropical or subtropical areas, such as in Lunan Stone Forest(路南石林), where the stone buds are as high as 10 m to 100 m.

5.2 Sinkhole

With the action of downward percolation of water, a nearly vertical cave will be formed. This is the sinkhole. Sinkholes usually grow with time and can be quite deep. The diameter ranges from a few to many hundreds of meters. Sinkholes are deep, steeply-sided holes in the ground, produced especially in jointed carbonate rocks. Sinkhole develops above the water table.

5.3 Karst Caves

Sinkholes collect surface water running off the surrounding land, and this groundwater may flow laterally within the aquifer. The moving groundwater chemically dissolves the rocks along the horizontal structural surface such as bedding, joint and fracture, creating subsurface cavities and caves. The horizontal karst caves develop beneath the water table. Some of the karst caves actually are or were the passage of underground river.

Due to the tectonic uplift, or the lowering of water table, a very complex multi-level cave system will be developed. The new caves will form just beneath high-level former caves.

When the caves rise to new level high above water table, the formation of stalactites and stalagmites happens.

When traveling through the vadose zone, the water will become acid and dissolve limestone. When the limestone-bearing solution seeps down into an airfilled cave, the carbon dioxide will leave from the water. The water can no longer hold the limestone (calcite) in solution. So it leaves small particles (crystals) of it on the wall of the cave. If the water is dripping it will create a stalactite hanging from the roof. If the water doesn't lose the carbon dioxide until it hits the floor of the cave, it will create a stalagmite growing upward (Figure 12.14). Stalactites and stalagmites usually occur together in caves and are collectively known as dripstones(滴石).

The world's longest cave is the Mammoth Cave of U. S. , extending over 560 km.

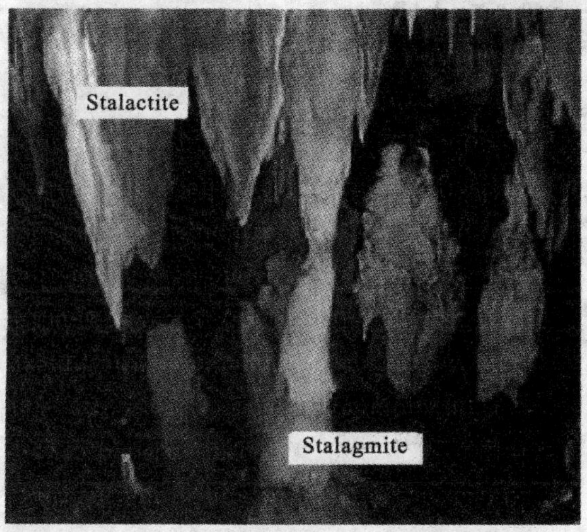

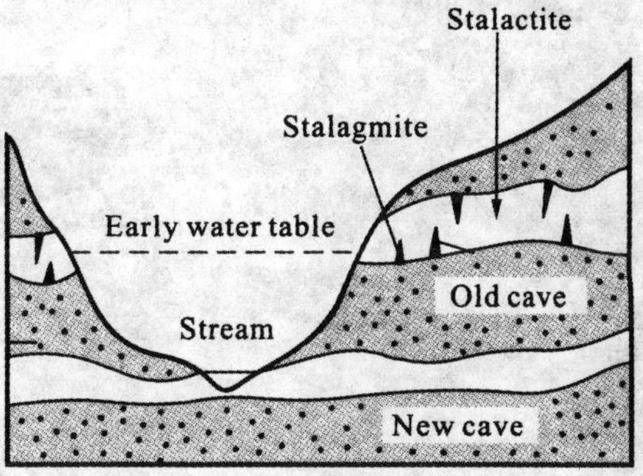

Figure 12.14 Water flowing down through cracks has resulted in this row of stalactites and stalagmites. Image credit: National Park Service (NPS), U.S. Public Domain

5.4 Karst Funnel

On the surface, sinkholes may develop progressively as bowl-shaped or steeply sided depressions due to the collapse. This is known as karst funnel. At the bottom of depression, there often is a sinkhole for surface water to enter the aquifer and replenish the groundwater supply. Sometimes water will fill the karst funnel and result in the karst lake.

5.5 Natural Bridge, Dry Valley and Blind Valley

Natural bridge(天生桥) is arches that are the remnants of collapsed caves and usually fairly short. Dry valley is the valley without a surface stream. Blind valley is a valley that is closed off at its downstream end, all its water sinking at or near that point. Or blind valley is a valley that end abruptly where stream vanishes underground (Figure 12.15).

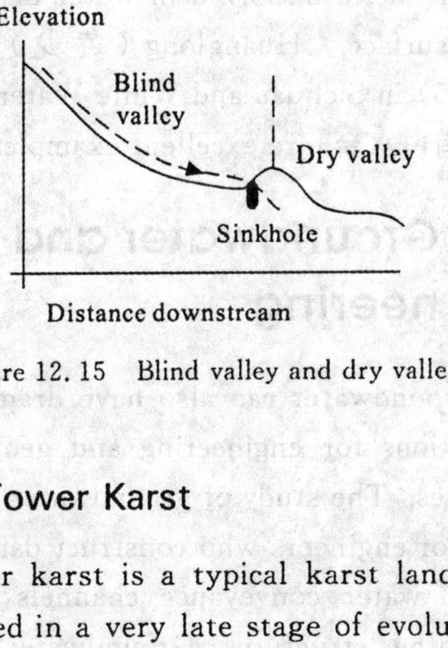

Figure 12.15 Blind valley and dry valley

5.6 Tower Karst

Tower karst is a typical karst landform developed in a very late stage of evolution. The formation of the towers is a combination of tectonic uplift and tropical karst erosion. Tower karst is characterized by isolated hill(孤峰), clusters of hills(峰丛) and hill forest(峰林). All of the spectacular hills are 30～300m high towers with vertical or overhanging sides. Tower karst occurs throughout Southeast Asia. By far the most extensive and best developed tower karst is in the Guilin(桂林).

A very special karst topography forms

Figure 12.16 Special karst topography in Huanglong (黄龙). Photo by Chen Fuxing(谌福兴)

when calcite dissolved in water deposits on the surface. Huanglong (黄龙) (Figure 12.16) in Sichuan and White Water Terrace (云南白水台) are excellent examples.

6 Groundwater and Engineering

Groundwater can also have dramatic implications for engineering and geotechnical studies. The study of groundwater is essential for engineers who construct dams, tunnels, water conveyance channels, mines, and other structures. Groundwater must be considered whenever the stability of slopes is important, whether the slope is natural or constructed. Groundwater must also be taken into account when devising measures to control flooding. In all of these situations, groundwater flow and fluid pressure can create serious geotechnical problems.

Groundwater, for example, may create structural weaknesses in dams, or it may flow underground right around the structure as it did at the Thirteen Mausoleums Dam (十三陵大坝). Water flowed so efficiently through the rock formations surrounding the reservoir that the dam would hold no water, even though it was structurally sound.

In another case, when geological exploration was being carried out in preparation for the construction of the Three Gorges Dam (三峡大坝), geologists and engineers were concerned about an ancient landslides on the banks of the proposed reservoir. They suspected that the water held in the reservoir could increase groundwater pressures enough to make the slide unstable. One of the solutions was to increase drainage around the slide to ensure that groundwater pressures did not increase.

Other problems result from the excessive use of groundwater. Overdrafting (过度开采) occurs when people draw water out of an aquifer faster than nature can replenish it. The most obvious problem created is a shortage of water. Overdrafting, however, can also create significant geotechnical problems. At many locations around the world overdrafting has caused land subsidence (地

面沉降). This can produce severe engineering difficulties. Parts of Shanghai City(上海), for instance, have subsided as much as 2m in the past 80 years, resulting in a host of problems in its water supply and sewerage system.

Land subsidence is a gradual settling or sudden sinking of the Earth's surface owing to underground movement of earth materials. Common causes of land subsidence from human activity are groundwater or oil extraction, dissolution of limestone aquifers (sinkholes) and collapse of underground mines(Figure 12.17).

In some areas where groundwater pumping has caused subsidence, the subsidence has been stopped by switching from groundwater to surface-water supplies. If surface water is not available, then other means must be taken to reduce subsidence.

Saltwater intrusion(海水入侵)is the movement of salt water into a non-salt water environment, such as a freshwater marsh. In coastal areas, the aquifers are very vulnerable to the saltwater intrusion through the overdraft of groundwater exploitation or insufficient recharge from upstream. When fresh water leaves the aquifier, salt water takes its place. Saltwater intrusion leads to the loss of freshwater vegetation and the spread of saline mudflats into previously vegetated areas. Saltwater intrusion also affects the high conservation values of coastal freshwater wetlands, as it reduces the diversity of flora and fauna.

Saltwater intrusion (Figure 12.18) caus-

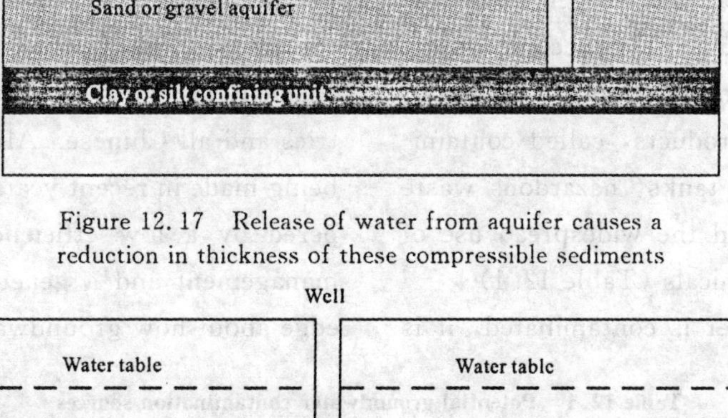

Figure 12.17 Release of water from aquifer causes a reduction in thickness of these compressible sediments

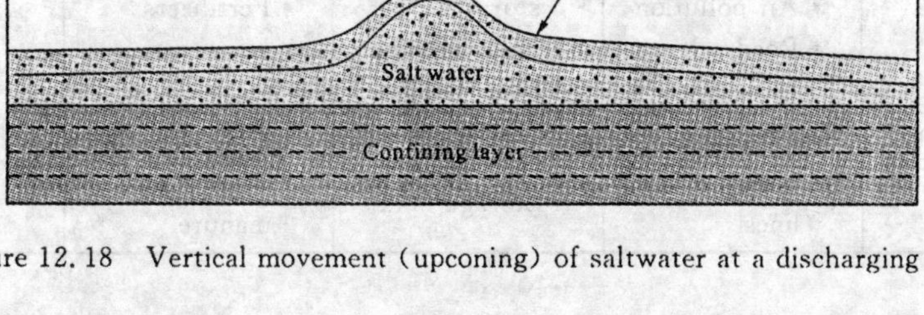

Figure 12.18 Vertical movement (upconing) of saltwater at a discharging well

ed by excessive pumping of the groundwater existed in coastal areas, such as, Shanghai, Tianjin and Ningbo. In these coastal cities, both the drink water supplies and industrial water usages have been affected seriously. One of the effective ways to prevent saltwater intrusion is pond surface runoff or river water to recharge the groundwater table.

7 Groundwater Contamination

Groundwater is one of our most valuable drinking water resources. It is also very important to many agricultural areas. Since it is filtered through the ground, it is often fresh and cold, and usually cleaner than surface water. Unfortunately, groundwater is threatened every day by people who do not even realize what it is or how they are affecting it.

Groundwater contamination occurs when man-made products such as gasoline, oil, road salts and chemicals get into the groundwater and cause it to become unsafe and unfit for human use. Some of the major sources of these products, called contaminants, are storage tanks, hazardous waste sites, landfills, and the widespread use of road salts and chemicals (Table 12.1).

Once groundwater is contaminated, it is an extremely costly operation to remove the contaminant. Sometimes it is impossible to clean it up to drinking water standards.

Proper land use management is the key to groundwater quality protection. Land use activities that result in contaminated groundwater can often be traced to lack of understanding of the potential for contamination, carelessness or negligence.

Groundwater contamination is a serious problem in China. Contamination problems are increasing in China primarily because of the large and growing number of toxic compounds used in industry and agriculture. Municipal landfills and industrial waste disposal are the main causes of groundwater contamination in some cities of China. Scientists suspect that many household wells in rural China are contaminated by substances from fertilizer, pesticides, and livestock wastes.

Preventing contamination in the first place is by far the most practical solution to the problem. This can be accomplished by the adoption of effective groundwater management practices by governments, industries and all Chinese. Although progress is being made in recent years, efforts are hampered by a low efficiency of government management and a general lack of knowledge about how groundwater behaves.

Table 12.1 Potential groundwater contamination sources

Place of Origin	Municipal	Industrial	Agricultural	Individual
At or near the land surface	• Air pollution • Road salt	• Chemicals storage • Fuels storage • Mine tailing	• Pesticides • Fertilizers • Livestock waste	• Detergents • Battery • Cleaners
Below the land surface	• Landfills • Leaky sewer lines	• Oil pipelines • Storage tank	• Storage tank • Storage manure	• Abandoned wells

1. Answer the Following Questions

(1) What is mean by the term artesian? Under what circumstance do artesian wells form?
(2) How do porosity and permeability differ?
(3) What is porosity? How is it calculated?
(4) Describe in sketch the formation of caves and relate it to water table.
(5) Define groundwater and relate it to the water table.
(6) Differentiate between stalactites and stalagmites.
(7) What is karst topography? How does it form?
(8) Sketch two different origins for springs.
(9) How does artesian well works its way to the Earth's surface?
(10) Describe in sketch the perched aquifer.

2. Multiple Choice

(1) Which one is the best groundwater reservoirs?
A: Sandstone B: Shale
C: Gneiss D: Granite
(2) Where do stalactites form?
A: Roof B: Floor
C: Head D: All of them
(3) The stalagmites only form ______.
A: Below the water table
B: Above the water table
C: All of the above
D: None of the above
(4) A zone where pore spaces are only partially filled with water is called ______.
A: Aquifer B: Aquiclude
C: Saturated zone D: Vadose zone
(5) The groundwater can be discharged by ______.
A: Well B: Spring
C: Stream D: All of them
(6) Which of the following statements about aquifer is not true?
A: Groundwater in perched aquifer is separated from the main groundwater below it.
B: The hydrostatic pressure on the surface of the body of unconfined groundwater is equal to atmospheric pressure.
C: The hydrostatic pressure on the surface of the body of confined groundwater is same everywhere.
D: Most confined aquifers are not directly recharged by vertical infiltration.
(7) The top of zone of saturation is ______.
A: Water surface B: Water table
C: Ground surface D: All of them
(8) The percentage of a rocks total volume that is taken up by pore space is called the ______.
A: Porosity B: Permeability
C: Volume ratio D: Pore ration
(9) The groundwater can be recharged by ______.
A: Precipitation B: Lake
C: Stream D: All of them
(10) The most extensive and best developed tower karst is found in ______.
A:Guilin(桂林)
B: Ha Long Bay in Vietnam(越南下龙弯)
C: Former Yugoslavia(前南斯拉夫)
D:Stone Forest of Lunan(路南石林)
(11) A periodical jet of hot water and steam is called ______.
A: Artesian well B: Geyser
C: Sinkhole D: Spring
(12) Which of the following rocks has the lowest porosity?
A: Sandstone B: Shale
C: Conglomerate D: Massive granite
(13) The removal of water from the saturated zone is called ______.
A: Recharge B: Discharge
C: Percolation D: Infiltration
(14) A layer of geologic material that impedes the movement of water into and out of an aquifer is called ______.
A: Sandstone B: Aquiclude
C: Water table D: Permeable layer
(15) Which of the following statements about groundwater is not true?
A: If a well is drilled into a confined aquifer, sometimes the water will flow out to surface of the Earth due to higher hydrostatic pressure.

B: Even in the baked desert, a thin water film surrounds the soil particles.

C: Gravity spring flow results directly form the force of gravity.

D: All the water beneath the ground surface is groundwater.

(16) Which of the following statements about groundwater is not true?

A: Water film surrounding soil grains in vadose zone is held by molecular attraction.

B: Only those highly fractured igneous and metamorphic rocks could have relatively higher porosity.

C: Sinkhole develops below the water table.

D: A perched aquifer sits above main water table.

(17) Confining layer is ______.

A: Sandstone B: Aquiclude

C: Aquifer D: Permeable layer

(18) The groundwater stored in gneiss is ______.

A: Pore water B: Fissure water

C: Karst water D: All of them

(19) The principal karst rock is ______.

A: Sandstone B: Granite

C: Sediment D: Limestone

(20) Clay is often served as ______.

A: Water table B: Aquiclude

C: Aquifer D: Permeable layer

(21) Karst topography results from the chemical weathering of limestone underground.

A: True B: False

Chapter 13 Ocean and Coastal Processes

Key words

Wave	Fringing reef	Barrier reef	Atoll	Breaker	Wave period	Salinity
波浪	岸礁	堡礁	环礁	破浪	周期	盐度
Tide	Longshore bar	Low tide	Flooding tide	Ebbing tide	Neap tide	Spring tide
潮汐	沿岸堤	低潮	涨潮	退潮	小潮	大潮
Waveform	Tidal current	Ocean current	Surface ocean current		Longshore current	
波形	潮流	洋流	表层洋流		沿岸流	
Amplitude	Barrier islands	Surf	Sea cave	Sea cliff	Headland	Bay
波幅	障壁岛	激浪	海蚀穴	海蚀崖	海岬	海湾

An ocean is a large body of salt water that surrounds a continent. The oceans cover about 71% of the Earth's surface. The geological process of ocean plays an important role in the evolution history of the Earth. The ocean is the end of the line for the weathered debris of continental erosion.

The term sea is used in two different ways. It is often used interchangeably with ocean in referring to bodies of salt water. In geology, seas are smaller branches of an ocean and often partly enclosed by land. In geology, however, an ocean is referred to as an area of oceanic crust covered by water.

The largest seas are the South China Sea, the Caribbean Sea, the Red Sea and the Mediterranean Sea. The popular division of ocean is the North Atlantic, South Atlantic, North Pacific, South Pacific, Indian, Arctic, and Antarctic. Actually, all limits of oceans are arbitrary, as there is only one global ocean. The greatest ocean depth of 11 035m occurs in the Mariana Trench and the average depth is 3.8km. More than half the ocean water resides in the Pacific basin.

The ocean is salty. Sea water is composed of, by weight, a mixture of approximately 96.5% pure water molecules and 3.5% of salts, dissolved gases, organic materials, and undissolved particles. Salinity refers to salt content of a body of water and is the total amount of dissolved material in grams in one kilogram of seawater. The average salinity of seawater is about 3.5%. At least 72 chemical elements have been identified in seawater. The most common salts in ocean are listed in the following Table 13.1:

The density of seawater varies with temperature and salinity of the water. As temperature and salinity increases, density de-

Table 13.1 Principal constituents of seawater [After Xia(夏邦栋) et al, 1995]

Constituent	NaCl	$MgCl_2$	$MgSO_4$	$CaSO_4$	K_2SO_4	$CaCO_3$	MgBr	Total
(%)	2.72	0.38	0.17	0.12	0.09	0.01	0.01	3.50

creases. Although the density of seawater varies at different points in the ocean, a good estimate of its density at the ocean's surface is about 1 027 kg/m^3.

The temperature of the surface seawaters varies mainly with latitude. The high latitude sea can be as cold as −2℃ while the low latitude can be as warm as 36℃. The average temperature of the ocean surface waters is about 17℃. On the other hand, the temperature decreases with increasing depth.

The study of ocean is very important both in science and economy. Nearly 60% of the world's population lives in the coastal zone. There are also some very unique processes such as coastal erosion and sediment loading on the continental shelves.

1 Ocean Movements

Ocean never stops moving and changing. There are three basic movements of the ocean, that is, up down movements of waves (波浪), steady movement of ocean currents (洋流) and the rise and fall of tides(潮汐).

1.1 Waves

Waves are the undulatory motion of a water surface. Waves are caused mainly by winds. When the wind blows across the water, it changes the water's surface, first into ripples and then into waves. Earthquakes and submarine landslides may also generate waves. A tsunami is a seismic sea wave caused by an earthquake or submarine landslide.

For a wave, the portion that is displaced above the still water line is called wave crest (波脊、波峰) and the part of the ocean wave that is displaced below the still water line is called wave trough (Figure 13.1). Or the top of wave is crest and the bottom of wave is trough(波谷、波槽). The wave height(波高) is the vertical distance between wave crest and wave trough. The wavelength (波长)is defined as the horizontal distance between two successive crests or troughs. Wave period(周期) is the time it takes two consecutive wave crests to pass a stationary point.

Ocean wave heights range from 0.3 to about 5m and the maximum-recorded ocean wave height is 34m. The amplitude is equal to 1/2 the wave height or the distance from the crest or trough to the still water or equilibrium surface. Wave period is dependent upon the speed of movement of the wave

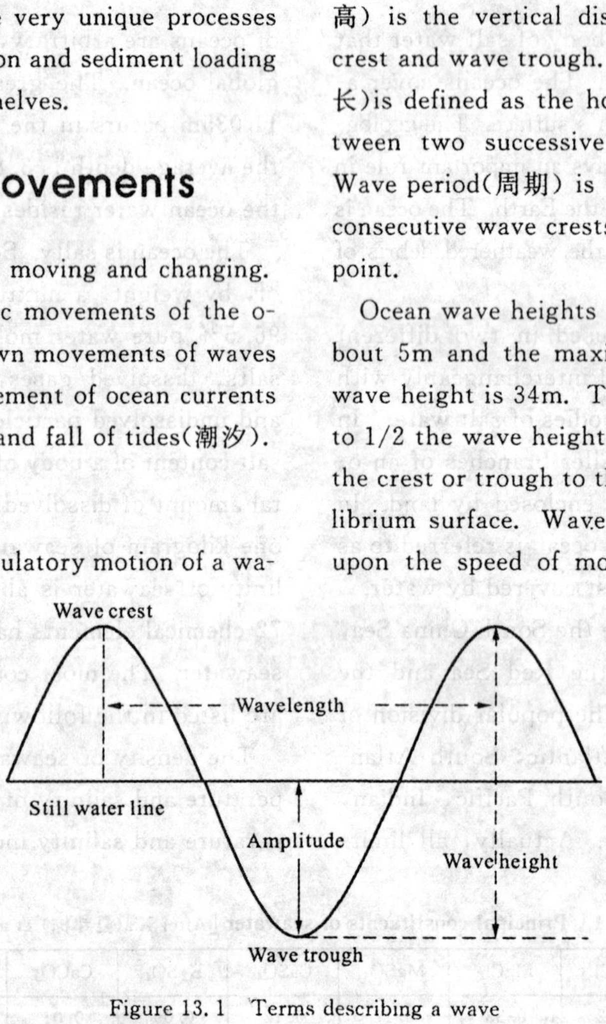

Figure 13.1 Terms describing a wave

across the surface. Ocean waves travel at speeds of 25 to about 90km/h.

A wave is an energy transport phenomenon which transports energy along a medium without transporting matter (Figure 13.2). As waves pass, it is the waveform (波形) and wave energy move rapidly forward, not the water. Waves do not carry water from one place to another, the water actually moves vertically in an orbital motion. The water moves up as a crest passes, and moves down as a trough passes. This can be demonstrated easily by observing the movement of floating objects beyond the breakers. The floating objects move towards the shore on the crest of a wave, and move the same distance backward with the trough of the wave. The amount of energy carried by a wave is related to the amplitude of the wave. A high-energy wave is characterized by a high amplitude.

Water waves are surface waves, a mixture of longitudinal and transverse waves. The surface water particles trace an orbit whose diameter is equal to the wave height as the wave passes. These vertical circles are more obvious at the surface. With the increasing depths, the energy of motion will dissipate gradually and the diameter of the orbits will become smaller and smaller. At a depth of 1/2 the wavelength (20 to 200m), the orbital motion has decreased to almost zero. So, wave base (波基面) is the depth at which wave motion ceases (Figure 13.3). It equals about 1/2 wavelength. Below that depth the water remain stationary as the wave passes. Thus, waves cannot erode the bottom or move sediment in water

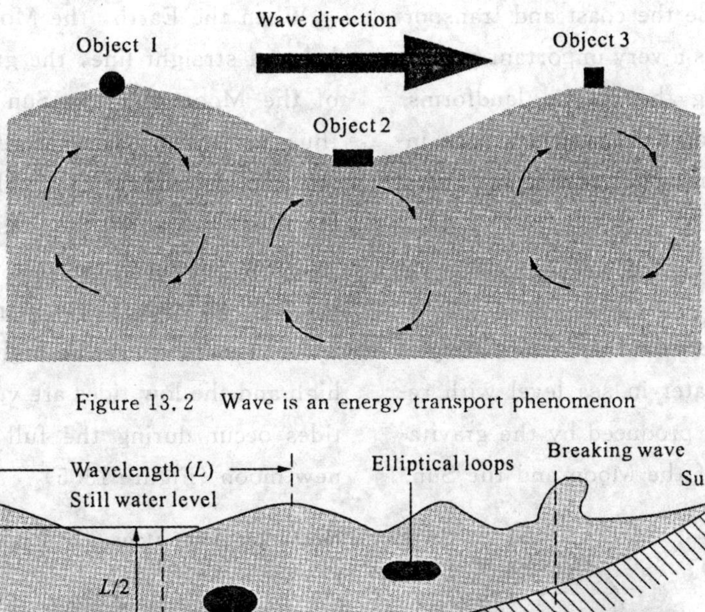

Figure 13.2 Wave is an energy transport phenomenon

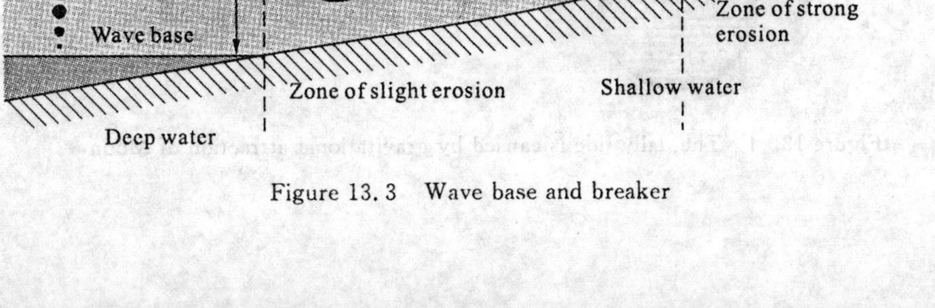

Figure 13.3 Wave base and breaker

deeper than $L/2$ because beneath which there is no wave movement.

As wave approaches shallow water (The water is shallower than wave base), they slow down by friction,and the circular loops of water motion change to elliptical shapes, and become increasingly flattened towards the bottom. Wave height increases, and the crests become more peaked. The orbit speed of the water near the crest is not slowed too much so water move faster toward the shore than the wave itself. The crest can curl and eventually break (fall over); this is the breaker(破浪). So, breakers are caused by the crest overtaking the trough. When breakers approach the coast or hit against rocks, it is called surf(拍岸浪、激浪、碎浪). The area where waves start breaking is called the surf zone. The water in surf zone is moving back and forth.

Waves can erode the coast and transport the sediment. It is a very important geological agent shaping the coastal landforms. The geological process of waves will be introduced in the following section of coastal process.

1.2 Tides

Tides are the periodic rising and falling of large bodies of water in sea level with respect to the land, produced by the gravitational attraction of the Moon and the Sun. When the tide is rising, it is called a flooding tide(涨潮)because it floods the channel and when its falling it is an ebbing tide(落潮、退潮). The maximum water level is called high tide (高潮) and the minimum level is low tide(低潮).

- **Daily Tide**(日潮)

The Sun's gravitational force on the Earth is only 46 percent that of the Moon, making the Moon the single most important factor for the creation of tides (Figure 13.4). The gravitational attraction of the Moon causes the oceans to bulge out in the direction of the moon. At the same time, a bulge occurs on the opposite side of the Earth due to centrifugal force of the Earth－Moon system. Since the Earth is rotating while this is happening, two tides occur each day.

- **Spring Tide** (大潮、朔望潮)

When the Earth, the Moon and the Sun lie on a straight line, the gravitational pull of the Moon and the Sun are combined, thus resulting a very large difference between high and low tides. This phenomenon is called spring tide. It has nothing to do with the season Spring. It is derived from the German verb *springen*, meaning"to leap up". In such case, the high tides are very high and the low tides are very low. Spring tides occur during the full moon and the new moon (Figure 13.5).

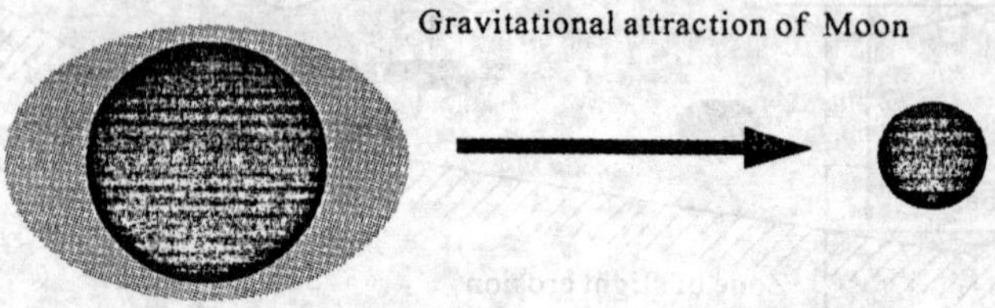

Figure 13.4 The daily tide is caused by gravitational attraction of Moon

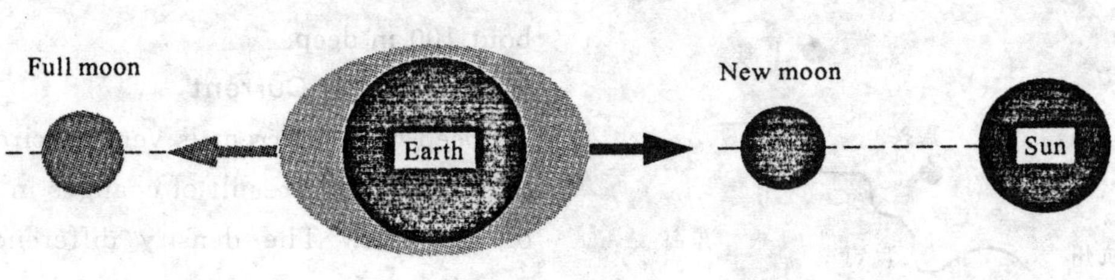

Figure 13.5 Spring tide

- **Neap Tide**(小潮)

When Sun and Moon are perpendicular to one another with respect to the Earth, the gravitational pull due to the Sun partially cancel out those of the Moon, thus resulting in very little differences between high and low tides. This especially weak tide is called the neap tide (Figure 13.6), every two weeks in between spring tides. In such case, a higher low tide and lower high tide occur.

Besides the Sun and Moon, the shape of the coastline, depth of the water and features on the ocean floor all affect tides.

Currents associated with the vertical rising and falling of the tide in coastal waters is a tidal current. Tidal current (潮流) moves horizontally. It can be described either as flood flow or ebb flow. The effects of tidal currents on the movement of water in and out of bays and harbors can be substantial and dangerous. The tidal current significantly affects the size, sorting, and distribution of sediment over most of the sea floor. Tidal currents along coasts can be effective in moving eroded material.

Estuaries (港湾、河口湾、三角湾、三角港) are semi-enclosed coastal bodies of water which are affected by tides and the interaction of fresh and seawater. In tide-dominated estuaries (also known as macrotidal estuaries and open estuaries), it is tidal currents, rather than waves, being the dominant force shaping the gross land shape.

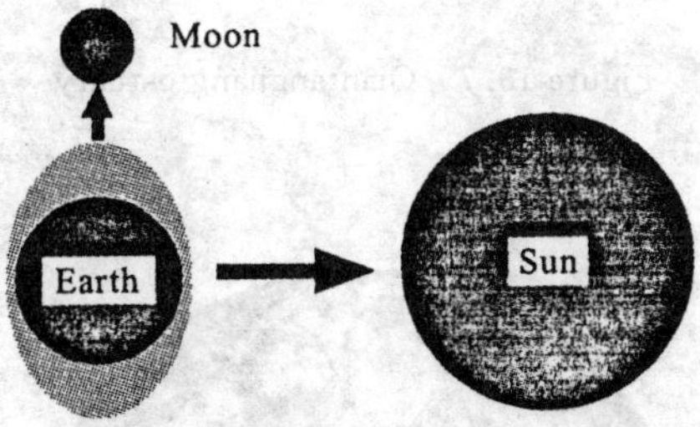

Figure 13.6 Neap tide

Qiantangjiang estuary(钱塘江河口湾) is virtually a tide-dominated estuary consisting of a landward-tapering funnel shaped channel. When the tide enters Qiantangjiang estuary, the tidal energy become increasingly strong due to the constriction by the funnel-shaped entrance, thus resulting the world famous Qiantang tidal current (钱塘江潮).

In some tide-dominated estuaries, such as Qiantangjiang estuary, the tidal current is so strong that any sediment will be washed away. So no delta can survive under such extreme conditions.

1.3 Ocean Current

The large-scale steady movement of the seawater in a prevailing direction is called ocean current (洋流). They are very important in determining the climates of the continents, especially those regions bordering on the ocean. There are two types of ocean currents: surface circulation and deep-water

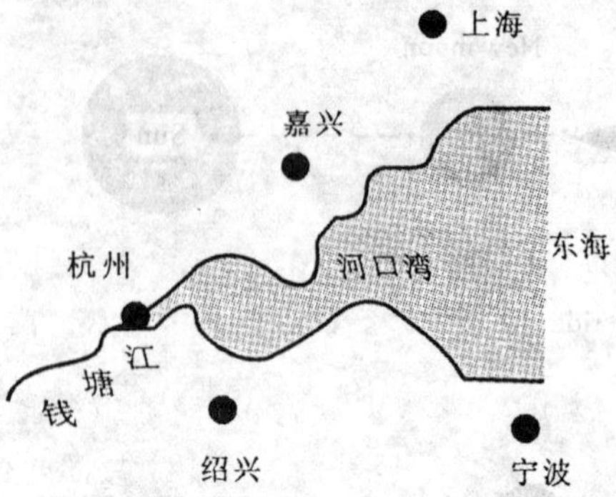

Figure 13.7 Qiantangjiang estuary

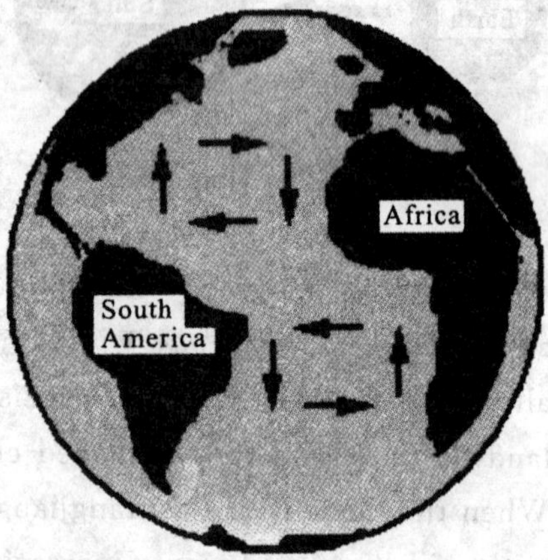

Figure 13.8 Simplified ocean surface currents in the Atlantic

currents.

• **Surface Ocean Current**

Surface ocean current is a horizontal movement of seawater at the ocean's surface. Large surface currents are mainly driven by winds that blow year round. Frictional stress at the interface between the ocean and the wind causes the water to move in the direction of the wind. Because of the Coriolis Force, the surface ocean currents have a typical clockwise spirals in the northern hemisphere and counter-clockwise rotation in the southern hemisphere.

The water of surface ocean current is about 100 m deep.

• **Deep-water Current**

Deep-water current is vertical circulation of seawater as a result of changes in density of seawater. The density difference is a function of temperatures and salinity. These deep waters sink into the deep ocean basins at high latitudes where the temperatures are cold enough to cause the density to increase. The water involved in the deep-water current accounts for 90% of the total ocean water.

The main geological process of the ocean current is the transportation of sediments from the bottom of the ocean to the shallow areas. The speed of ocean current is usually less than 0.5～1.5m per second.

2 Coast

The ocean can be largely divided into three different geographical units. They are coast region, continental margin and ocean basin.

Coast represents the boundary between sea and land, which is all the land near the sea, including the beach and inland areas. Coastal processes include all the geological processes acting on or near coast.

Shore (or sea shore, 海滨、滨海) is the land on the side of sea, or any other large body of water. Strictly speaking, however, when the water does not ebb and flow, in a river, there is no shore. The beach (沙滩) is a kind of special shore which is sandy or pebbly. The term "shore" and "coast" usually can be used interchangeably.

Off shore (外滨) (Figure 13.9) zone is the region between wave base and low tide level. Foreshore (前滨) zone lies between the

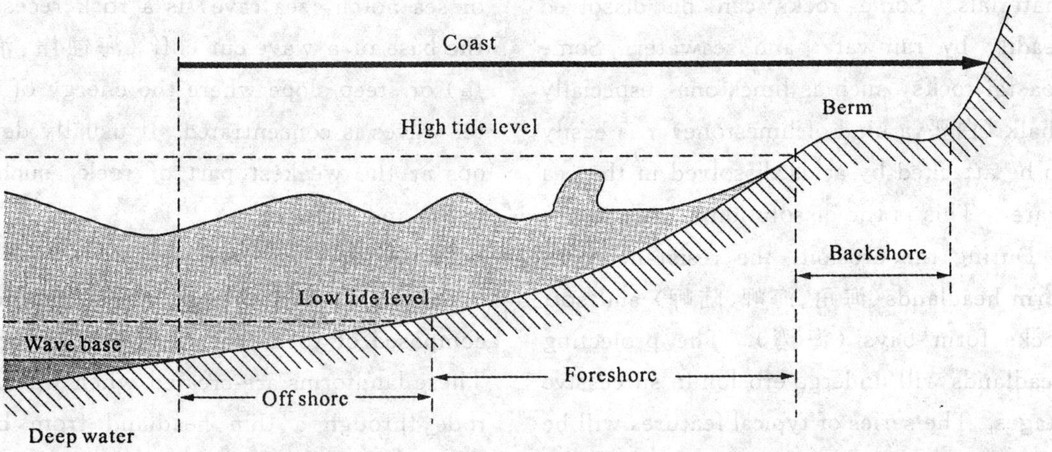

Figure 13.9 Schematic diagram of coast structure

average high tide mark and the average low tide mark. And backshore(后滨)is the region extending from high tide level to wave-cut platform（海蚀平台）. Only storm events reach this high region. The berm（后滨阶地、滩肩）is the raised land above water (depending upon tide). It is composed of deposited materials.

Coastal region is a region where sediment is being delivered from the land, moved around by the waves, and exported out to the shelves. There are many different classification schemes of coast. In this textbook, we classify the coasts into erosional and depositional coasts depending on whether their primary features were created by erosion of land or deposition of eroded material.

2.1 Erosional Coast

Erosional coast（侵蚀性海岸）is the one that has been undergoing erosion. Erosional coast is featured by steep rocky cliffs. Jagged, irregular shapes dominate in hard, rocky coast. The forces that shape coasts are essentially the same as those that form other topographies: the various erosional processes operating in conjunction with waves, tidal current and ocean current. At the coastline, the major erosive agent is wave action.

Wave erodes the coast by hydraulic action, abrasion and dissolution. The surf hits again the cliff and smashes the rock into pieces. Seawater can compress air into cracks in rocks. When the air escapes it does so under pressure and shatters the rock. It is estimated that the pressure exerted against the cliff may be up to several tons per square meter. This is the hydraulic action. Large waves throw beach material against the cliffs. And waves also make rocks and pebbles to crash into each other, causing them to break up and reduce in size. Wave action wears down rugged, sharp stones into smaller rounded pebbles. This is the abrasion. In fact, the abrasion is probably more intense in the surf zone than in any other environment. The physical destructiveness of the waves is further enhanced by the chemical weathering action of seawater. Coastal shorelines differ markedly in rock materials. Some rocks can be dissolved readily by rainwater and seawater. Some coastal rocks, such as limestone, especially chalk (白垩,a kind of limestone), is easily to be attacked by acids dissolved in the sea water. This is the dissolution.

During the erosion, the resistant rocks form headlands(海角、海岬、陆岬) and soft rocks form bays (海湾). The projecting headlands will undergo erosion in successive stages. The series of typical features will be formed in orders of sea notch, stack, wave-cut platform and finally wave-cut cliff, as a result of progressive erosion during the evolution of erosional coast (Figure 13.10). All of the above-mentioned features develop from the initial cracks in less resistant rock.

• **Wave-cut Notch**

Wave-cut notch (海蚀穴、浪蚀洞、浪蚀龛, or sea notch, sea cave) is a rock recess at the base of a wave-cut cliff (海蚀崖、浪蚀崖) or steep slope where the energy of water waves is concentrated. It usually develops at the weakest part of rock, such as faults and joints.

• **Sea Arch**

Sea arch (海蚀桥) is a coastal landform composed of rock that resembles an arch. These landforms are created when waves erode through a thin headland from both sides. When the sea cave is deepened and enlarged until it passes all the way through the headland, the sea arch forms.

• **Sea Stack**

Sea stack (海蚀柱) (Figure 13.11) is a steep pillar of rock located in the ocean a short distance from the coastline. These landforms are created due to the collapse of

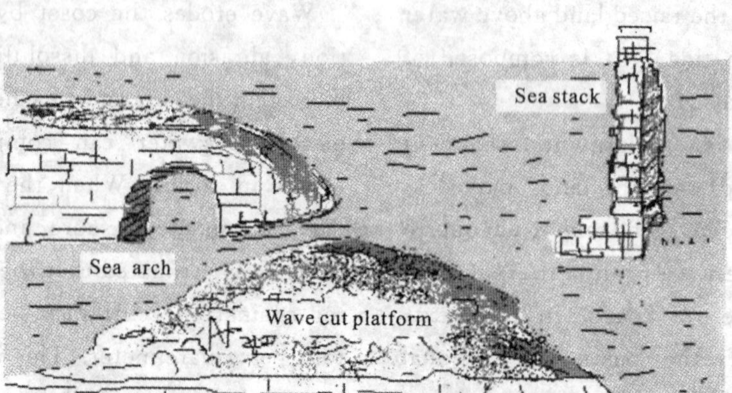

Figure 13. 10 Schematic sketch of principle features along erosional coast

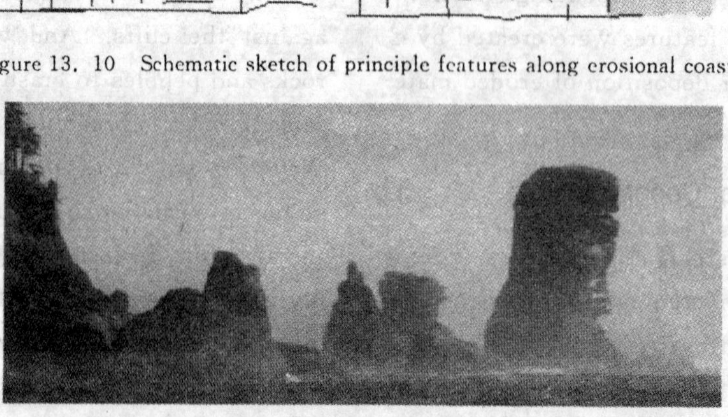

Figure 13. 11 Sea stacks. Courtesy of NOAA, public domain

top of sea arch. If the upstanding stack is lowered further by erosion, it is called sea stump (海蚀墩). The famous stone block, named Tianyahaijiao(天涯海角) in Hainan Island (海南岛), is a good example of stump.

- **A Wave-cut Platform (or Terrace)**

A wave-cut platform (海蚀平台、波切台、海蚀阶地) is a narrow horizontal surface at the base of sea cliff. It is formed beneath the surf zone as a coast retreats by wave action (Figure 13.12). Wave-cut platform is covered by water at high tide but exposed at low tide. It actually represents the last remnant of an eroded headland. With the further development of wave-cut platform, eventually sea cliff forms. It is important to note that wave-cut platform is the erosion of bedrock (Figure 13.13). Steep cliffs, such as these in the Fujian Province(福建省) and eastern Taiwan(台湾), occur when the sea water removes all the material eroded by the strong waves.

Due to the continually erosion, an irregu-

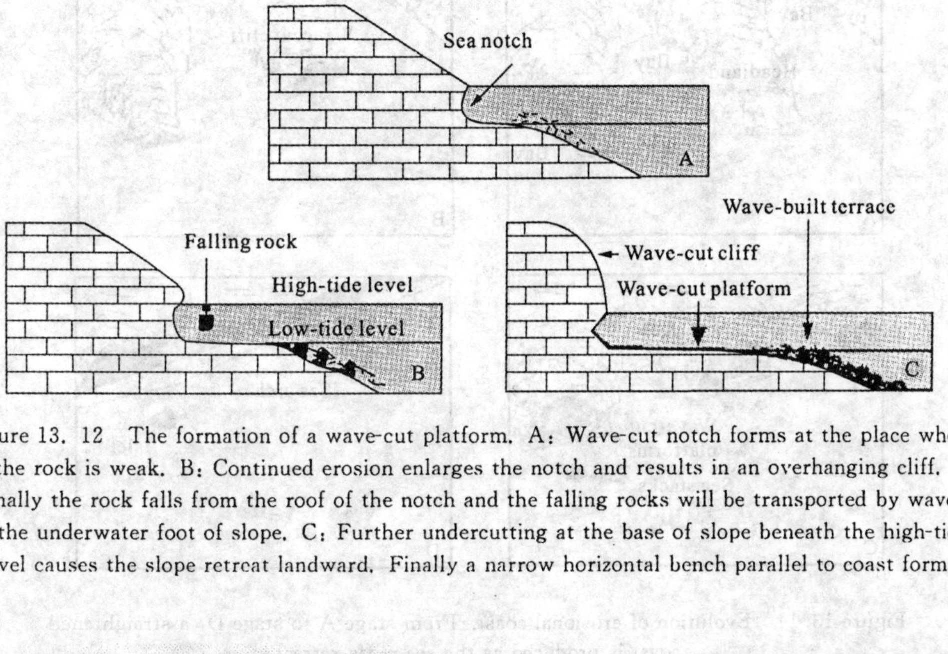

Figure 13. 12 The formation of a wave-cut platform. A: Wave-cut notch forms at the place where the rock is weak. B: Continued erosion enlarges the notch and results in an overhanging cliff. Finally the rock falls from the roof of the notch and the falling rocks will be transported by waves to the underwater foot of slope. C: Further undercutting at the base of slope beneath the high-tide level causes the slope retreat landward. Finally a narrow horizontal bench parallel to coast forms

Figure 13.13 Wave-cut platform and beach. Courtesy of NOAA. Public domain

lar coast with bays and headlands can be gradually straightened. This is the evolution of erosional coast (Figure 13.14).

At the initial stage, the coastlines are rugged and irregular, with numerous bays being separated by headlands. This kind of bedrock coast is usually formed by drowning of a former valleys due to the rising of sea level (Since the last Ice Age (~10 ka) (冰期), sea level has risen approximately 130m) or regional depression of land. Then due to the wave action, the coast will be eroded to form various stacks, arches and wave-cut platforms. At the same time, a series of seaward cliffs form due to the continuous retreating of coastline. And finally, with the connection of all the truncated headlands, a nearly straight coastline is produced (Figure 13.15). The former wave-cut platform may be covered by a veneer sand deposition, thus forming a beautiful ever-stretching beach.

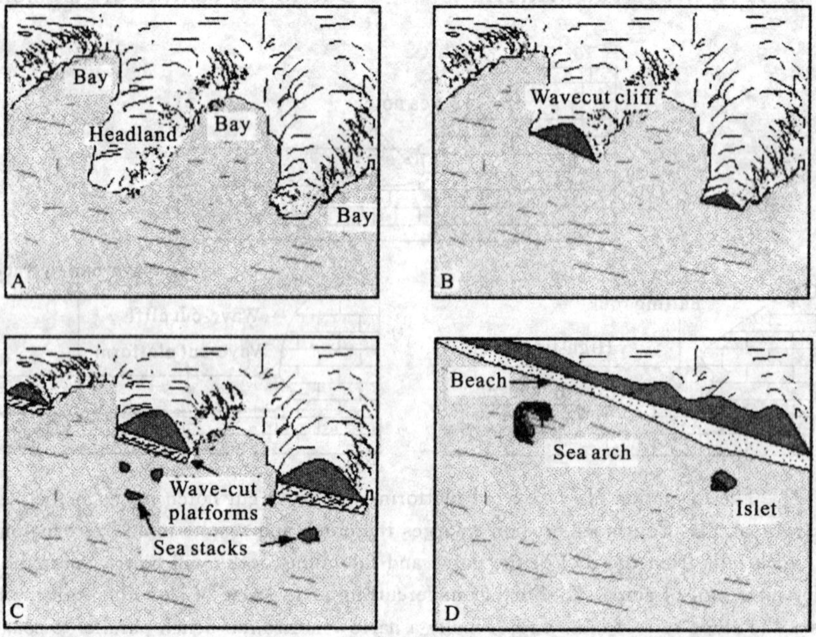

Figure 13.14 Evolution of erosional coast. From stage A to stage D, a straightened coast is produced as the sea cliffs retreat

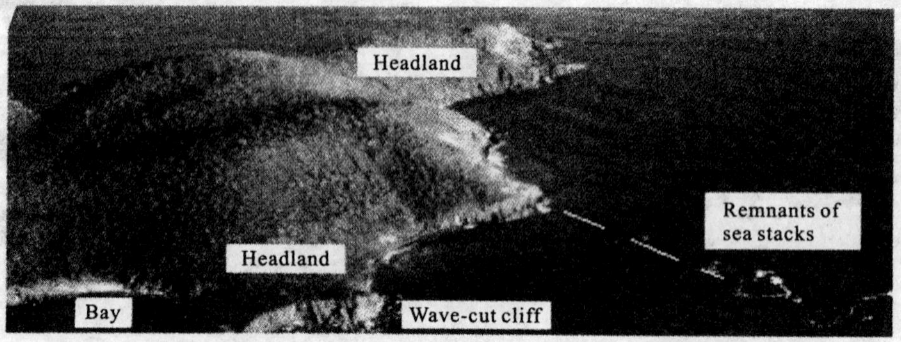

Figure 13.15 Straightened coast. Courtesy of NOAA. Public Domain

2.2 Depositional Coast

Depositional coast is the place where the major geological process is the deposition and accumulation of sediment due to the wave action. The most striking features on the depositional coast include deltas, lagoons (泻湖), barrier islands, and beach-sand dunes and ever-stretching beach.

The sediment carried by wave comes from many sources, including creeks and rivers, cliff faces, sand bars, seagrass and shells. Due to the high energy of waves, the sediments on the short will be transported. These materials can be transported and deposited at someplace near the shore, or redistributed by a wave action called longshore drift (沿岸漂移). Longshore drift transports sediments along shorelines by way of longshore currents(沿岸流).

When a wave hits the shore, we say that the wave move as swash(进流). Swash is actually the turbulent sheet of water from collapsing breakers. When the water retreats and flows back down the beach, we say that wave moves as backwash(回流).

Swash is the turbulent sheet of water from collapsing breakers and moves up the slope of the beach. When the energy of the swash has been expended, the water flows back down the beach toward the surf zone as backwash. The grains are lifted as the waves in the swash zone move onto the beach, and then the grains are deposited again as the water retreats. If breakers (waves) are approaching perpendicular to the shoreline, the sand will be lifted and then redeposit at the some position. Therefore no real net movement of sand takes place. In this case, the sand moves in a continuous cycle (Figure 13.16).

Forturenately, it is rare for waves to hit the land"head on". Longshore currents are formed when incoming waves approach the shore at an oblique angle. The energy of the breakers has both a perpendicular and a parallel component to the shoreline. The parallel component of energy cause part of the water mass to be transported along the shoreline. The parallel movement of water mass along the shoreline is called longshore current (Figure 13.17).

The longshore current not only moves water in the surf zone, it also moves sediment parallel to the shoreline. Longshore transport of sand is caused by the longshore current in combination with the stirring up of sand by wave action. Figure 13.17 illustrates how beach drift operates. The water moves onto the beach in the same direction

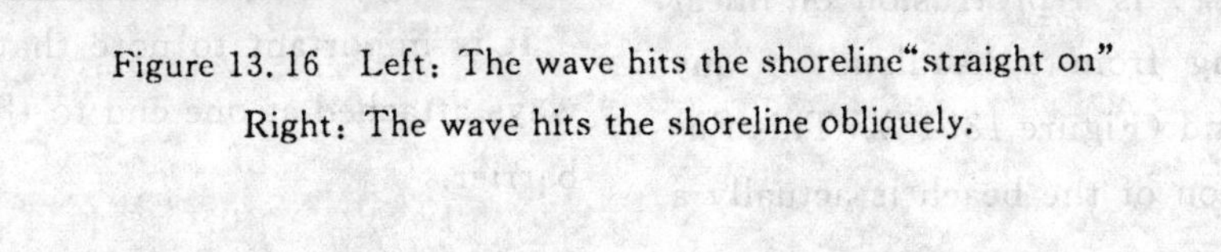

Figure 13.16 Left: The wave hits the shoreline"straight on"
Right: The wave hits the shoreline obliquely.

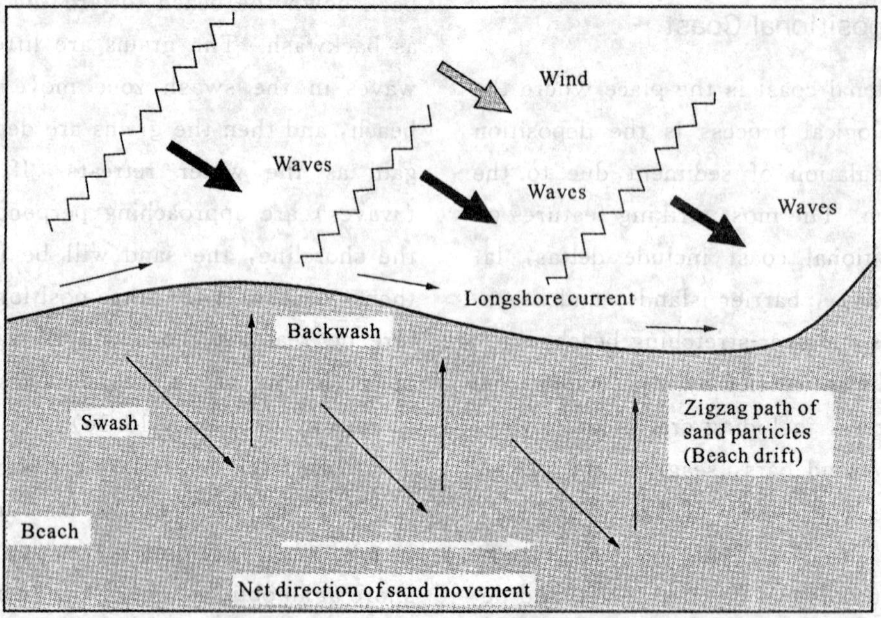

Figure 13.17 Longshore current. Note the zigzag path followed by sand grains moved by the swash and backwash of the waves

of waves. Therefore the sand grains are being carried in movement direction of water. However, when the water retreats, the backwash will take the sand grains straight down the slope of the beach, which is perpendicular to the shoreline, due to the gravity. Eventually the sand grains are deposited further down the beach. So, the actual path taken by sand grains is a zigzag way and the net overall movement of sand is down current. This net movement of the beach sand is known as"beach drift".

The major depositional features produced by wave and longshore current are listed here. It is important to note that some of the features can be found either on erosional coast or depositional coast.

• **Spit and Baymouth Bar**

A spit (沙嘴) is a protrusion or linear beach extending from a headland or any point on the land (Figure 13.18). This fingerlike projection of the beach is actually a peninsular, built by longshore current. They are usually found to extend into open water or form at the mouth of a bay. In the latter case, they often curve inward towards the bay due to refraction of the waves around the mouth of the bay.

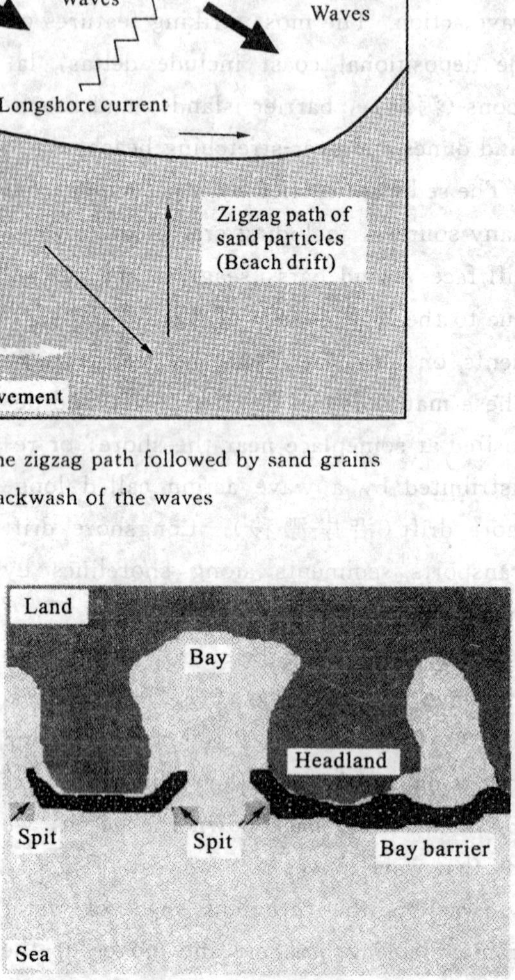

Figure 13.18 Spit and bay barrier

It is important to note that spits are always attached at one end to the land or bay barrier.

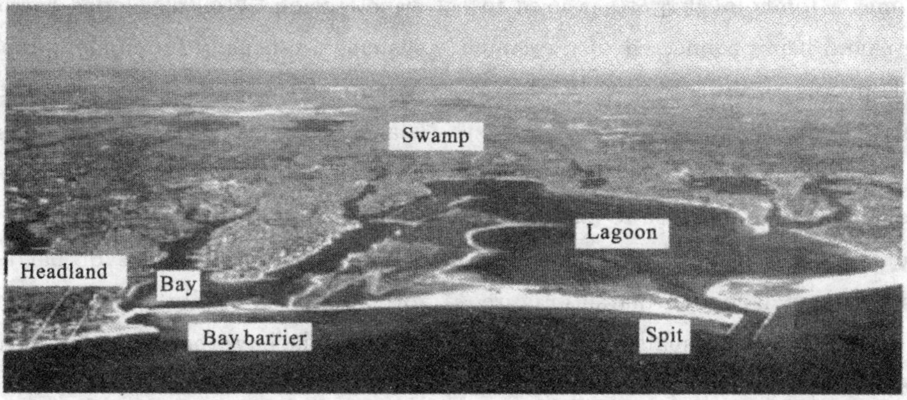

Figure 13. 19 Bay barrier and spit Courtesy of NOAA. Public Domain

When the spit on either side of a bay meets together, the baymouth bar or bay barrier(拦湾坝,湾口坝) forms. Baymouth bar separates a bay from the open ocean (Figure 13. 19). Spit and baymouth are commonly smaller in size.

• **Longshore Bar**

Longshore bar(沿岸堤) or sandbar is the straight submerged ridge of sand running roughly parallel to shoreline. It generally extends along the shore near the breakpoint of the waves (Figure 13. 20).

Longshore bar forms where the waves are breaking. The swash lifts the sands to attack the beach berm. And the backwash then take them back to place where the waves break. Eventually the sands deposit near offshore. The longshore bar is strongly influenced by seasonal storms. During storms, the sandbar will migrate seaward because of the seaward migration of break point of larger waves.

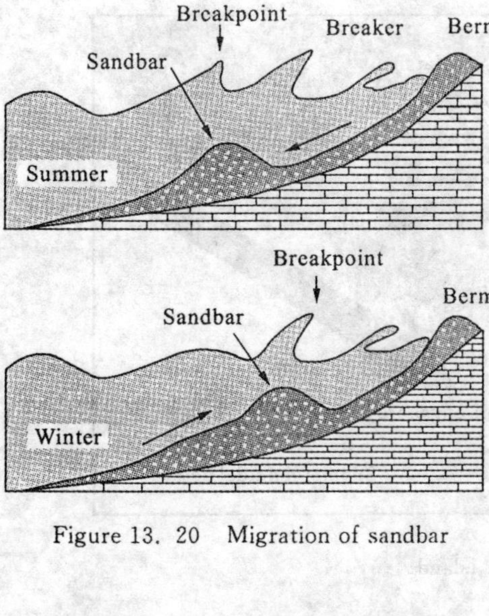

Figure 13. 20 Migration of sandbar

The sandbar is never long-lasting. It frequently migrates from place to place along the coastline due to the longshore current. Sometimes it can be exposed on the surface of seawater.

• **Tombolo**

A tombolo(连岛沙洲)(Figure 13. 21) is the sandbar that connects the island to mainland or another island. They are usually arcuate in shape and formed by longshore drift resulted from wave refraction. The bending of the wave is called wave refraction. The island bends the incoming waves around it so that their energy sweeps sand onto the tombolo from both sides. Once the stack erodes down to the waterline, the tombolo will disappear. Stacks don't last long, and that's why tombolos are uncommon.

Sometimes a tombolo(陆连岛) is used to refer the island being connected to the land by a sand spit.

- **Barrier Island**

Barrier islands(障壁岛、沙洲岛)are long thin sand islands that are close to and parallel to the shore and are separated from the mainland by a bay, lagoon or marsh (Figure 13.22). Barrier islands usually stretch for hundreds of kilometers along a coastline.

Barrier islands sit along the gently sloping coast plains which show few effects of wave erosion. Well-developed barrier island systems are usually found along most of the Atlantic Ocean. For example, the barrier island stretches almost continuously from Long Island in New York to the scenic Miami Beach in Florida (Figure 13.23).

The exact mechanisms of barrier island formation aren't fully understood yet. Several models have been put forward to explain the barrier island development. One of theories proposed that the barrier islands are greatly elongated sand spits formed by

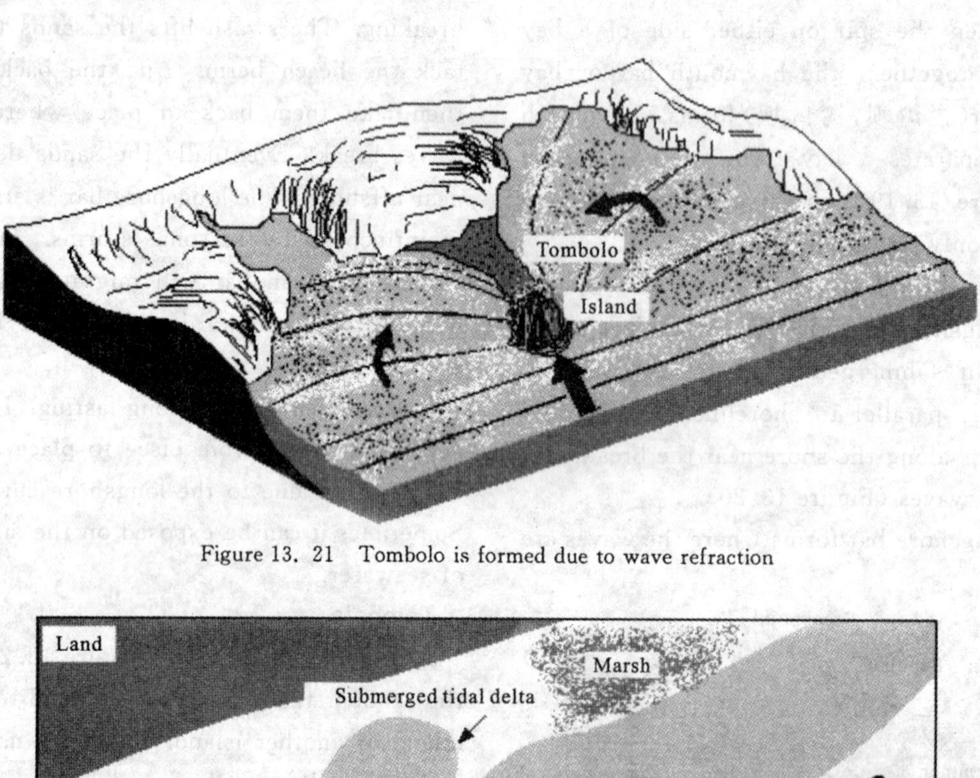

Figure 13.21 Tombolo is formed due to wave refraction

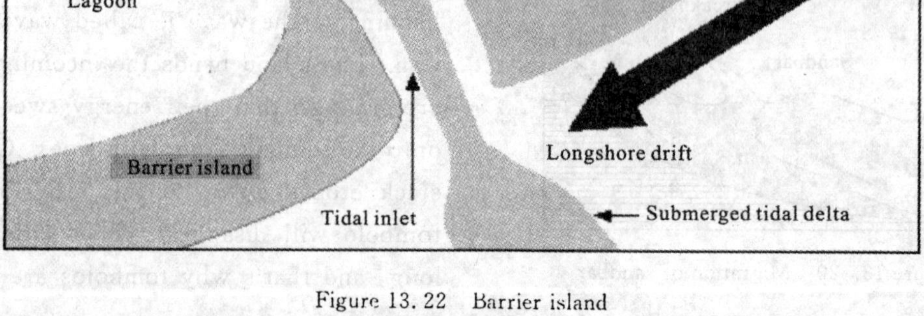

Figure 13.22 Barrier island

Figure 13.23 Barrier islands

Image credit: US National Oceanic and Atmospheric Administration , Public Domain

longshore drift. However, the rapid rise in sea level, associated with melting glaciers from the last ice age, may have been an important factor in their development.

Once formed, barrier islands will be reformed by storms, waves, tides and longshore currents. At some places, the gaps in barrier islands may allow tidal movement to form tidal deltas. Tidal deltas are submerged sand bodies located inland and outboard of barrier gaps. Furthermore, the ongoing global warming will cause the rise of sea level and the rising sea level will make the existing islands to migrate shoreward. Large storms contribute significantly to an island's landward migration.

- **Lagoons**

Lagoon (泻湖) is enclosed or semi-enclosed water body which is situated at the coast, but separated from the sea by land spits, barriers islands or other similar land features. Coastal lagoons are unusually open to the sea in restricted spaces.

The exact geological process of lagoon will be discussed in Chapter 17.

3 Continental Margin and Ocean Basin

Continental margins are the boundaries between the oceans and the continents. They are dynamic regions for the production, recycling, and deposition of sedimentary particles. Continental margin generally consists of the continental shelf, slope, and rise. Active continental margin is featured by various tectonic activities. Shelf is narrow to non-existent, dropping off quickly into the depths of the trench. Passive continental margin is characterized by minor or local tectonic activities. These margins are marked by smooth relief due to tectonic inactivity and major sediment accumulation.

The deep ocean basin covers 30% of the Earth's surface and has features, such as abyssal plains(深海平原), deep-sea trenches and seamounts.

Coral reefs are among the most picturesque features in the ocean. By definition, a reef is a rigid, wave-resistant structure and coral reef is island consisting chiefly of com-

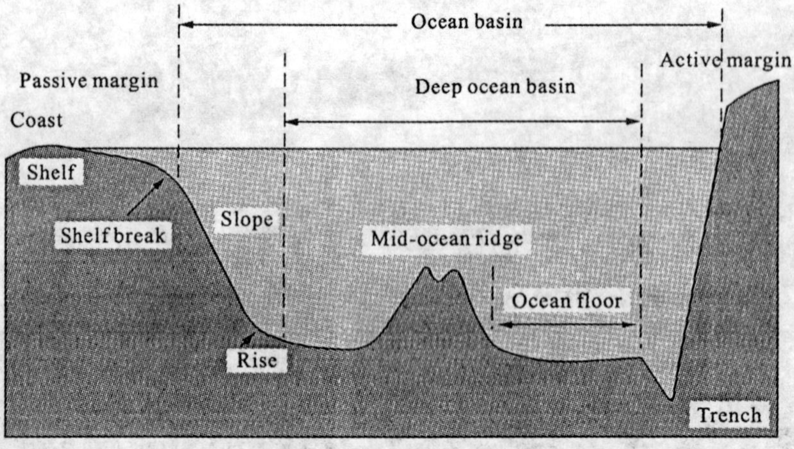

Figure 13.24 Schematic representation of the ocean basin and continental margin

pacted coral together with algal material and sediments. The coral reefs, often of great extent, are mainly found in the clear waters of the Pacific and Indian oceans. They are found only in tropical waters between 30°N and 30° S latitude. Australia's beautiful Great Barrier Reef（大堡礁）is the world's longest coral reef. It extends from near Papua New Guinea down Australia's east coast for about 2 000 km.

Reef-building corals（造礁珊瑚，often very erroneously called coral insects）secrets calcareous skeletons（钙质骨骼）as it grows, cementing itself to the carbonate remains of dead coral below. Reef building corals require warm water conditions to survive. If possible, temperatures should be between 18℃ and 30℃. On the other hand, reef corals require clear water because of their dependence on light. Thus, the limiting depth of active reef growth is about 50m under the surface. It is important to note that corals are not the only, and in some cases not even the major, reef-forming organisms. Some algae, mollusks and other marine organisms also deposite calcium carbonate and form reefs. However, any reef formed by a biological community is usually called a coral reef.

The presentation of coral reefs on the Pacific islands sitting on the deep ocean floor is a paradox. Drill cores from coral reefs go down into coral without reaching bedrock for a very long ways（often kilometers）. The problem is how can corals ，which need light to live, build islands that reach to the surface from the floor of the deep sea thousands of meters below?

During his voyage on the H. M. S. Beagle in the 1830s, Charles Darwin（达尔文）studied Pacific and Indian Ocean reefs. He noted that reefs were of three basic types: fringing reefs（岸礁）, barrier reefs（堡礁）and atolls（环礁）and postulated a single sequence of events which could cause each type to evolve.

The essence of Darwin's theory was that: reef-building corals would continue to build reefs upward toward the sunlight even if the bedrock below was sinking. As long as the rate of subsidence of the bedrock was equal to or lower than the growth rate of the corals, a coral reef could maintain its position in the ocean for thousands of years.

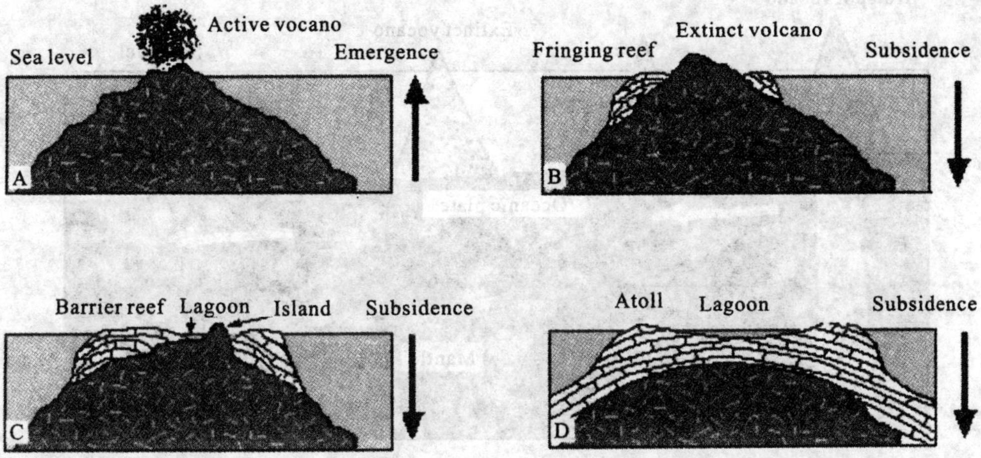

Figure 13. 25 Evolution of a coral reef from a subsiding volcanic island

Here is the model of coral reef formation originally proposed by Darwin and modified by modern plate tectonics (Figure 13. 25).

• Stage A: Active Volcanoes

The fist stage is the formation of a volcano built up from the sea floor. The volcano can be produced by activity of hot spots (热点) beneath the ocean floor, or by the plate movement along the mid-ocean ridges.

Hot spots are caused by plumes(地幔柱) that rise from the lower mantle. Mantle plumes are columns of bouyant molten rock material which rise through the Earth's mantle. They burn through the lithosphere plates to erupt as "hotspot" volcanoes. For example, the Hawaiian Islands, perhaps the best known hotspot, have formed in the middle of the Pacific Ocean more than 3 200km from the nearest plate boundary.

• Stage B: Fringing Reef

Fringing reef is the reef formed along the coastline of island. Fringing reefs occur close to land and often extend out to sea for long distances. This happens when the volcanoes become extinct and begin to subside.

The subsidence of volcano is caused by the plate movement. For example, due to the continuing plate movement, the active volcano island will eventually be taken away from the hotspot. At this moment, the magma source is cut off and the volcano becomes extinct. A new active volcano will form over the hotspot. The oceanic plate will sink due to the cooling (thus becoming heavier) on its way to trench. Therefore, the extinct volcano carried by plate will subside simultaneously. The gradual subsidence of cooling plate also occurs as it moves away from mid-ocean ridges.

• Stage C: Barrier Reef

With the continuing subsidence along with erosion, the central volcano may then be lowered almost to sea level. This is stage of barrier reef. Barrier reefs are reefs around a central island with a protected lagoon in between.

• Stage D: Atoll

If the entire island is submerged, an atoll forms. An atoll is a ring-shaped coral reef enclosing or nearly enclosing a shallow lagoon.

A seamount is a submerged mountain ris-

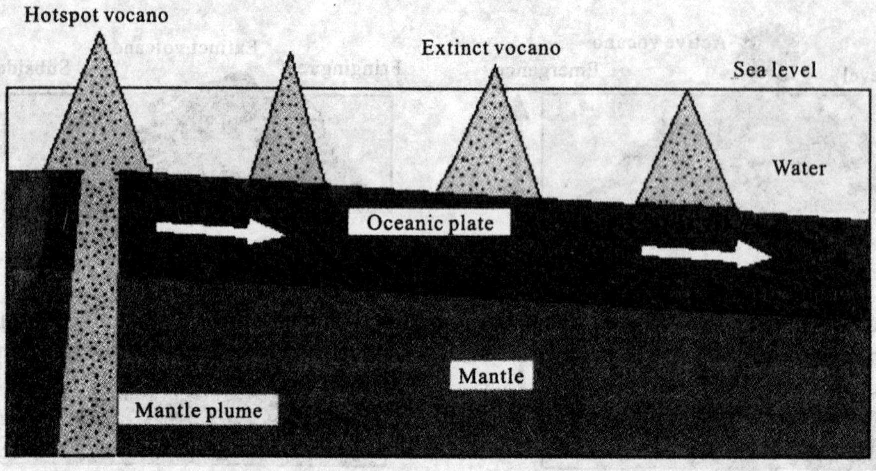

Figure 13.26 Hotspot volcano is caused by mantle plume

ing from the seafloor. A guyot(海底平顶山), named after the geographer Arnold Henry Guyot, is a flat-topped seamount.

Guyot is actually the end product of evolving coral reef islands. When the coral reef can not keep up with the sinking of the island, the entire island will be submerged. During its subsidence, the top of guyot will be planed by waves.

1. Answer the Following Questions

(1) Describe and explain the changes a wave undergoes as it approaches shore and breaks.

(2) What is wave base?

(3) What is the difference between spring tides and neap tides.

(4) What is longshore current?

(5) List the shoreline landforms created by erosion.

(6) Describe various landform features produced during the straightening of coastline.

(7) Show in sketch how longshore drift of sand can form a baymouth bar.

(8) Describe longshore current and their significance to longshore drift of sand.

(9) Describe wave-cut platforms (and marine terraces) and how they are formed.

(10) What is the difference between wavelength and wave period?

(11) Describe in sketch the transforming of coral reef from fringing reef to atoll reef.

(12) Why do most volcanoes on the ocean floor subside?

(13) List and describe the various tides.

(14) Differentiate between longshore bar and barrier islands.

2. Multiple Choices

(1) If you followed a single drop of water during a passing wave, in which motion would it move?

A: Horizontal circle　B: Vertical circle

C: Forward　D: None of the above

(2) What is the horizontal distance, either between the crests or troughs of two consecutive waves?

A: Wavelength　B: Wave height

C: Wave period　D: Wave amplitude

(3) The lowest point between two waves is ______.

A: Wave crest　B: Wave height

C: Wave trough　D: Wave amplitude

(4) A good example of an erosional coast is ______.

A: The coast of Bohai (渤海湾)

B: The eastern coast of Taiwan (台湾东海岸)

C: Tianjin (天津)

D: Shanghai (上海)

(5) Low ridges of sand that parallel coastlines and rise above sea level are called ______.

A: Spit　B: Barrier island

C: Longshore bar　D: Baymouth bar

(6) The tide occurs during full moon is called ______.

A: Neap tide　B: Spring tide

C: Daily tide　D: None of them

(7) A wave base for a wave with wavelength of 80m and period of 4s will be ______.

A: 50m　B: 160m

C: 0m　D: 80m

(8) Longshore bar is formed by breaker in conjunction with longshore current.

A: True　B: False

(9) Spit is always attached to land or baymouth bar.

A: True　B: False

(10) The sediments under the wave base are not disturbed by waves.

A: True　B: False

(11) Spring tides are usually in Spring.

A: True　B: False

(12) The swash takes back the sands into offshore.

A: True　B: False

(13) Longshore bar forms beneath the wave base.

A: True　B: False

(14) Which of the following statements about tide is not true?

A: Tides are the alternating rise and fall of sea level with respect to land, as influenced by the gravitational attraction of the moon and sun.

B: Tide occurs only one side of the Earth at any moment.

C: Spring tides occur when the Earth, the Sun, and the Moon are in a line.

D: Neap tides occur during quarter moons.

(15) Which of the following statements about coral reefs is not true?

A: Most coral reefs lie between the latitudes of 30° north and south where sea temperatures are warmest.

B: Fringing reefs are relatively young. They can

develop in shallow waters along the coast of tropical islands or continents.

C: Corals atolls are on the top of submarine mountains. These mountains are remnants of volcanoes.

D: Barrier reefs are featured by extensive lagoon without central island.

(16) Long Island in New York is an example of ______.

A: Atoll B: Barrier reef

C: Barrier island D: Longshore bar

(17) Which of the following statements about longshore current is not true?

A: The longshore current not only moves water in the surf zone, it also moves sediment parallel to the shoreline.

B: Longshore currents are formed by waves striking the shore at an oblique angle.

C: Sand is carried along the shore in a zigzag pattern by waves rushing onto the beach and back to the ocean.

D: Longshore current transport sand from beach to offshore.

(18) A wave can not transport matter.

A: True B: False

(19) Spring tide occurs ______.

A: Once a month B: Twice a month

C: Once a year B: Twice a year

(20) Backwash is always perpendicular to the coastline.

A: True B: False

(21) Guyot is a remnant of atoll.

A: True B: False

Chapter 14 Glacier

Key words

Glacier	Snow line	Continental glacier	Valley glacier	Till	Cirque
冰川	雪线	大陆冰川	山岳冰川	冰水沉积物	冰斗
Lateral moraine	Horn	Terminal moraine	Ground moraine	Drumlin	Arête
侧碛	角峰	终碛堤	底碛	鼓丘	刀脊
Glacial erratic	Drift	Roche moutonnée	U-shaped Valleys	Esker	Kettle
冰川漂砾	冰川沉积物	羊背石	U形谷	蛇形丘	锅穴

Glacier is the large permanent mass of ice that forms on land and moves downhill under the force of gravity. The term glacier does not include icebergs, sea ice, or immobile snowfields in mountainous areas.

As a very important part of hydrologic cycle, glacier ice is the largest reservoir of fresh water on the Earth. Although about 2% of the Earth's water is contained in glaciers, but nearly 85% of total fresh water on the Earth is stored in the form of ice in glacier. Many of the great rivers originate from glaciers.

Glaciers are found on every continent except Australia. Glaciers cover about 10% of the Earth's land surface, with the Arctic and Antarctica ice sheets accounting for about 99% of the land covered. Other glaciers are found in mountainous regions with high elevations. The western China has the highest mountains and plateaus in the world. Because of the high elevation and cold weather, glaciers are distributed on the mountains regions of Western China with elevations above 3 500m.

Like great rivers of ice, glaciers have sculpted mountains and carved out valleys. They continue to flow and shape the landscape in many places today. A number of geologic features are associated with glaciers. Glaciers are powerful grinding machines for turning solid bedrock into sediment and shaping the surface of the Earth. And their distinct deposition provides a valuable geologic clue for inferring where ancient glaciers flowed. Glaciers are indicators of climate. The retreating and advancing of glaciers can dramatically change the sea level. During times when glaciers covered millions of square kilometers of the land areas in the Northern Hemisphere and world sea level dropped more than 150m, the climate was generally cooler than it now is.

In this chapter you will learn about how glaciers form, how they are classified, and what types of landforms they create.

1 The Formation of Glacier

A glacier is formed by multi-year ice accretion in mountainous or sloping terrain. Glacial ice forms through a process of compaction, recrystallization of snow, and growth.

Ice is a mineral and glacial ice is a rock. Glacial ice is essentially a metamorphic rock for which the principal agent of metamorphism was pressure. There are two requirements for the formation of a glacier: ①The temperature must be below freezing. ②There must be an adequate amount of snow accumulation.

The conversion of snow to ice involves several steps: Snow →Firn →Glacial ice.

1.1 Accumulation of Snow

Snow and ice are both minerals — they are both naturally occurring, solid, crystalline chemical compounds. Recalling the hexagonal geometry of snowflakes, it is apparent that snow is"mostly air", and therefore, is less dense than ice (Figure 14.1).

Glaciers form in areas where more snow falls in winter than melts during summer. Snow line(雪线) is an irregular line marking the minimum elevation of snow lying on the ground or glacier surface above which the snow can persist throughout the year and below which the annual snow cover is lost in summer.

The snowline is the product of the combined effect of climate, snow fall as well as topography. In general, weather is hot and humid in the equator and cold and dry in the poles. So, the elevation of snowline is highest near the equator and lowest near the poles. For examples, the elevation of snowline in equatorial areas of Africa is 5 700～6 000 m and in Alps it is about 2 400～3 200m.

1.2 Compaction and Recrystalizaiton

When a snow pile becomes thick, the great overlying weight causes the fluffy accumulated snowflakes to recrystallize into smaller, thicker, and more spherical granular snow. This recrystallized snow is called firn(粒雪).

1.3 Glacial Ice

As more snow is deposited and becomes firn, the pressure on underlying grains increases. When the thickness of the snow and ice exceeds tens of metres, the weight is sufficient to cause the firn to grow into even larger ice crytals. Thus the glacial ice forms. Glacial ice is a mono-mineralic type of rock, composed of crystals of the mineral ice, formed through metamorphism of snowflakes. On average, the transformation of firn into glacial ice may take 25～100 years. Glacial ice contains minute air bubbles (<20% air) as a result, giving it a distinctive blue tint due to Rayleigh scattering.

Snow is analogous to sediment, firn to a sedimentary rock , and ice to a metamorphic

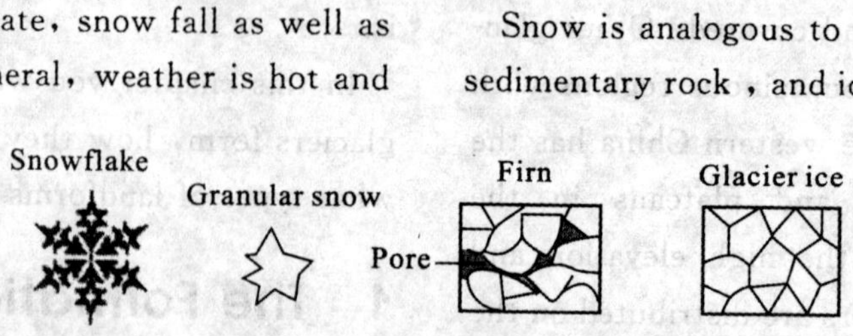

Figure 14.1 Transformation of snowflake to glacier ice

rock. If the thickness of accumulated ice increases so high that the entire ice mass can be moved by its own gravity, it is said that the whole procedure for the formation of a glacier is completed.

2 The Budget of a Glacier

The size and extent of glacier is determined by the relationship of accumulation and ablation.

Accumulation zone(积累区) is the area of the glacier where material is only added, no mass is lost. Usually this area is near the top of the glacier. Zone of accumulation is covered by perennial snow (i. e. above the snow line) (Figure 14. 2).

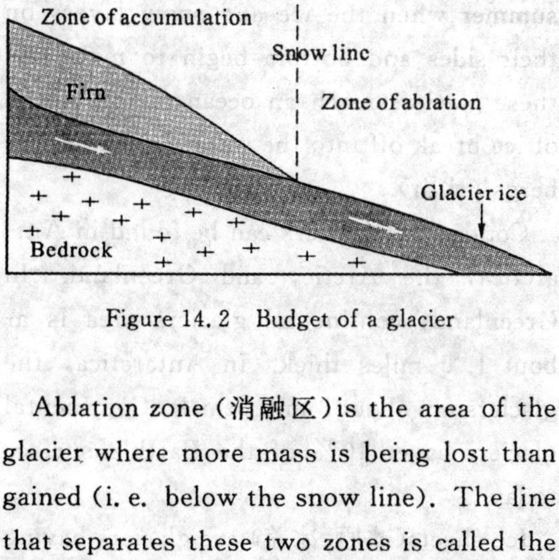

Figure 14. 2 Budget of a glacier

Ablation zone (消融区)is the area of the glacier where more mass is being lost than gained (i. e. below the snow line). The line that separates these two zones is called the equilibrium line. The balance between what accumulates high up on the upper part of the glacier, and what melts near the glacier's foot is called the glaciers mass balance. A glacier may advance when accumulation exceeds ablation; it is called advancing glacer(前进冰川). And when ablation exceeds accumulation, it is called receding glacier(后退冰川). Bear in mind that the glacier ice moves down valley whether the galcier is advancing or receding. The glaciers mass balance is controlled by climate.

3 Movement of Glacier

Glaciers are always in movement. It is a living entity, not a fixed mass. One can compare a glacier to a river, although the rhythm of the flow is certainly slower. Valley glaciers flow down valleys, and continental glaciers flow outward in all directions from a central point. Glaciers move due to a combination of internal flow in the solid ice and basal sliding along the base of the glacier.

Internal flow occurs when the weight and mass of a glacier causes it to spread out. Glacier flows plasticly by internal flow. Basal sliding (底滑) means that the base of the ice sheet is near the pressure melting point, and that a layer of water a few centimeters thick may be present at the glacier base (Figure 14. 3). Water reduces friction between the glacial ice and the ground surface, and allows the ice to move faster.

The flowing speed of glacier ranges between a few centimeters and meters daily, depending on the slope of the mountain, the amount of snow that falls, and the size of the glacier. It is difficult for our eyes to figure out that it is moving. The moving speed of glacier in Rongbu Glacier (绒布冰川,西藏) is 64m yearly. As for the glacier in other regions, the annual moving range is about a few meters to tens of meters.

The center of a glacier moves more rapidly than the sides and the surface more rapidly than the bottom, because the sides and bottom are held back by friction. Due to the

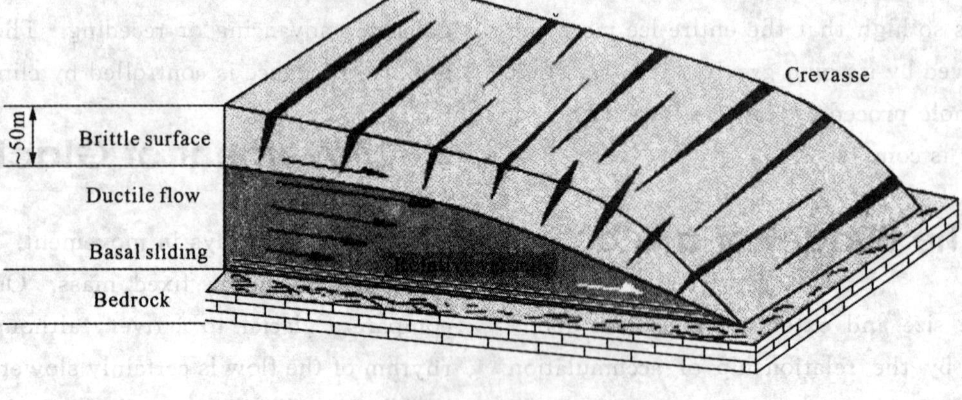

Figure 14.3 Movement of glacier

differences of speeds, the upper parts of glaciers will break rather than undergoing gradual flow and thus crevasses (fissure or crack) develop. Crevasses can be up to 20m wide, 45m deep, and hundreds of meters long. Their orientation may be in any direction with respect to the glacier flow. Crevasses can not exceed 50m deep because they are closed by plastic flow below that depth. A serac(冰塔、冰雪柱) is a large, jagged pinnacle or tower of glacier ice formed by intersecting crevasses. The glacier in castle-like seracs is divided at steep points by crossing crevasses.

4 Glacial Classification

There are three basic types of glaciers, valley glaciers (山岳冰川) which form in high mountain valleys, piedmont glacier (山麓冰川) and continental glaciers(大陆冰川) which form on ice caps in frigid polar regions.

4.1 Continental Glaciers

Continental glacier is a glacier that spreads out from a central mass of ice and it is a slow moving sheet of ice may lie over land or water. Continental glaciers (Figure 14.4) form at high latitudes where temperatures remain cold enough during the summer to keep the previous winter's snow from melting, allowing snow and ice to accumulate. Continental glaciers move in the summer when the ice and snow layers on their sides and bottom begin to melt. As these glaciers reach an ocean, huge blocks of ice break off into the water, creating icebergs (冰山).

Continental glaciers can be found in Antarctica, the Arctic, and Greenland. In Greenland, continental glaciers area is about 1.5 miles thick. In Antarctica, the glaciers are about 3 miles thick. Continental glaciers cover 10% of the Earth's surface today.

Ice sheets(冰原、冰盖) are glaciers having considerable thickness and being more than 50 000km^2 in area. It forms a continuous cover of ice and snow over a land surface. Extensions of ice sheets over the ocean are known as ice shelves(冰架). An ice cap is a dome-shaped cover of perennial snow and ice. Ice caps(冰盖,冰帽) are actually small ice sheet, generally less than 50 000 km^2 in size. Today, they are found at the poles.

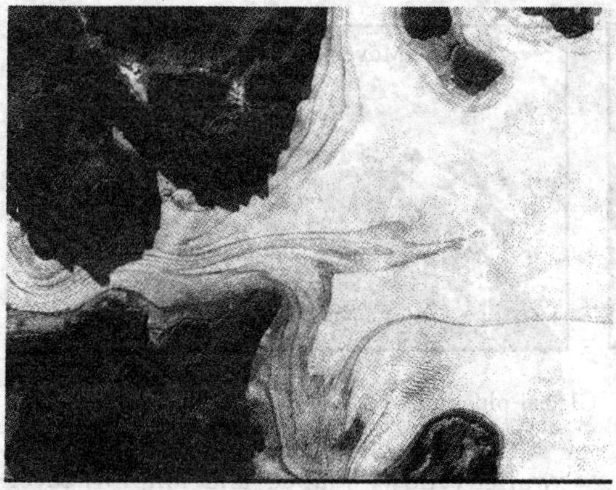

Figure 14.4 Left: Valley glaciers in Chile (satellite image). Right: Continental glacier
Image left: NASA-JSC-ES&IA. Image right: NOAA. Public domain

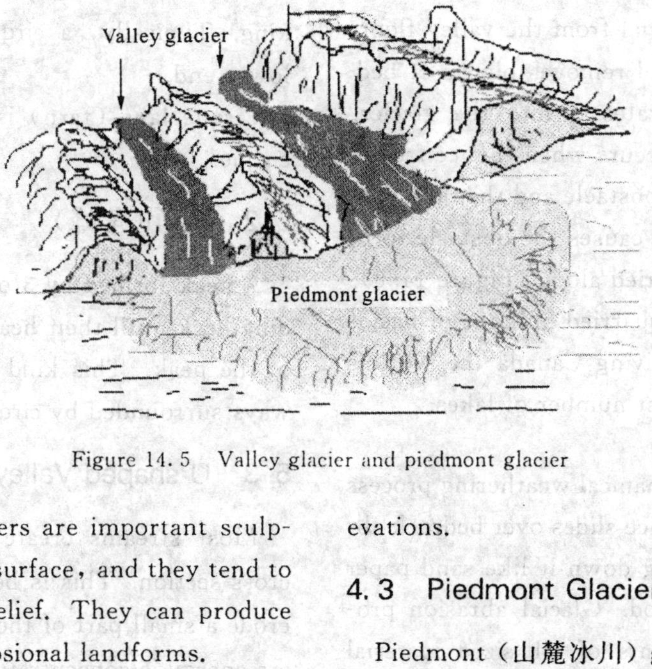

Figure 14.5 Valley glacier and piedmont glacier

Continental glaciers are important sculptors of the Earth's surface, and they tend to lower the overall relief. They can produce depositional and erosional landforms.

4.2 Valley Glacier

Valley glacier, also known as Alpine glacier(阿尔卑斯式冰川)(Figure 14.5), form at high elevations where temperatures remain cold enough during the summer to keep the previous winter's snow from melting allowing snow and ice to accumulate. This glacier is dominantly confined by the surrounding topography and slowly makes its way down the mountain toward lower elevations.

4.3 Piedmont Glacier

Piedmont (山麓冰川) glacier is the large lobe-shaped, expanded, terminal part of a valley glacier spreading out over broad lowlands at the base of mountains. It forms where valley glaciers exit from the mountains and meet flat ground. So piedmont glaciers commonly are a merger of several valley glaciers.

5 Glacial Erosion

The glacier erodes the lands in two ways.

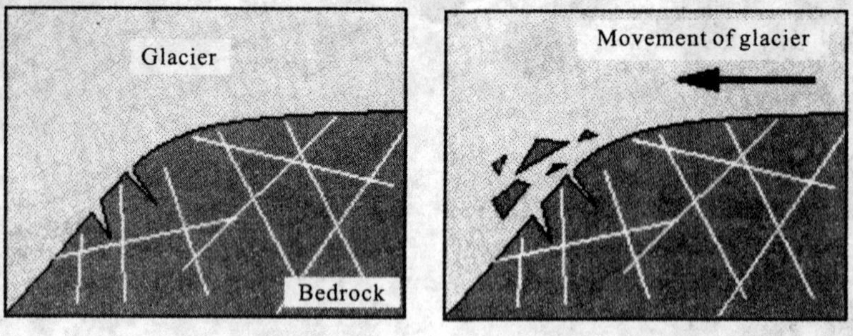

Figure 14. 6 Glacial plucking

• **Quarrying (plucking)(拔蚀作用)**

Quarrying (plucking) is a common mechanical weathering process where glacial ice frozen into cracks in the bedrock literally "pluck" rock material from the valley floor. The ice wedges and removes blocks of bedrock and incorporates it into the glacier. Glacial plucking occurs when the ice freezes around a bedrock obstacle and then the subsequent movement causes the obstacle to be pulled out and carried along (Figure 14. 6).

The ice sheets quarried millions of basins in the bedrock, giving Canada the world's largest and greatest number of lakes.

• **Abrasion(磨蚀)**

A common mechanical weathering process where debris-rich ice slides over bedrock abrading and wearing down it like sand paper on a block of wood. Glacial abrasion produces large amounts of silt-sized material during this process. The very fine particles of pulverized rocks that are often carried away from the front of the melting glacier by sediment-charged streams are also called rock flour. It is also called glacier flour or glacier milk.

A range of glacial landforms is produced by processes of glacial erosion:

5. 1 Cirque(冰斗)

Cirque is a French term for a steep-walled, bowl-shaped and amphitheater-like basin occurring at the upper end of a mountain valley (Figure 14. 7). It is carved out of mountain terrain by a glacier due to plucking. Typically, a cirque has a lip at its lower end.

Cirque lake (tarn) is a lake found in a cirque.

5. 2 Horn(角峰)

A peak formed by 3 or more cirques eroding back until their headwalls are the sides of the peak. This kind of rock peak is always surrounded by cirques.

5. 3 U-shaped Valleys(U 形谷)

Most streams create a V-shaped valley cross-section. This is because streams only erode a small part of the valley and the rivers energy is concentrated only some special points of bed and bank. When the stream-carved valley is filled entirely by a glacier, the glaciers may erode all sides of the valley. Bit by bit, the glacier eats away at the meandering curves of the original stream valley, carving a wider, straighter valley. This is a U-shaped valley (Figure 14. 8).

Valley glaciers tend to erode valleys that have broad, relatively flat floors and steep sides.

Figure 14. 7 Cirque. Image credit: Richardson et. al. U. S. Gelogical Survey. Public Domain

Figure 14. 8 U-shaped hanging valley. Image credit: Left: form NOAA. Public Domain. Right: Richardson et. al. U. S. Gelogical Survey. Public Domain

5. 4 Arête(脊)

Arête is a sharp mountain ridge between U-shaped glacial valleys. The knife-edged arête is formed from the plucking action of two glaciers on opposite sides of the ridge (Figure 14. 9).

5. 5 Hanging Valleys (悬谷)

Hanging valley(悬谷) is a U-shaped tributary valley suspended above main glacial valley floor. When small tributary glaciers join larger glaciers, a hanging valley will be produced . After retreat , or melting , the

Figure 14.9 A well-developed arête in Jiuzhaigou(九寨沟). Photo by Chen Fuxing(谌福兴)

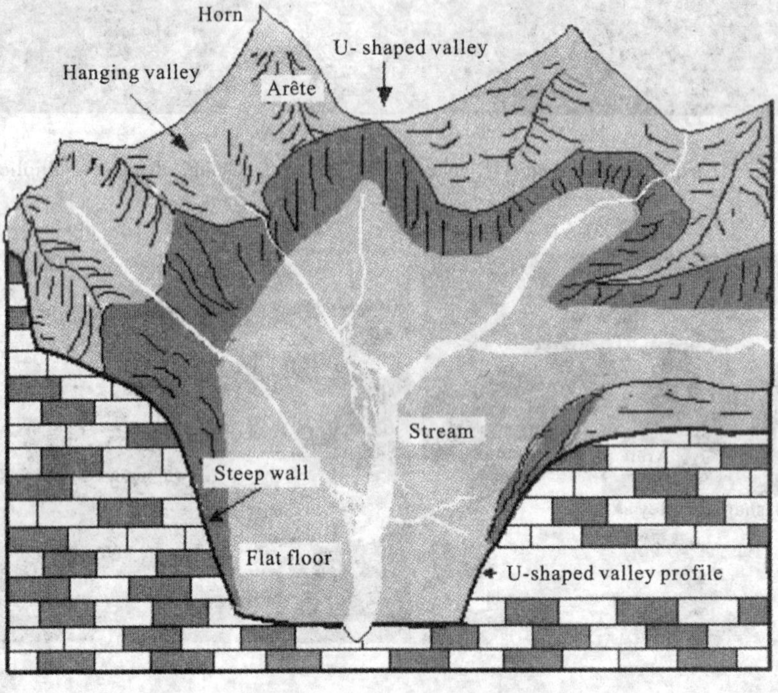

Figure 14.10 Erosional landforms of glacier. Note that from the shape of the valley, we know that glacier eroded this valley, not the river presently existing here

hanging valley floor will be above the main valley. Waterfalls are often present, cascading from the hanging valley into the larger valley below (Figure 14.10).

5.6 Roche Moutonnée(羊背石)

A roche moutonnée (moutonnée, from the French, shaped like a sheep, being assigned to the shape of the structure as similar to a resting sheep,羊背石) is an eroded and smoothed bedrock hill of mesoscale (1m to 1km of length). The gentle side faces up glacier and was smoothed, rounded, grooved, striated and polished by glacial abrasion.

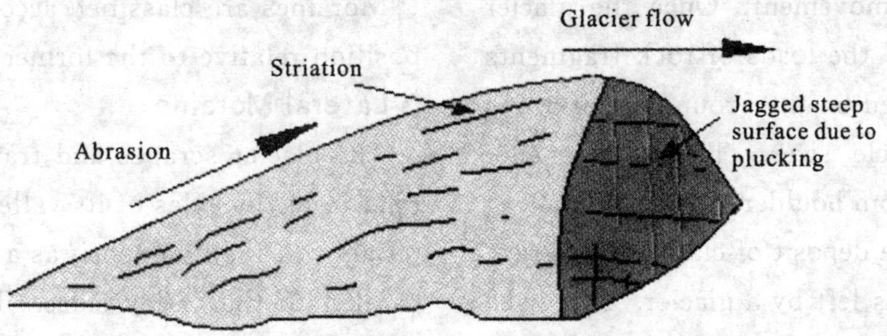

Figure 14.11 Asymmetrical profile of a roche moutonnée

Glacial striation(冰川擦痕) is a grooves in the bedrock due to grinding action of rocks embedded in the glacier. The down glacier side is a more irregular, steeper face, where pieces of bedrock have been plucked out by glacier ice. Because of the two different processes an asymmetric profile is developed (Figure 14.11). Roche moutonnée is regarded as a hallmark of glacial erosion and can be used to determine the flow direction of glacier. Roche moutonnees often appear as isolated"hills"in glacial valleys.

6 Glacial Transportation

Glacier ice is a much more powerful erosive and transportation agent than water or wind because its greater viscosity enables it to carry much larger particles. The glaciers pick up the plucked rock fragments and transport them during movements. All of the materials transported by glacier are debris.

Rock material eroded by a glacier, or weathered debris that has fallen onto a glacier, is transported in one of three different positions: debris carried on (supraglacial, 表碛), within (englacial, 内碛), and dragged along the bottom (subglacial, 底碛). An englacial debris is the one that has fallen down crevasse and becomes embedded in the glacier.

7 Glacial Deposition

When glaciers recede, deposition of material begins. Glacial deposits differ greatly from rivers and wind deposits. In many areas once covered by the ice sheet of the recent Ice Age, the bedrock is rarely seen because it is deeply buried by the tens of even hundreds of meters thick glacial deposits.

The term drift (冰川沉积物) is used to refer all materials of glacial origin, no matter how, where, or in what form they were deposited. Long before there was a glacial theory, much of the soil and rock debris covering parts of Europe was thought to be deposited by drifting icebergs (floating ice), perhaps in Noah's flood! That is why the term"drift"is used to refer to sediments deposited by glacier. In general, glacial drift can be classified into two different types: till and stratified drift.

7.1 Till

Till (冰碛物、冰砾泥) is deposited directly from the glacier. Because the sediments carried by glacier are relatively fixed in their positions, they can not be sorted or rounded during glacial movement. Once the glacier begins to melt, the loads of rock fragments will be deposited on the ground as mixtures of many particle sizes. Those fragments range in size from boulders to clay particles.

So, till is the deposit of clay, mud, gravel, and boulders left by a glacier. It is usually described as massive (块状的, not layered, that is, it does not form clear layers or beds), poorly-sorted, and composed of multiple types of angular to sub-rounded rocks (Figure 14.12). The materials in till vary greatly with source material.

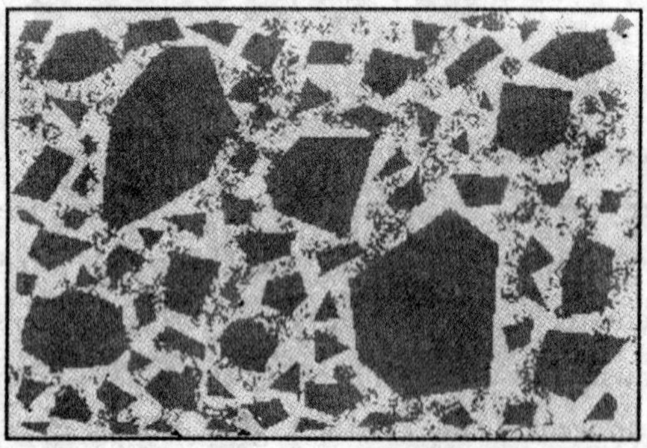

Figure 14.12 Nonstratified and nonsorted

If the surrounding temperature is higher than normal, glaciers may melt early. As a glacier melts, it deposits large boulders and other debris that it has collected over time. Along the edges of a glacier, mounds of dirt and smaller debris are deposited. These areas are called moraines. They are composed of tills. Moraine is a French word that refers to any glacier-formed till accumulation. The late Pleistocene (晚更新世, "latest") glaciers left huge moraines throughout northern North America and much of Europe. The rolling hills, fertile fields and small lakes are common country scenes in northern North America. They are all products of those glaciers.

Moraines are classified according to their position relative to the former glacier.

- **Lateral Moraine**

The glacier scrapes and transports materials from the sides of its valley. These materials will finally deposit as a ridge running parallel to the valley sides. This is lateral moraine (侧碛堤). When two or more valley glaciers unite, their lateral moraines form a medial moraine (中碛堤) running down the center of the glacier.

Both lateral moraine and medial moraine are common in valley glaciers.

- **End Moraine**

End moraines (终碛堤), as the name implies, mark the end of a glacier.

At the end of glacier, if the glacier's forward progress stops, a balance is maintained between the accumulation and ablation. Or if the terminus of a glacier remains stationary for a few years, the material carried by glacier is dumped at that point and eventually a linear, slightly curved ridge of rocky debris piles up along the front edge of the ice. This is the end moraine, which represents the furthest point of advance of a glacier. They often form a series of crescents, corresponding with the lobes of the glacier (冰舌).

Similar moraines deposited during a temporary halt in the retreat of glacial ice are called recessional moraines. Recessional moraine (后退碛) marks the temporary edge of a retreating glacial. It is important to note that when the glacier retreats, the ice is always flowing forward actively, but the edges melting back. A glacier has only one terminal moraine but may have several recessional moraines.

Advancing glaciers push the end moraine

further, perhaps smearing it out and running it over, but retreating glaciers leave the end moraine behind.

- **Ground Moraine**

Ground moraine (底碛、冰碛丘陵) is widespread thin sheet of drift formed during retreat of a glacier. They consist mainly of till. The ground moraine of continental glacier is usually in the form of undulating plains having gently sloping swells, sags, and enclosed depressions. The predominant material is till, but some stratified drift is present.

Rolling-to-flat landscapes that form under the ice sheet are also referred to as till plain. It is usually formed by the rapid retreat of ice sheet. In such case, there is no chance for the formation of end moraines, instead, the till will deposit irregularly across the landscape.

- **Drumlin**

Drumlins (Gaelic druim, the crest of a hill) are streamlined mounds of till that form at the base of moving glaciers. It is produced by a glacier overriding and shaping either a bedrock hill or a ground moraine left by an earlier glacial advance. So some are composed of entirely of till while others are made up of till plastered to bedrock cores. The long axis of these features is usually parallel with glacier flow, and the steep end faces the direction from which the glacier advanced. Drumlins usually have a teardrop shape and the tails of the teardrop shapes point in the direction of glacier movement (Figure 14.13).

Drumlins range in height from 15 to 60m and may reach almost 1km in length. Drumlin may occur singly or in groups.

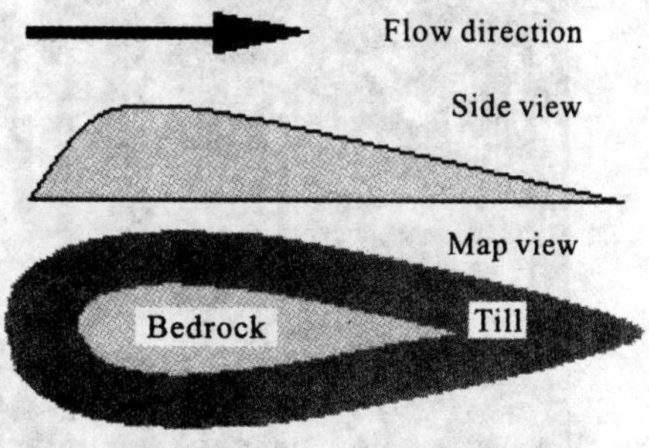

Figure 14.13 Teardrop-shaped drumlin

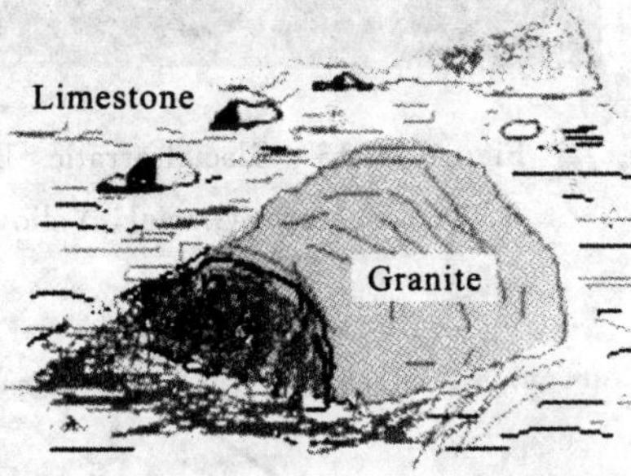

Figure 14.14 Granite erratic on limestone ground

- **Glacial Erratic**

Glacial erratics (冰川漂砾) (Figure 14.14) are boulders have been brought from greater distances. These are large boulders or rocks, which "do not belong" in the locations where they are found in present times. Erratics have sometimes been moved 50km from their mountain of origin and frequently placed in a very unusual and precarious position (Figure 14.15).

7.2 Stratified Drift

Stratified drift is deposited by running water from melting ice. Landforms of stratified drift are listed here (Figure 14.16):

- **Outwash Plain** (冰水平原)

Outwash plain is a flat or gently sloping surface composed of stratified drift. Sediments were washed out by flowing water

Figure 14.15 Glacial erratic. Image credit: From Jim Peaco, September 1998; Geology/Glacial, Yellowstone Digital Slide File. Public domain

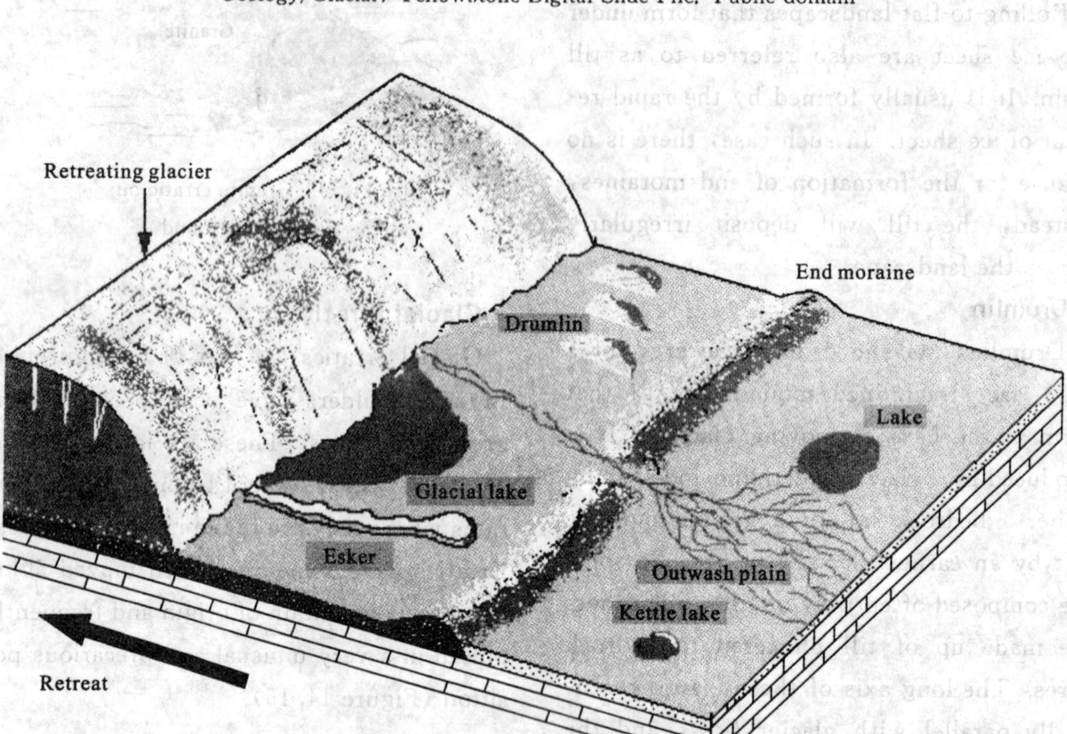

Figure 14.16 Schematic diagram of various depositional landforms of continental glacier

beyond the glacier and laid down as stratified drift in thin foreset beds. The particle size may vary from boulders to silt.

• **Kame**

Kames(冰碛阜、冰砾阜) are steep-sided mounds or hills created when drift fills a hole in a glacier. When the ice eventually melts away, the mound consisting of stratified drift is left behind.

• **Kettle**

As the ice retreated, big chunks were broken off of the main glacier. The large blocks of ice may be buried in the outwash plain and slowly melt away. When these blocks of ice subsequently melted, they left a closed depression, known as a kettle(锅穴), above the former buried blocks of the glacial ice. If the depression is filled with water, it is called kettle pond or lake(锅穴湖). The Southwestern Sichuan (四川) and Northwestern Guizhou (贵州) are dotted with many small scenic kettle lakes. Some of the so-called Haizis (海子) in those regions are actually kettle lakes.

• **Eskers**

An esker(蛇形丘) is a glacial outwash landform that appears as a long sinuous ridge made up of sand and gravel, formed in a meltwater tunnel under a glacier. Esker may be several meters high and can stretch for tens of kilometers.

• **Proglacial Lakes**

Lakes that receive meltwater from a glacier are named proglacial lakes. Proglacial lakes are found in a variety of environments and in considerable numbers. Most proglacial lakes(冰前湖) are created when glacial meltwater is dammed by terminal or lateral moraines during the retreat of a melting glacier. Another type of proglacial lake, an ice-dammed lake(冰堰湖), forms if an ice mass blocks a river or stream valley.

The deposition of sediments in such still water is strongly seasonal. In the summer, a large amount of sediments from the glacier meltwater is dumped into the lake, resulting in deposition of a light-colored layer composed of coarser particles. On the contrary, nothing at all enters during the winter because the lake is frozen. Only the finest sediment, such as clay, sinks down slowly after the lake surface freezes, resulting in the deposition of thin dark-colored layer under the reduction environment. This kind of seasonal stratification in proglacial lake sediments is called varve(季候泥、纹泥). Because each pair of light and dark layers represents a year's deposition, just like the tree ring dating, the varves can be used to determine how long a glacial lake was in existence.

8 Glaciation and Its Effects

Glaciation(冰川作用) is the formation, advance and retreat of glaciers and the results of these activities. Or glaciations are the general effects on landforms produced by the presence and movement of a glacier.

Ice age (glacial period) is a period of geologic time interval when large parts of the surface of the globe are covered with ice sheets (large continental glaciers). The term is used to describe time intervals on two very different scales. It describes long, generally cool intervals of the Earth history (tens to hundreds of millions of years) during which glaciers advance and retreat. If "ice age" is used in this way we are still in one today. The term also describes shorter time periods (tens of thousands of years) during which glaciers were near their maximum extent. In this sense, the last ice age ended about 10 000 years ago. In this textbook, we use the ice age in the former sense, and use the term "glacial periods" (冰期,冰川期) for colder periods when glaciers advance during ice ages and interglacial (间冰期) for the warmer periods when glaciers retreat.

In addition, the term "Ice Age" is sometimes used to refer to the last time that glaciers extended over a large portion of the Earth's surface. When used in this way, the first letters of both words are often capitalized.

The existence of ice ages during geological history can be proved by tillite(冰碛岩) and striations on the rock fragments. There are three major glaciations occurred during the past 800 million years. The first one occurred during the late Proterozoic (晚元古代, between about 800 and 600 million years ago), the second one during the Carboniferous and Permian (between about 350 and 250 million years ago), and the last one late Neogene(晚第三纪)to Quaternary (the last 4 million years). It is important to realize that the ice never covered more than one third of the Earth's land surface, even at its greatest extent.

The glacial periods during late Neogene to Quaternary is said by geologists to have started about 2 million years ago and ended about 11 000 years ago. It was punctuated by relatively warm interglacial periods, which lasted about 10% of the time. There are supposed to have been advances and retreats of the ice sheet during many thousands of years, with the last of these ending about 11 000 years ago. During the Ice Age, mountain glaciers formed on all continents, the icecaps of Antarctica and Greenland were more extensive and thicker than today, and vast glaciers, in places as much as several thousand feet thick, spread across North America (as far south as the northern United States) and Eurasia (Scandinavia to Germany and England, Siberia). The most recent period of glaciation, which many people think of as the"Ice Age", was at its height approximately 20 000 years ago. During the present ice age, glaciers have advanced and retreated over 20 times. Our present relatively warm period actually represents a brief interglacial between glacial advances and the glaciers may again advance in the future. However, the presence of massive continental ice sheets on Greenland and Antarctica, along with numerous smaller glaciers in mountainous regions throughout the world, indicates that the Earth is still in the grip of an ice age.

9 Causes of Ice Ages

Although the Great Ice Age has been accepted as a proven fact, but the origin of the ice age has greatly perplexed scientists. Much lower temperature and copious snowfall are required, but they are inversely related, since cooler air is drier. Although the exact causes for ice ages have not been proven, they are most likely the result of a complicated dynamic interaction between such factors as solar output, distance of the Earth from the Sun, position and height of the continents, ocean circulation, and the composition of the atmosphere. These different factors, at many different time scales, probably combine to result in glaciations of the Earth's surface.

9.1 Solar Output

The Sun experienced an extended quiet period in which its intensity fell dramatically. This leads to less radiant heat emitted from the Sun to neighboring bodies, therefore a cooling occurs. It is estimated that a fluctuation of only a few tenths of 1% in energy output would be sufficient to produce significant changes in climate. Amount of sunspots is correlated to the activeness of the Sun. More sunspots mean more emission of radiant heat from the Sun.

9.2 Milankovitch Cycles

In the 1920s Yugoslav scientist Milutin Milankovitch proposed the Milankovitch Astronomical Theory, which states climatic fluctuations and the onset of glaciation can be caused by variations in the Earth's position relative to the Sun. Important parameters that vary include the eccentricity of the orbit around the Sun, the tilt of the Earth's axis, and precession of equinoxes. Variation in these three factors changes the amount and distribution of incoming solar radiation, resulting in 100 000 year ice age cycles of the Quaternary glaciation over the last few million years.

- **Eccentricity of Orbit**

Eccentricity(偏心率) refers to the changing shape of the Earth's orbit around the Sun, which varies from nearly circular to elliptical over a cycle of about 100 000 years. Eccentricity changes the Earth's average annual distance from the Sun and slightly alters the amount of sunlight hitting the Earth.

- **Tilt of Rotation Axis**

The tilt of the Earth's axis changes through time. The amount of tilt in the Earth's rotation affects the amount of sunlight striking the different parts of the globe (Figure 14. 17). The greater the tilt, the stronger the difference in seasons. The time scale for changes in tilt is about 41 000 years.

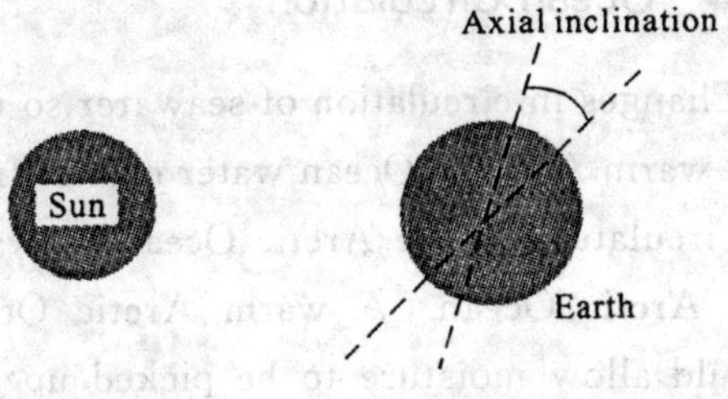

Figure 14. 17 Tilt of rotation axis

- **Precession of Equinox(岁差)**

Precession of equinox is the phenomenon whereby the Sun on the day of the equinox [that day when the day and night are of equal length 分点,包括 Vernal Equinox(春分点)和 Autumnal Equinox (秋分点)] rises in different constellations, changing by about 50″ per year. Precession completes a cycle every 21 000 years.

9.3 CO_2 Variation

A general reduction in amount of CO_2 in the atmosphere may contribute to the development of ice ages. Carbon dioxide is an important greenhouse gas. Decreases in the amount of CO_2 in the atmosphere may lead to global cooling. Air bubbles in ancient glacial ice suggests that CO_2 and CH_4 are much higher in the atmosphere during interglacial periods than during glacial periods.

Many processes can cause a long-term decrease in the amount of CO_2 in the atmosphere, including many complex interactions among organisms, ocean currents, erosion, and volcanism. But many scientists believe that the most important reason may be the volcanic activity.

Global warming due to fossil fuels may be catastrophic in many ways, but will probably not much affect these longer-term cycles. We will have run out of fossil fuels long before the duration of a typical interglacial.

9.4 Ocean Circulation

Changes in circulation of seawater so that the warm Atlantic Ocean water cannot freely circulate into the Arctic Ocean to warm the Arctic Ocean. A warm Arctic Ocean would allow moisture to be picked up and precipitate heavy amounts of snow on continents and thus cause continental glaciation.

9.5 Tectonic Movement

Plate tectonics probably contributes to the development of long periods with many glaciations in a second, more subtle way. Tectonic changes can cause changes in atmospheric circulation and oceanic circulation. These changes in turn initiate the ice ages. Plate movement can lead to the ① Changing of the positions of the continents; ② Large scale uplifting of land mass; ③ Formation of mountain chains; ④ Open and close of ocean. Those events except for the last one can initiate dramatic decreasing of temperature. Therefore some scientists hypothesize that climatic changes caused by uplift are critical to the development of ice ages. The last event may create continents surrounded by oceans like the present Antarctica.

1. Answer the Following Question

(1) What is a glacier and name the two types of glaciers?
(2) Is ice within a glacier a mineral? Is a glacier a rock?
(3) How does material deposited by glaciers differ from material deposited by stream?
(4) How do recessional moraines differ from terminal moraines?
(5) What is roche moutonnée? How to determine the flow direction of glacier by roche moutonnée?
(6) Explain briefly the causes of Ice Ages.
(7) What is drumlin? How to determine the flow direction of glacier by roche drumlin?
(8) Differentiate between the zone of ablation and zone of accumulation.
(9) What is the stratified drift? List the main landforms consisting of stratified drift.
(10) What is plucking?

2. Multiple Choice

(1) Which of the following features is NOT composed of till?
A: Lateral moraine B: Drumlin
C: Kame D: Terminal moraine
(2) Plucking occurs when ice flows into or refreezes in fractures in the bedrock.
A: True B: False
(3) The moraine that marks the end of a glacier is ______ moraine.
A: Lateral B: Ground
C: Recessional D: Terminal
(4) Boulders having been brought from greater distances is called ______.
A: Drift B: Till
C: Moraine D: Erratic
(5) Which one of the following is not a feature of valley glacier?
A: Cirque B: Horn
C: Hanging valley D: Drumlin
(6) A roche moutonnée is composed of ______.
A: Bedrock B: Drift
C: Erratic D: Till
(7) The followings are depositional landforms created by glaciations except ______.
A: Moraines B: Eskers
C: Drumlins D: Roche mountanée
(8) When ice accumulation equals ice wastage, the form of the glacier will ______.
A: Remain stationary B: Advance
C: Retreat D: None of them
(9) If the glacier dumps the material in an unsorted mixture, this is called till.
A: True B: False
(10) Larger grooves are usually created by abrasion on the back of ______.
A: Drumlin B: Roche moutonnée
C: Moraine D: Erratic
(11) Eskers are sinuous ridges of sand and gravel deposited in or under ice in a meltwater tunnel.
A: True B: False
(12) Glaciers are such a powerful force that they can carve valleys, deposit moraines, or lead to the creation of landforms such as deltas.
A: True B: False
(13) Crevasse in the upper surface of a glacier that cannot from at depth because of ______.
A: Drumlin B: Roche moutonnée
C: Moraine D: Erratic
(14) What is the difference between plastic flow and basal slip?
A: Plastic flow is dry glacial movement, and basal slip is wet glacial movement.
B: Plastic flow is made of small movements of individual crystals, and basal slip is movement of the entire glacier.
(15) What is glacial ice?
A: A metamorphic rock
B: Recrystallized snow
C: A mineral
D: All of the above
(16) The bowl-shaped bedrock depression from which an alpine glacier flows at the top of a glacial valley is called a ______.
A: Drumlin B: Hanging valley
C: Moraine D: Cirque

(17) In what ways can a glacier effect the climate?

A: It may cause rising sea temperature.

B: It may cause the Earth to be a large ball of ice.

C: It may cause the sea levels to rise and fall.

D: All of the above.

(18) How is a kettle formed?

A: A kettle is formed when an ice block left behind by a retreating glacier melts and forms a depression.

B: A kettle is formed when till is deposited and a streamlined hill forms parallel to the direction of the ice movement.

C: A kettle is formed when you are in a metal forging plant.

D: None of the above.

(19) Which type of glaciers are the largest?

A: Piedmont glaciers B: Alpine glaciers

C: Ice sheets D: Piedmont glaciers

(20) Which type of glaciers occupy the lowland at mountain fronts?

A: Piedmont glaciers B: Alpine glaciers

C: Ice sheets D: Piedmont glaciers

(21) What is the term for sediment that is sorted and deposited by the meltwater of a glacier?

A: Till B: Drift

C: Outwash D: Moraine

(22) Glaciers are found on every continent except ______.

A: African B: Australia

C: South America D: Asia

(23) What term refers to glacial melting below the snowline?

A: Plastic flow B: Plucking

C: Abrasion D: Ablation

Chapter 15 Wind Action and Desert

Key words

Aeolian erosion	Deflation	Desert pavement	Ventifact	Aeolian landform	Blowout
风蚀	风蚀、吹蚀、吹扬	沙漠砾石滩	风棱石	风成地貌	风蚀洼地
Yardangs	Aeolian erosional landforms		Aeolian depositional landforms		Gobi
雅丹地貌	风蚀地貌		风积地貌		戈壁
Playa	Pediments		Abrasion	Barchan Dunes	Transverse dunes
干盐湖	麓原、山麓面、山前侵蚀平原		磨蚀	新月型沙丘	横向沙丘
Star dunes	Longitudinal dunes	Aeolian sediment	Dust storm, Sand storm		Loess
星状沙丘	纵向(线状)沙丘	风积物	尘暴、沙暴、沙尘暴		黄土

Variations in pressure lead to the development of winds, which in turn influence weather. Erosion, deposition, and transport of sand caused by the flow of air over the Earth's surface are denoted as aeolian, which is derived from *Aeolus*, the Greek god of the winds. Wind is caused by the movement of gas molecules in the Earth's atmosphere, and has the ability to erode, transport and deposit sediment in ways similar to the action of running water. However, wind is not as effective an agent of erosion as are ice and water, but over time it can transport much material and finally shape a landscape.

1 Wind Erosion

The erosion caused by wind is called aeolian erosion (wind erosion). Wind is an important agent of erosion and deposition in any climate, as long as sediment particles are loose and dry. Although many of the geologic processes in arid areas are related to the work of the wind, but wind erosion and deposition can be very significant in other climates as well. The two major aeolian erosion processes are deflation and abrasion.

• Deflation

Deflation is an erosive process of lowering the land surface due to the removal of sand and dust and loose rocks by wind. The wind picks up and carries away the weathered loose materials, thus leaving the grains larger than sand on the surface. Although the effects of deflation in deserts or arid areas may be difficult to notice, but the lowering of entire eroded area may be significant (Figure 15.1).

When this happens in a desert, it usually leaves the larger pebbles and boulders which are called desert pavement(沙漠砾石滩,沙漠砾石盖层).

• Abrasion

Abrasion is the scouring of exposed surfaces by the sandblasting action of windborne material. The"sandblasting"impact of windblown particles on exposed surfaces will wear away of the surface of particle as a

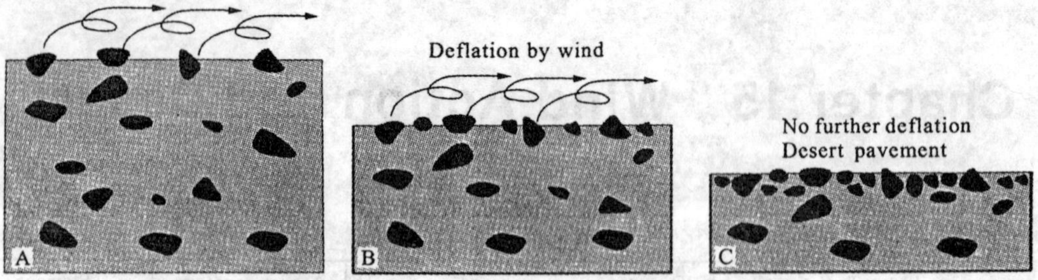

Figure 15.1 Successive stages in development of desert pavement

A: The loose, dry sediments consisting of clay, silt, sand, gravel, and cobbles.

B: Sand-sized particles or smaller are removed by deflation, thus lowering the ground surface.

C: Concentrated larger particles form a continuous layer which protects the sediments below from further deflation

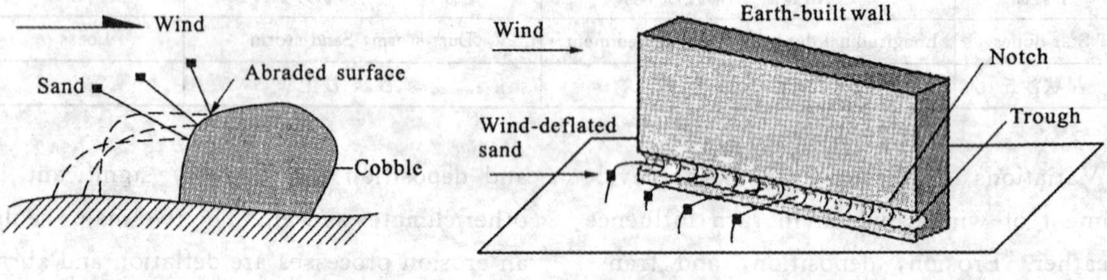

Figure 15.2 Left: Sand-blasting action of wind. Right: Erosion at the foot of wall

result of applied impact force. Abrasion can reduce the size of the particles that are being moved.

In Northwestern China, wind is so strong that numerous notches or even troughs are often scoured at the foot of earth-built wall (土筑墙) (Figure 15.2). The notches and troughs are due to the ever-lasting sand-blasting action of sand-sized particles deflated by wind. The sandblasting action is also responsible for the erosion and collapse of earth-built Great Wall in Northwestern China.

Aeolian landform is a feature of the Earth's surface produced by either the erosive or depositional action of the wind. The followings are some of the landforms produced mainly by the erosive process of wind, called aeolian erosional landforms. Remember that even in arid areas, the wind is a relatively insignificant erosional agent compared to running water and glaciers. Most of the landforms in deserts and arid areas are sculptured by intermittent streams, not by wind.

1.1 Blowout

In some areas shallow depressions are formed where all of the loose rock particles have been carried away, called blowouts. The diameters of blowouts range from a few meters to 1 000m. Oases(绿洲) can even be formed from deep blowouts that hit the water table. When the low-lying land among the dunes in the desert is lower than the groundwater level, the water gathers there and forms the aeolian lake(风蚀湖泊) (Figure 15.3).

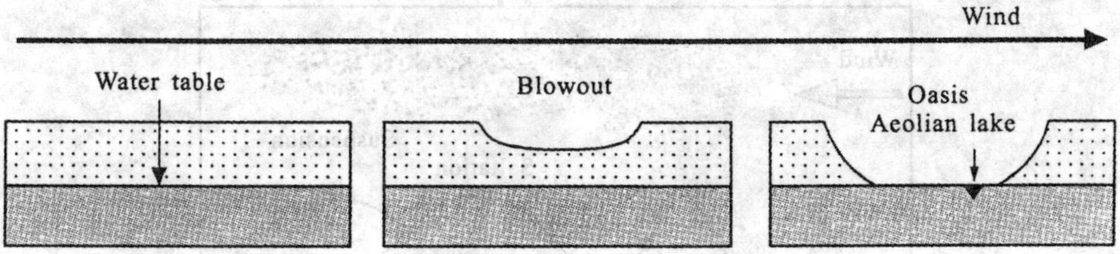

Figure 15.3 Deflation by wind stops at the water table. Oasis can be formed by wind deflation

Blowouts are common in the Xingjiang Uygur Autonomous region and the Northern Gansu province.

1.2 Desert Pavement

When fine material is removed from the surface, a cover of pebbles, cobbles and boulders will be left. This is desert pavement. The coarser material protects the underlying fine material from further erosion.

1.3 Ventifacts

Sand grains impacting rocks will eventually erode the rock surface, polishing the exposed surface to form a flat face. Such rocks are called ventifacts. It is the most common product of wind erosion. Ventifacts are characterized by a variable number of sheen facets on which pits and gouges can be found. The numerous facets meet with sharp ridges.

1.4 Yardangs

This term derived from the Uygur (维吾尔) word, meaning ridge or steep bank from which material is being removed. Yardangs are streamlined, sharp-crested elongate hill oriented parallel to dominant wind direction. The remnant hills are often strongly grooved or furrowed as the wind erodes away weakly cemented material.

They are commonly found in arid regions especially in Northwestern China and believed to be formed by abrasion. Some of the Yardangs landforms are known as the Devils' Town (魔鬼城) to the locals in Northwestern China.

2 Wind Transportation

Transportation processes in wind are similar to those in streams, including suspension, saltation, and surface creep. About 80% of aeolian transportation is in the form of saltation. Because of the light density of air, the wind usually cannot move particles larger than 2mm.

2.1 Suspension

The smallest dust particles may be held in the atmosphere through suspension.

Suspension occurs when surface materials are lifted into the air and upward air currents are strong enough to support the weight of the suspended particles and hold them indefinitely in the surrounding air (Figure 15.4). Typical winds speeds (5m/s) near the Earth's surface will suspend particles less than 0.2mm in diameter. Only severe windstorms can hold large particles. Once lifted, suspended particles can remain aloft for many hours or even weeks. Under strong wind conditions, suspended dust particles may be lifted thousands of meters

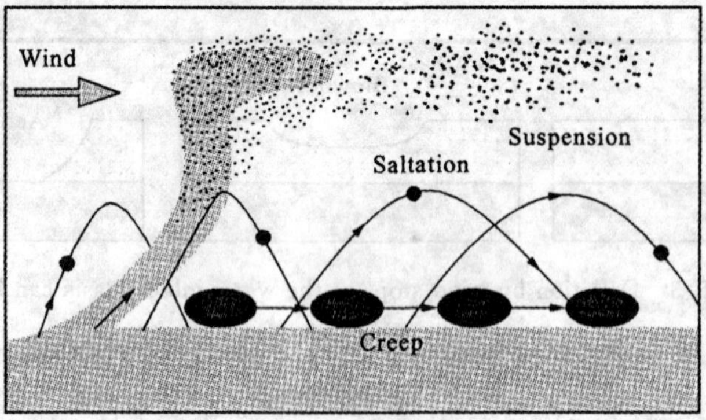

Figure 15.4 Suspension, saltation and creep

upward and thousands of kilometres downwind, held in suspension by turbulent eddies.

A strong, steady wind may hold large amount of suspended particles, thus forming the dust storms. The dust-filled clouds can be as much as several kilometers thick and swept all the way along the wind direction for about several thousands kilometers. Smaller particles can be carried far further than coarser ones, so wind erosion is an excellent sorter of sediments.

2.2 Saltation

Just as in the bed load of streams, saltation (Latin for leaping) is downwind movement of small particles near the surface forward through a series of short jumps or skips. A saltating grain may hit other grains as it returns to ground that will jump up to continue the saltation. Saltation normally lifts sand-size particles no more than one centimeter above the ground. Saltation process accounts for 75% to 80% of total wind transportation. The particles bounce onto the surface lifting other particles and causing damage to the surface and to the vegetation.

Sandstorms are the sand-filled low clouds holding a large amount of sand-sized particles lifted by saltation.

2.3 Creep

Creep is a sliding and rolling movement of particles which are too large to be lifted by wind, whose motion is produced by impact from saltating grains. When saltating particles hit larger particles, they nudge the larger grains (up to six times larger than saltating particles) slightly and slowly forward, this can result a sliding and rolling movement known as creep. Creep accounts for 20% of total wind transportation.

3 Wind Deposits

Material deposited by the wind is called aeolian sediment.

3.1 Characteristic of Aeolian Sediment

Compared to stream and glacier sediments, the aeolian sediment often has the following characteristics:

• Components

Aeolian sediment is composed of sand, silt and small amount of finer clay-sized particles. The particles greater than 2mm are rarely found in aeolian sediment.

Sand grains carried by the wind may be of almost any minerals, but most are quartz. Unlike stream sediment, some unstable minerals, such as pyroxene, amphibole and biotite can occur in aeolian sediment with a relatively high content. That is because chemical weathering is weak in arid and desert regions.

- **Well-sorted and Well-rounded**

Fine-grained silt and clay are carried much further than sand, and grains coarser than sand are left behind when sand moves. The final depositions of the sediment at different places are composed of particles with uniform grain sizes. The finest material is sometimes picked up on land by wind and deposited in the deep sea floor.

Due to the abrasion, the particles are frequently attacked with each other, thus resulting in well-rounded particles.

- **Large-scale Cross Bedding**

Due to the constant movement of sand of dunes from one side to another side, wind-deposited rocks are often featured by well-developed cross bedding.

- **Variable Colors**

Aeolian sediment may have different colors, but the dominant color is red as a result of oxidation.

3.2 Dune Formation and Migration

Deposition by the wind creates the most striking landforms. The most commonly found aeolian depositional landform of wind is dune (sand dune). Dunes are asymmetrical mounds and/or sand ridges that have been deposited by the wind. The extensive fields of dunes occur on Taklamakan Desert (塔克拉玛干沙漠) which contain vast sand seas.

Dunes occur where there is a large supply of sand, a wind to move it, and a place where it can accumulate. Dunes are found on coasts, rivers, and desert basins. These places all have in common a ready supply of loose sand.

Dunes begin to form when moving sand grains encounter an obstacle, such as a rock (Figure 15.5). The sand will accumulate in

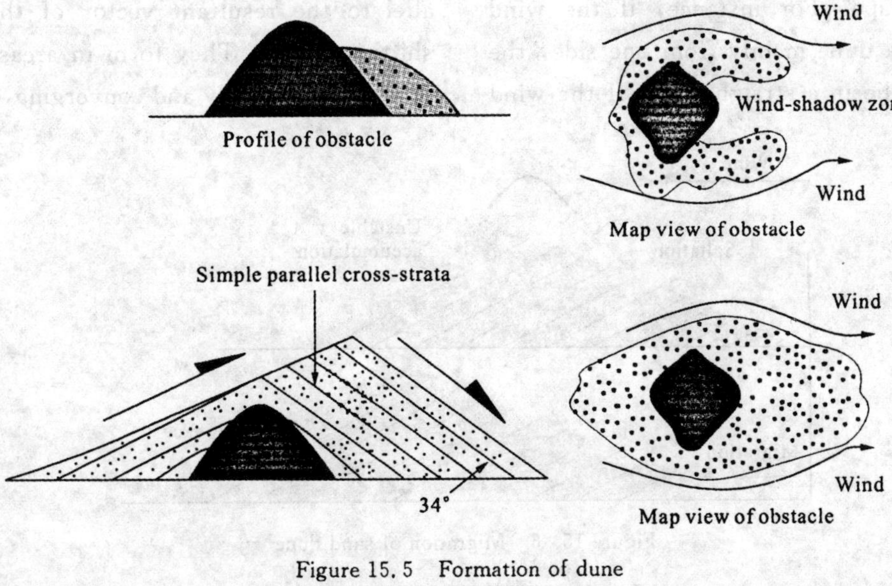

Figure 15.5 Formation of dune
Left: Side view. Right: Map view

the leeward side of the obstacle. Once started, this sand pile will grow. The sand moves up the windy side and is deposited on the leeward side of the dune, building that slope up to the angle of repose(休止角). Angle of repose is the maximum degree of gradient on which sand or loose rock remain stable. The angle of the leeward side of a dune is about 34 °, the angle of repose of dry sand. The growth of dune caused by deposition of sand on the leeward side can form cross-strata. If the wind has uniform direction and velocity, it would build a simple, parallel cross-strata.

Once the dune reaches certain height, it will migrate due to the continuous movement of sand on the dune surface (Figure 15.6).

3.3 Dune Types

The shapes of dunes are diverse. The direction and velocity of winds, in addition to the local supply of sand, result in a variety of dune's shapes and sizes. The direction of the wind is the determining factor of the dune's shape. For instance, if the wind comes at a dune mainly from one side, the dune will lie in a straight line. If the wind changes directions alternating from side to side, the dune will be crescent-shaped, like a crescent moon. If the wind comes from all sides, a dune will be star-shaped.

• Barchan Dunes

Barchan dunes are crescent-shaped dunes with a curved slip face on the downwind side of the dune. They form in areas where there is a hard flat ground surface, a limited supply of sand, and a constant wind direction.

• Transverse Dunes

Transverse dunes are asymmetrical narrow linear dunes that form at right angles to the prevailing wind direction, in which the downwind slope is at the angle of repose of sand and the upwind slope is comparatively gentle. Transverse dunes form in areas where there is abundant supply of sand and a constant wind direction. If more sand is available, the barchan dunes may merge into transverse dunes.

• Longitudinal Dunes

Longitudinal dunes are elongated straight ridges of aeolian sediments, which run parallel to the resultant vector of the sand-shifting winds. They form in areas with a limited sand supply and converging wind di-

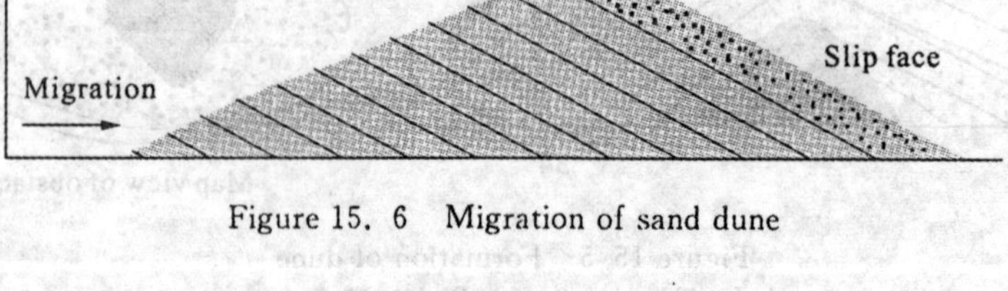

Figure 15. 6 Migration of sand dune

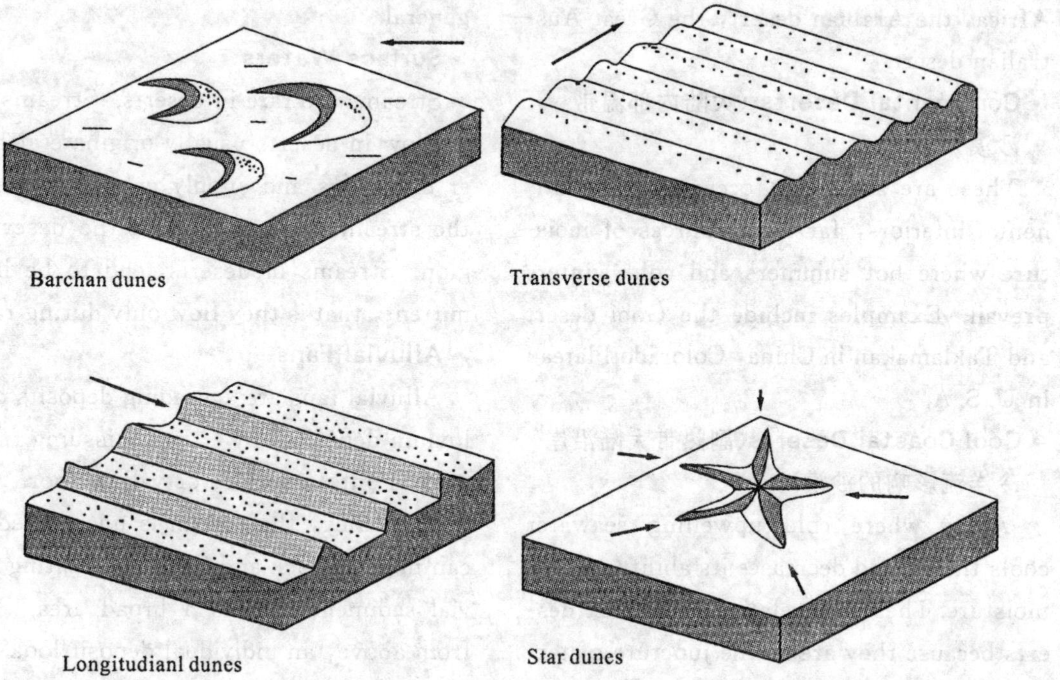

Figure 15. 7 Dune types

rections. Smaller types of longitudes dunes are less than 3m high, but some larger dunes can reach 100m in height and extend for 100km.

• Star Dunes

Star dunes are dunes with several radial arms extending in many directions, which develops in an area where there are abundant sand and variable wind directions over the course of a year.

4 Deserts

The word desert derives from the Latin word for abandoned or forsaken. Thus, formally defined, a desert is an abandoned place. In geology or geography, a desert, also called arid region, is a landscape form or region with an average annual rainfall of rain (or its equivalent as snow) of 25cm, or regions with evaporation exceeding precipitation.

Deserts tend to have sparse vegetation due to low moisture, and high relief due to short duration catastrophic rain with no vegetation to hold soils. Nearly 30% of the Earth's land area is covered by desert. Deserts cover about 11 percent of the China's territory.

The characteristics and classification of deserts can be summarized as follows:

4.1 Classification

Generally deserts can be classified into four types: subtropical, continental, coastal and polar.

• Subtropical Deserts（亚热带沙漠）

These are deserts located in belts of descending dry air between latitudes 20° and 30° N and S of the equator. Subtropical deserts are the hottest, with rapid evaporation. Examples include the Sahara of Northern

Africa, the Arabian desert, the Great Australian desert.

• **Continental Deserts**（大陆内部温带沙漠）

These are areas that occur in the continental interiors, far from sources of moisture where hot summers and cold winters prevail. Examples include the Gobi desert and Taklamakan in China, Colorado Plateau in U. S. A..

• **Cool Coastal Deserts**（副热带大陆沿岸冷洋流影响的沙漠）

Areas where cold upwelling seawater cools the air and decreases its ability to hold moisture. These are relatively complex deserts because they are at the juncture of terrestrial, oceanic, and atmospheric systems. A coastal desert, the Atacama of South America, is the Earth's driest desert.

• **Polar Deserts**（极地沙漠）

Polar regions are also considered to be deserts because nearly all moisture in these areas is locked up in the form of ice. Examples include the dry northern Greenland, arctic Canada, and in the ice-free valleys of Antarctica.

4.2 Surface Processes in Deserts

• **Desert Varnish**

Desert varnish（沙漠漆）is a thin red to dark coating found on exposed rock surfaces in arid regions. The coating is composed dominantly of fine-grained clay minerals along with black manganese oxide and red iron oxide. Desert varnish is caused by chemical and biochemical actions. The color of the varnish varies from shades of brown to black.

A more general term is rock varnish which applies to dark coatings on rocks in general.

• **Surface Waters**

Streams are rare in deserts. Streams that do flow in deserts usually originate at higher elevations and supply enough water for the stream to pass through the desert region. Streams in deserts tend to be intermittent, that is they flow only during rains.

• **Alluvial Fans**

Alluvial fans are aggrading deposits of alluvium deposited by a stream issuing from a canyon onto a surface or valley floor. Once in the valley, the stream is unconfined and can migrate back and forth, depositing alluvial sediments across a broad area. View from above, an individual deposit looks like an open fan with the apex being at the valley mouth.

Where a long, linear mountain range has many adjacent canyons, alluvial fans will build and coalesce with neighboring fans, producing a broad apron of sediments at the base of the mountains known as a bajada（山麓冲积扇）.

• **Pediments**（山麓面）

A pediment is a gently sloping erosion surface of bedrock cutting into the solid rock of a mountain at the base of a receding mountain front. A pediment is typically covered by a thin layer of gravel derived from erosion of the pediment.

• **Playa**（干盐湖）

Playa is short-lived lakes that form where water drains into basins with no outlet to the sea and quickly evaporates, or a lake that is dry for most of the year. Lakes that form during the rare periods of rainfall in desert area will quickly evaporate, leaving a dry lakebed covered by salts that were carried in by streams and precipitated during evaporation. These precipitated salts give the dry lakebed a white color resembling a beach (Playa means beach in Spanish).

5 Desertification and Dust Storm

Desertification is the degradation of land in arid, semi-arid areas and is subject to increasing interest due to the rapid increase of deserts at the expense of arable land.

Desertification is caused primarily by climatic changes and human activities. At present, human impacts, such as overgrazing, over-exploitation and inappropriate land use, contribute much to desertification. The result is the destruction of soil followed by loss of the land's ability to sustain crops, livestock or human activity. More than 27% of our country comprises useless sand (Just 7% of Chinese land feeds about a quarter of the world's population). Between 1994 and 1999, the Gobi Desert, one of the main sources of dust storm sand, grew by 52 400km^2 due to farming, grazing and deforestation. Desertification costs China about \$2~3 billion annually, while 800 km of railway and thousands of kilometers of roads are blocked by sedimentation. Combating desertification is essential to ensuring the sustained development of our country.

A dust storm is a meteorological phenomenon and created whenever strong winds travel over dry, arid land with little vegetation and pick up tiny particles such as sand, dust, and other debris in such quantity as to reduce visibility seriously. The dust and sand storms that have plagued Asia for millennia are becoming more frequent and intense, with some areas experiencing a five-fold increase in the last 50 years. According to Chinese government estimates, roughly one million tons of desert dust and sand blow into Beijing each year. These sands come from the degraded pastureland and dry land of the area around the deserts of Inner Mongolia.

The dust afflicts humans with respiratory problems, disrupts communication and transportation, and causes the loss of crops, livestock and human life. In May 1993, dust storm killed 85 people in north-western China, causing damage worth 550 million yuan. On March 20, 2002, a most severe sand storm in a decade hit a vast area of 1. 4 million km^2 in China. The sand storm reached into upper atmosphere, dumped more than 40 000t of sand per hour on the Korean peninsula, and finally arrived in North America in a week.

China has already noticed the problem of desertification and dust storm, and adopted a series of remedial measures. Currently such measures as the withdrawal of land from cultivation and pasturage and its return to forestland and grassland have been taken. Only by decreasing the damage caused by human and restoring the natural environment will it be possible to hold in check effectively the sources of the dust and to improve the situation fundamentally.

6 Loess

Loess is a geologic term (of German derivation, meaning loose) that refers to a soft, porous, pale yellowish in color, loose deposit of silt (sediment with particles 2~64μm in diameter) that has been laid down by wind action. Most loess deposits formed during the Quaternary period.

Loess is composed of angular unweathered grains of quartz, mica, feldspar, hornblende and other rock-forming silicates that are unconsolidated, but usually held together by calcareous cement. Bedding is absent from typical loess depositions. The angular grains of loess have little rounding and polishing. One characteristic property is its capacity to retain vertical, or even over-hanging, walls in the banks of streams for many years without slumping because of its angular grains. That is why we can construct innumerable cave dwellings in the loess slopes. On the other hand, loess is highly erosion-prone soil that is susceptible to the forces of wind and water. Deep gullies develop very easily during a single rainy season, so the presence of the loess should be considered in all construction operations.

Loess plateau in our country covers an area of some 640 000km^2 in the upper and middle parts of Yellow River. Thousands of years deforestation and over-grazing, exacerbated by population increase, have resulted in degenerateds ecosystems of loess plateau.

1. Answer the Following Questions

(1) How does a pediment differ from bajada? Discuss the differences and similarities between the two features, particularly in regard to appearance and origin.

(2) Name four types of sand dunes and describe the conditions under which each forms.

(3) Sketch a cross section of an idealized dune, labeling the slip face and indicating the wind directions. Why does the dune move?

(4) List and describe various dune types.

(5) Describe in sketch the ways by which the sediments are transported by wind.

(6) Describe the characteristic of aeolian sediment.

(7) What is Yardangs?

(8) Describe the classification of deserts.

(9) What is alluvial fan?

(10) What is blowout?

(11) Differentiate the desert pavement and pediment.

2. Multiple Choice

(1) A dry lakebed in the desert is known as a(n) ______.

A: Yardang B: Ventifact

C: Desert varnish D: Playa

(2) The term "saltation" refers to which of the following processes?

A: The sliding of boulders along the floor of gently sloping basins

B: The deposition of salt in the pores between sand grains

C: The hopping of sand grain as they are blown by wind across the desert floor

D: The sudden change in temperature that occurs in deserts from day to night

(3) What is the most effective agent of erosion in the deserts?

A: Wind B: Ice

C: Wind D: Gravity

(4) A sand dune that is convex in the downwind direction is a ______.

A: Star-shaped dune B: Transverse dune

C: Barchan dune D: Longitudinal dune

(5) A surface layer of closely packed pebbles found in desert areas is ______.

A: Ventifact B: Bajada

C: Desert varnish D: Desert pavement

(6) Which type of desert erosion consists of the lifting and removal of loose material?

A: Ventifact B: Abrasion

C: Varnish D: Deflation

(7) As the original deposit has sand, silt and clay particles eroded away, only the coarser particles remain, forming desert pavement.

A: True B: False

(8) Dunes occur where there is a large supply of sand, a wind to move it, and a place where it can accumulate.

A: True B: False

(9) Dune migrate upwind.

A: True B: False

(10) Playa is an ephemeral lakes in arid area.

A: True B: False

(11) Which of the following constitutes most of the bed load carried by wind?

A: Pebbles B: Dissolved ions

C: Sand D: Fine dust particles

(12) What is composed of fine-grained, windblown sediment?

A: Till B: Moraine

C: Drift D: Loess

(13) Which of the following statements about the wind action is not right?

A: Sediment hold in dust storm is hold in suspension.

B: The dune shape is determined mainly by wind direction.

C: Aeolian sediment is generally well-sorted.

D: Creep is caused by plastic movement.

(14) The rounded particle of sand is the result of ______.

A: Suspension B: Abrasion

C: Saltation D: Deflation

(15) What is the term for wind erosion that leaves behind coarse sediments and moves the finer grained material?

A: Suspension B: Abrasion
C: Saltation D: Deflation

(16) The streamlined, sharp-crested elongate hill in some arid area is called ______.
A: Yardangs B: Bajada
C: Blowout D: Desert pavement

(17) Which statement about dunes and desert is not right?
A: Sand moves up on the windy side and deposit on the leeside.
B: The wind direction is parallel to general ridge crest in longitudinal dunes.
C: If the slope is greater than the repose angle, the slope will collapse.
D: Barchan dunes are formed by abundant supply of sand.

(18) The merging of adjacent alluvial fans forms a sloping depositional surface called a ______.
A: Pediment B: Bajada
C: Blowout D: Desert pavement

(19) Ventifact is the result of ______.
A: Saltation B: Suspension
C: Abrasion D: Deflation

(20) Sand dunes migrate in the predominant wind direction.
A: True B: False

(21) Which one is not the aeolian erosional landform?
A: Ventifact B: Bajada
C: Blowout D: Desert pavement

Chapter 16 Mass Wasting

Key words

Mass wasting (movement)		Slump	Rock slide	Rock fall	Landslide
块体运动、重力作用		滑坡	岩滑	岩落	滑坡
Angle of repose		Earthflow	Grain flow	Talus	Creep
休止角		土流	颗粒流	倒石锥	蠕变
Sediment flow	Mudflow	Debris flows	Slurry flow	Debris	Shear strength
沉积物流	泥流	泥石流	泥浆流	碎屑	抗剪强度

Mass wasting (also called mass movement) is a general term for a variety of processes by which large masses of earth material are moved by gravity from one place to another. It is the downslope movement of rock and soil, under the influence of gravity. The definition excludes material transported downslope by streams, winds, or glaciers. Mass wasting is a form of erosion and plays a vital role in transferring the products of weathering from their original sites to lower lying places where the agents of erosion can pick them up for transporting a longer distance. The general non-technical term"landslide"is used to describe all rapid forms of mass wasting.

The"mass"part of the name implies that a somewhat coherent grouping of sediment rock begins moving downward due to the force of gravity, and usually in combination with some triggering mechanism such as an earthquake or rapid erosion of the base of a slope. Wasting refers to the destructive process. The"wasting"part of mass wasting means that a cliff or mountain slope is diminishing in size, or wasting away. "Mass wasting"turns out that we call this"mass" wasting because the fragments generally tend to stay together during transport and move as a unit. This can occur suddenly with tremendous destructive force, or very slowly with only a gradual alteration of the Earth's surface over a period of many years.

Mass wasting is a natural process that continuously shapes the landscape, and it occurs without human involvement; however, it's important for us to recognize that our decision, such as where we build our roads and structures, and the ways in which we alter natural landscapes can affect this process and sometimes trigger mass wasting events (Figure 16.1). Given enough time and repetition, the different types of mass wasting can play significant roles in reducing a tall mountain to a mound of low rolling hills, or in widening a narrow canyon into a broad stream valley.

Mass wasting affects humans in many ways. The force of rocks, soil, or other debris moving down a slope can devastate anything in its path. Its defect ranges from the devastation of a killer landslide to the

Figure 16.1 Mass wasting: earthflow. Photo by Dr. Jian Wenbin (简文彬)

nuisance of having a fence lowly pulled apart by soil creep. The cost in lives and property from landslide is surprisingly higher. Whether slow, fast or somewhere in between, the various kinds of mass wasting play an important role in the wearing away of the Earth's surface.

The basic physics controlling the stability or instability of landforms is relatively simple and well understood, but the hazards are not always recognized, even when geological deposits indicate past slope failures in the region. In many cases, surface instabilities of this type are compound events, associated with earthquake or volcanic processes, which enhance their catastrophic potential. From the surface geology perspective, landslides and debris flows are important landscape modifying agents, and play as large of a role in eroding topography and depositing debris as is played by other mechanisms such as rainfall and runoff.

1 Factors controlling Mass Wasting Hazards

Disaster occurs when mass wasting interacts with human society. In general, this means that certain factors conspire to cause a change in the normal conditions. A slope in nature evolves to an angle such that a balance is achieved so material moves downhill at some rate. If the conditions change, then this balance may be upset, and a sudden mass movement may take place. The factors that most influence the stability of slopes are listed here:

1.1 Gravity

Gravity(重力) acts on all objects on the Earth's surface. Gravity can be resolved into two components for objects resting on sloping surfaces. One component is parallel to the slope (G_t) and one is perpendicular to the slope (G_p). The force G_t is called the shear stress and the force G_p is called normal stress (Figure 16.2). The downhill movement of soil and loose unconsolidated sediments is due to the shear stress and is resisted by the force of friction. The force of friction, or the shear strength(抗剪强度) is the internal resistance of the rock to move. If shear stress is greater than shear strength, the rock moves downhill.

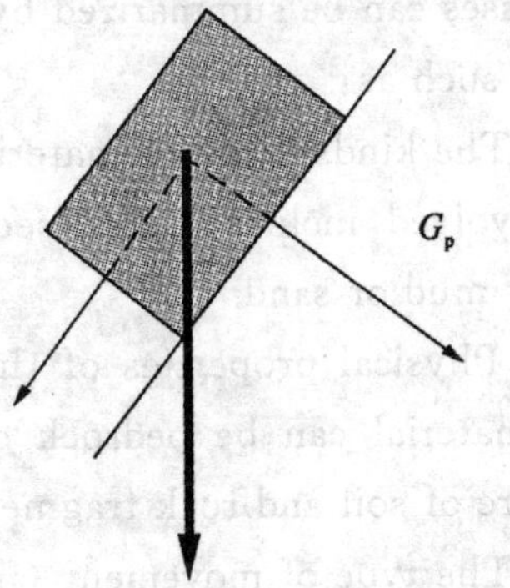

Figure 16.2 Components of gravity

Although many of landslides differ greatly from one another, they all share one characteristic: they occur on slopes. The forces of gravity and friction are in balance at the angle of repose which is the maximum slope angle that unconsolidated materials can maintain. The angle of repose, typically 25°~40°, is the angle at which loose material sits at rest. At angles steeper than the angle of repose friction is not sufficient to counter gravity and mass wasting occurs. At angles less than the angle of repose, gravity cannot overcome friction and sediments may accumulate to form steeper slopes.

Some slopes are naturally steep, such as those along glacial valleys, riverbanks, or at the shore where waves cut back the cliffs, but in many areas the primary force responsible for steeping the slopes is human activity.

1.2 Water

Water is extremely important in slope stability. It is a critical factor commonly associated with downslope movement.

The effects of water depend on how abundant it is. A small amount of water in soil can actually prevent downslope movement. On the contrary, if the soil is saturated with water, which makes it heavy, the force of gravity will overcome the resistance of the slope and thus result in landslide.

Damp sediments are cohesive as water coats the sedimentary grains and hold them together with its surface tension (Surface tension is the result of the dipolar nature of water.). The angle of repose increases (think of a sand castle).

But an excessive amount of water within a slope can actively assist slope failure by increasing the weight of the sediment or rock, or passively assist failure by reducing the friction and cohesion between particles (Figure 16.3). Both of these situations, often associated with heavy rainfall or rapid snowmelt, can decrease the angle of repose and lead to mass wasting. Water can also flow between a permeable layer and an impermeable rock layer, thereby decreasing friction and possibly causing mass wasting. Finally, the addition of water may promote instability by adding weight to a slope.

1.3 Other Factors

Finally, miscellaneous factors such as earthquakes, the presence or absence of vegetation, geological structure, and human

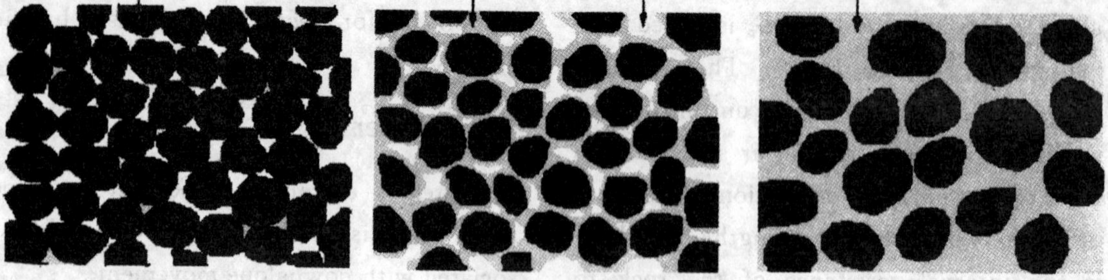

Figure 16. 3 Left: Dry sands. Middle: Unsaturated sands. Right: Saturated sands

activities can also influence the potential for mass wasting. For example, if the rocks dip in the same direction as the slope, or if there are joints which dip in the direction of the slope, mass wasting is more likely — water can more easily percolate along the bedding planes and lower the cohesiveness and friction between grains (Figure 16. 4).

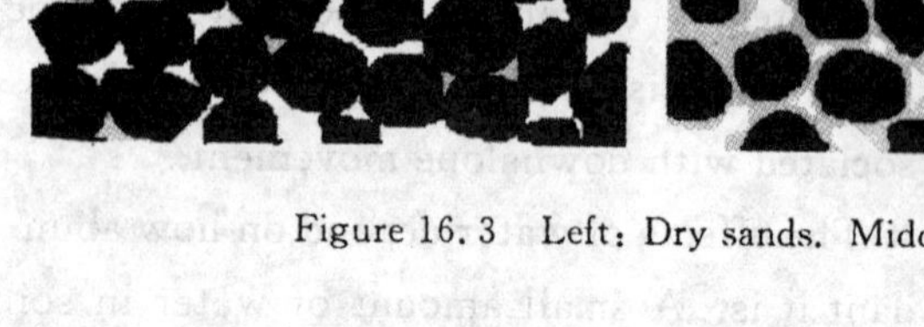

Figure 16. 4 Influence of bedding on the slope stability

Vegetation can absorb water which can prevent water from percolating into the unconsolidated material. The root systems help to stabilize a slope. Human impact, primarily in the form of building the road and structures, can seriously impact the integrity of a hillside and is capable of triggering mass wasting events.

2 Types of Mass Wasting: Slope Failures

There is a broad complex of activities that geologists call mass wasting. Mass wasting processes can be summarized by several criteria, such as:

(1) The kinds of earth materials: Materials involved include rock, sediment, ice, snow, mud or sand.

(2) Physical properties of the materials: The material can be bedrock or debris (a mixture of soil and rock fragments).

(3) The type of movement (mechanism): falling, sliding or flowing. It is the most common and unambiguous criterion.

(4) Speed of mass movement: Speed can be less than 1cm/a to greater than 5km/h.

Based on the above criteria, mass-wasting processes can be classified by many different ways. In fact, it is very difficult to classify them absolutely. For example, based on the type of movement, mass-wasting processes can be classified into four different types: fall, slide, flow, and creep. Flow is like a viscous fluid, it is turbulent, always mixing and tumbling. Slide is coherent and moving along on one well defined surface. Fall is material free falling or bounding down a slope. In this textbook, we group mass wasting processes into two broad categories: slope failure and sediment flow.

Be sure that the boundaries between the different types are arbitrary and all of these processes generally grade into one another.

Slope failure is a sudden failure of the slope resulting in transport of debris down hill by sliding, rolling, falling, or slumping.

2.1 Slump

A slump is the movement of a block of earth material through downward rotation along a curved surface of failure. The upper surface of each slump block remains relatively undisturbed, as do the individual blocks. Slumping typically involves unconsolidated sediment that becomes saturated with water. The curved sliding surface results in rotation of overlying slump blocks. A crescent-shaped cliff or "scarp" is left at the upslope end (sometimes more than one scarp) (Figure 16.5). At the bottom (or toe) of the slump, flow of soil occurs.

Slumps form where the base of a slope is removed by natural processes (stream or wave erosion) or by human efforts (road or building construction). Removal of the lower part of a slope effectively removes physical support for the upper part of a slope, causing the formation of a new fracture in the sediment/rock comprising the slope. Soon thereafter, the slope will begin sliding downward, often rotating along the curved surface of rupture. Heavy rains or earthquakes usually trigger slumps.

Slumps are common in hilly regions. They are the most frequent cause of property damage during period of heavy rainfall.

Table 16.1 Classification of mass wasting process

		High speed (~5km/h)— Intermediate — Low speed (~1cm/a)			
Slope failure	Massive rock	Rock falls	Rock slides	Slumps	
	Debris	Debris falls	Debris slides		
Sediment flow	Granular flow (dry flow)	Grain flows		Earth flows	Soil creep
	Slurry flow (Wet flows)	Mud flows		Debris flow	Solifluction

Figure 16.5 Slump. Courtesy of NOAA. Public Domain

2.2 Rock Falls and Debris Falls

Rock fall is the falling of pieces of rock from very steep portions of slope. The free fall of detached pieces of material can be of any size; and may fall directly downward or bounce and roll. Rock fall usually happens in steep mountains, especially with little vegetation as result of freeze-thaw, or the loosening action of plant roots. Debris falls are similar, except they involve a mixture of soil, regolith, and rocks.

Rock fall can cause the formation of talus (Figure 16.6). Talus is the accumulation of fallen boulders or weathered material at the base of the cliffs. The slope of the talus is controlled by the angle of repose for the size of the material. No matter how much dry sand is added to a pile, it can never form a slope that is steeper than 35° inclination (Figure 16.7). Since talus results from the accumulation of large rocks or masses of debris the angle of repose is usually greater than it would be for sand.

2.3 Rockslides and Debris Slides

Rockslide or debris slide occurs when downlsope movement occurs along a pre-existing plane of weakness in the rock. Such planes of weakness are typically either suitably oriented bedding planes or a fracture surfaces (fault or joint) within a body of

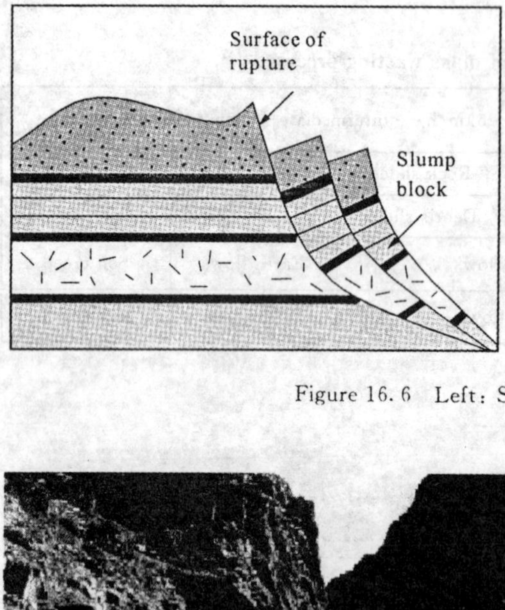

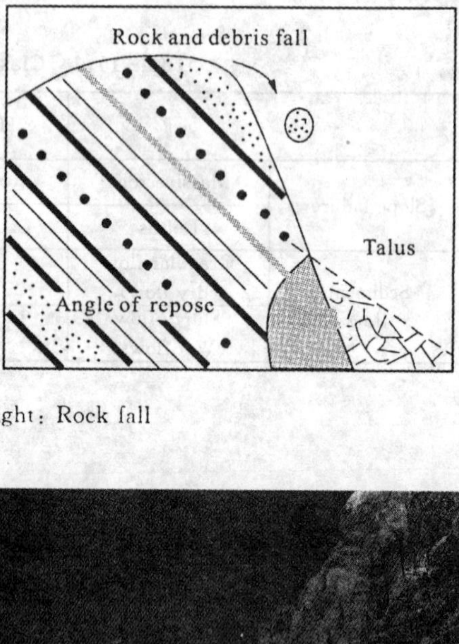

Figure 16.6 Left: Slumps. Right: Rock fall

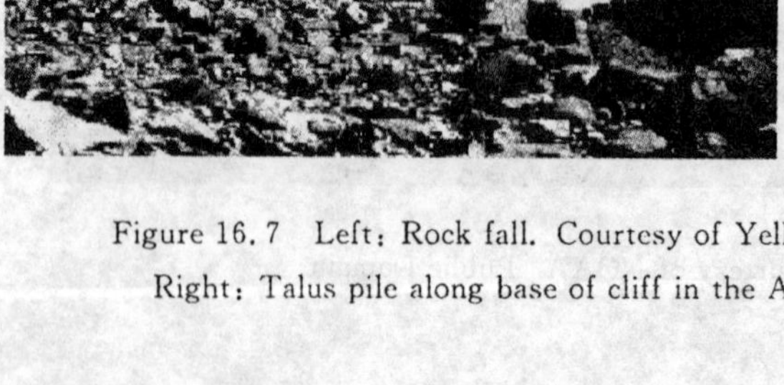

Figure 16.7 Left: Rock fall. Courtesy of Yellowstone Digital Slide File. Public domain. Right: Talus pile along base of cliff in the Arizonan. Photo by Dr. Yu Shan (俞姗)

rock. Debris slides are similar, except they involve a mixture of soil, regolith, and small pieces of rock.

The Figure 16. 8 shows layers of rock tilted downward to the right. The topmost rock layer is prone to sliding because it lacks support at the base of the slope. Gravity will eventually overcome the friction between the topmost rock layer and the rock beneath it. Once this occurs, the topmost rock layer slides downward as a coherent block. As it comes to a sudden stop the slide block may break apart and continue moving for some distance as a debris flow.

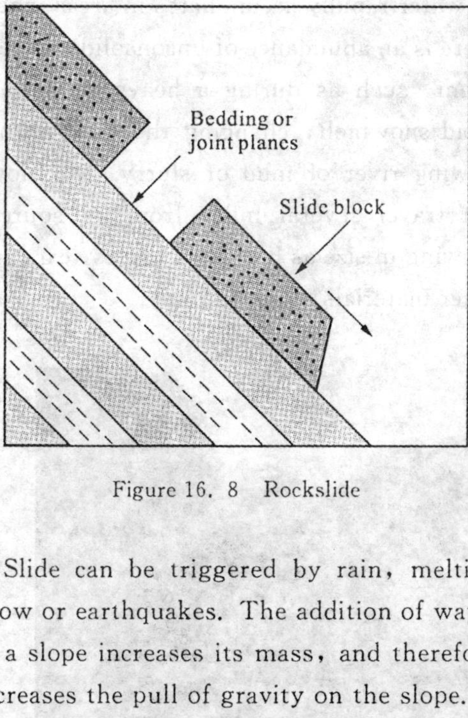

Figure 16. 8 Rockslide

Slide can be triggered by rain, melting snow or earthquakes. The addition of water to a slope increases its mass, and therefore increases the pull of gravity on the slope. In addition, water can lubricate a layer of clay or shale within a slope, which then serves as a slide surface for the rock above it. Slides differ from slumps in that there is no rotation of the sliding rock mass along a curved surface. Rockslides can be massive, occasionally involving an entire mountainside, making them a real hazard in areas where a surface of weakness tilts in the same direction as the surface of the slope.

Rockslides generate relatively thin sheets of rock that are broken into smaller blocks as they move downslope, and usually form extensive scar, massive pile of highly irregular debris at its bottom. Therefore, piles of talus are common at the base of a rockslide or debris slide.

3 Types of Mass Wasting: Sediment Flows

Sediment flows (沉积物流) are the downslope flow of mixtures of solid material, water, and air. Rather than moving downslope as a coherent mass (slump, landslide) the material flows downhill as a chaotic mixture. Water and air help promote flow, but the gravity remains the primary reason for their movement.

They can be broken into two types depending on the amount of water present or the sediment concentration.

3.1 Slurry Flow(泥浆流)

Slurry flow is a moving mass of water-saturated sediment. In general, the water content in slurry flow is between about 20% and 40%. As the water content increases above about 40%, slurry flows grade into streams. The poorly sorted sediment mixture in slurry flows is often so dense that large boulders can be suspended in it. Based on the velocity, slurry flow can be subdivided into following types:

- **Solifluction(融冻泥流)**

Solifluction is the very slow downslope movement of saturated soil and regolith. Solifluction occurs in areas with a permafrost. During the summer the surface layer of permafrost melts creating a water-saturated layer that becomes mobile. The underlying frozen ground acts as a sliding plane along which the mass of soil can slowly move downslope over. Rates of solifluction movement are measured on the order of centimeters per year, usually less than about 30cm/a.

• **Debris Flows(泥石流，碎屑流) or Debris Avalanches(泥石崩流)**

Debris flows are the downslope movement of unconsolidated regolith and often result from heavy rains causing saturation of the soil and regolith with water (Figure 16. 9). Like fresh concrete, the greater part of debris flow is sand and gravel. Even boulders bigger than cars ride long distances in debris flows. They usually have a very high velocity and may travel several kilometers before coming to rest. The velocities can vary between a few m/a and 100km/h. The bigger the avalanche the bigger its speed and thus its danger.

A debris flow often begins with a slump or debris slide, and then continues to flow downslope. Deposits commonly have a tongue-like front with an irregular surface consisting of ridges and furrows. They are frequently associated with intervals of extremely heavy rainfall that lead to saturation of the ground. They are often triggered by earthquakes and volcanic eruptions.

• **Mudflows(泥流)**

Mudflow is the downslope rapid movement of unconsolidated clay, silt and sand with a water content sufficient to make it highly fluid (Figure 16. 10). It develops when water rapidly accumulates in areas where there is an abundance of unconsolidated sediment, such as during a heavy rainfall or rapid snowmelt, changing the earth into a flowing river of mud or slurry. Mudflows can travel several miles from its source, growing in size as it picks up trees, cars and other materials along the way.

Figure 16. 9 Debris fliow. Photo by Dr. Jian Wenbin(简文彬)

Figure 6.10 Mudflow in Hongkong. Courtesy of Dr. Jian Wenbin(简文彬)

3.2 Granular Flow(粒流)

Granular flow is a flow of mixture of sediment, air, and water. Usually the water content varies between 20% and 0%. Note that granular flows are possible with little or no water. Due to the mixture of air in sediments, the granular flow can behave like fluid. Unlike slides, in which the material sticks together as a coherent mass as it moves downslope, granular flows are characterized by internal movements of individual grains (tiny like silt or sand up to large boulders and small blocks of crust) within the flow itself.

• Creep(蠕移)

The downslope migration of regoligh and broken-up rock at a very slow but more or less continuous rate is called creep. The rate of movement is usually less than a centimeter per year and can be detected only by observations taken over months or years. It is commonly seen as curved tree trunks, distorted bedding surfaces of weathered sedimentary strata and tilted utility poles (Figure 16.11).

Figure 16.11 Creep

Creep is a near-surface phenomenon and diminishes with depth to where movement is no longer discernible. Two factors that contribute significantly to creep are water in the soil and daily cycles of freezing and thawing. The alternate expansion and contraction of surface material caused by freezing and thawing or wetting and drying can move the material a short distance downhill each time.

• **Earthflows(土流)**

Earthflow is the flow of unconsolidated materials on an open slope. They generally tend to be narrow tongue-like features that begin at a scarp or small cliff. In a longitudinal profile from head (a steep scarp) to toe, an earthflow is concave upward near the head and convex upward near the toe.

Earthflows are usually associated with heavy rains and move at velocities between several cm/d and 100s of m/d. Earth flows (Figure 16. 12) differ from mudflows in that they ① tend to be slower; ② are not confined to channels; ③ are more common in humid areas than dry; ④ have a lower water content.

• **Grain Flow(颗粒流)**

The sediment of grain flows is largely dry sand. If the slope of sand dune exceeds the repose angle, a small disturbance will send the dry unconsolidated sand grains moving rapidly downslope.

Figure 7. 12 Aerial views of earthflow. Courtesy of USGS, public domain

1. Questions for Review and Thought

(1) How does a slump differ from a slide?

(2) What is the distinction between earthflows and mudflows?

(3) How does the water influence the mass movement?

(4) What is the difference between rock fall and a rockslide?

(5) Describe the characteristics of a slump.

(6) List the different types of slurry flows.

(7) How to classify the mass wasting events?

(8) What is mass wasting?

2. Multiple Choices

(1) What type of mass wasting involves block rotation and leaves scarps?

A: Rockslide B: Creep

C: Slump D: Mudflow

(2) The type of mass wasting involving a very high water content and fluid like flow is called a ______.

A: Landslide B: Debris flow

C: Rock fall D: Creep

(3) Which type of mass wasting occurs most slowly?

A: Rockslide B: Creep

C: Slump D: Mudflow

(4) All kinds of mass wasting occur on slopes.

A: True B: False

(5) What process builds talus cones?

A: Rockslide B: Creep

C: Slump D: Mudflow

(6) Which of the following factors can trigger mass wasting?

A: An earthquake

B: A heavy rainstorm

C: Removal of material at the base of slope

D: All of the above

(7) Which of the terms is the most important factor responsible for mass movement.

A: Pressure B: Temperature

C: Water content D: All of the above

(8) All kinds of mass wasting are caused by gravity.

A: True B: False

(9) A mudflow is formed by the rapid sliding of large masses of rock on steep slopes.

A: True B: False

(10) A mass movement in which the material in motion remains as a coherent block is called ______.

A: Slide B: Fall

C: Flow D: Slurry

(11) The term mass wasting implies a ______.

A: Movement of rock or soil in bulk or as a unit

B: Fast moving flow of mud

C: Slow wasting away of a hillside

D: Mass of broken rock at the base of a cliff

(12) Heavy rains can trigger rockslides.

A: True B: False

(13) The accumulation of rock fragments found at the base of many cliffs is referred to as the ______.

A: Earth flow B: Landslide

C: Rockfall D: Talus

(14) Which one is the most common mass wasting type?

A: Rockslide B: Creep

C: Slump D: Mudflow

(15) The form of mass wasting known as slump can be recognized ______.

A: As a flow of mud in a defined channel

B: By slow movement of a gentle slope

C: By a steep scarp or cliff at the head of a slide

D: By a pile of boulders at the base of cliff

(16) The slowest form of mass wasting is ______.

A: Rockslide B: Creep

C: Slump D: Mudflow

(17) Which type of mass wasting is associated with permafrost?

A: Creep B: Solifluction

C: Rockslide D: Earthflow

(18) A rockslide is formed by the rapid sliding of large masses of rock on steep slopes.

A: True B: False

(19) The two opposing forces involved in mass movement are ______.

A: Weight and mass

B: Gravity and buoyancy

C: Shear stress and shear strength

D: Friction and heat

(20) What type of mass wasting involves sliding on well-defined surfaces?

A: Creep B: Slump

C: Rockslide D: Earthflow

(21) A mass movement in which the material in motion mixes together is called ______.

A: Slide B: Fall

C: Flow D: Slurry

(22) Which of the following statements about aquifer is not true.

A: The material can include everything from great slabs of solid rock to loose unconsolidated rock debris or soil.

B: Earth flow can cause the formation of talus at the foot of slope.

C: Steep slope are generally more susceptible to mass wasting events.

D: Mass wasting events are one of the erosion agents of landscape evolution.

Chapter 17 Lake and Wetland

Key words

Tectonic lake	Crater lake	Fluvial lake	Endorheic lake	Exorheic lake	Dammed lake
构造湖	火山口湖	冲积残留湖	内流湖	外流湖	堰塞湖
Glacial lake	Karst lake	Impact crater lake		Aeolian lake	Wetland
冰川湖	喀斯特湖	陨石撞击湖		风成湖	湿地
Mire, Swamp, Bog, Marsh, Fen 总称沼泽,无标准翻译					

A lake is a large body of water surrounded by land on all sides. More than 98% of accessible surface water is lake water. Really huge lakes are often called seas. A mire is a type of freshwater wetland that has spongy, muddy land and a lot of water. Many trees and shrubs grow in swamps.

Finland is known as the Land of Thousand Lakes. The largest lake in the world is Caspian Sea and the deepest is the Lake of Baikal in Siberia. Lake Baikal is located north of the Mongolian border and is the largest freshwater lake in the world by volume. Baikal is also the deepest lake on the Earth, with a depth of 1 741m — more than one-mile straight down. The surface of Dead Sea (it's considered a lake but called a sea) is the lowest point on the Earth's surface at an elevation of 417m below the sea level. The Dead Sea is also the saltiest lake in the world. It is called "dead" because its high salinity means almost nothing can survive in it besides some special green algae. The highest lake in the world is Namu Cuo Lake (纳木错) in Tibet with an elevation of 4 718m above the sea level.

China is a country with numerous lakes. It has more than 2 800 natural lakes of varying types. There are also a large number of artificial lakes (reservoirs). Lake is one of our nation's greatest resources.

1 A Genetic Classification of Lakes

Lakes can be classified based on their origins:

1.1 Tectonic Lake

The lake basin is the result of tectonic movement. Most of the famous lakes in the world are of tectonic origin. For example, the big lakes of the African Rift Valley (Victoria, Tanganyika, Nyasa) are all graben lakes. Lake Baikal is one of the most ancient lakes in the world, created 20 to 30 million years ago by tectonic forces. Faults and rifts during different geological periods created mountainous ridges and deep valleys, and the tectonic hollow of Lake Baikal. Thermal vents in the depths of the lake are evidence of ongoing tectonic activity in the area.

The Sun-Moon Lake (日月潭), located in

the middle of Taiwan, was formed due to the fault activity in the areas between Yushan Mountains（玉山）and Alishan Mountains(阿里山). Earthquakes often occurred along this fault zone. Sometimes, the large-scale crustal subsidence can also form tectonic lakes. Poyanghu Lake(鄱阳湖) and Dongtinghu Lake(洞庭湖) are the typical subsidence lakes. This kind of lake often has an irregular shape and enormous size.

1.2 Crater Lake

A volcanic crater (Figure 17.1) lake is a cap of meteoric water over the vent of an active volcano. Only 12% of the world's active volcanoes have such a lake. The crater lake is very clean and clear because it contains few dissolved minerals and particles, and no streams run into or out of the lake.

Tianchi Lake in Changbaishan Moutains (长白山天池) is the deepest volcanic crater-lake in China. It is shaped like an enormous funnel. After the eruption of volcano, the water gradually accumulated and finally formed the huge crater lake — Tianchi Lake.

1.3 Fluvial Lake

Fluvial lake is formed due to the interaction of erosion and deposition of river.

The depression on the floodplain sometimes can accumulate water, forming fluvial lake. The discarded riverbed can change into oxbow lake during the evolution of meander.

Lake Saint-Pierre in Canada is the largest fluvial lake ($400\ km^2$) of the St-Lawrence River, one of the three largest rivers in North America. Lake Saint-Pierre has recently been classified by UNESCO as an Ecological Reserve of the Biosphere and is now officially recognized as a world heritage site.

The East Lake in Wuhan(武汉东湖) is a typical fluvial lake. It is situated on the floodplain of Changjiang and periodically flooded by Changjiang(Figure 17.2).

Figure 17.1 Crater lake and wizard island.

Image credit: USGS/Cascades Volcano Observatory. Public domain

Figure 17.2 The East Lake in Wuhan (武汉东湖). Photo by Chen Fuxing(谌福兴)

1.4 Dammed Lake

A lake basin may result from the damming of a river valley by a landslide. These lakes are also known as barrier lakes.

Some valleys with steep wall often produce rockfalls or landslides, and become mantled with talus. the talus will block a river or stream. On the evening of April 9, 2000, in 10 minutes, a giant dam measuring 2 500m long, 2 500m wide and 54m high was naturally formed by a great landslide to block the Yigong River (易贡藏布江) in Linzhi (西藏林芝). After being dammed, the river became a large barrier lake with an area of $33km^2$.

Damming or diversion of waterways by accumulation of volcanic products can also produce volcanic dammed lakes. Like the landslide dammed lake, large unstable tephra dammed lakes formed in a river and lake basin within a volcanic area may be the most dangerous water bodies.

1.5 Glacial Lake

Glacial lakes are ephemeral lakes created where glacier ice blocks a valley or embayment and ponds the drainage (Figure 17.3). During the glaciations, a series of pocket lakes can be formed at various elevations along both sides of the valley. The global climatic change during the first half of the twentieth century has brought a tremendous impact on the high mountainous glacial environment. Many of the big glaciers melted rapidly and gave birth to the origin of a large number of glacier lakes. The Tianchi in Xinjiang (新疆天池，古称瑶池) and Xinluhai Lake (新路海，川西甘孜) may be the most famous glacial lakes in China.

Some of the very famous lakes in the world are also closely related to glaciation activity. The Great Lakes in North America are considered to be the cumulative result of several phases of glaciations(冰川作用) that took place during the last 1.5 million years (the Quaternary Period). During the Pleistocene Epoch, the continental glaciers repeatedly advanced over the Great Lakes region from the north. As the glacier retreated, large volumes of meltwater occurred along the front of the ice. Because the land was greatly depressed at this time from the weight of the glacier , large glacial lakes

Figure. 17. 3 Glacier-dammed Lake in Jiuzhaigou(九寨沟冰川堰塞湖). Photo by Chen Fuxing(谌福兴)

formed.

1.6 Karst Lake

Karst is a type of landscape found on carbonate rocks (limestone, dolomite, marble) or evaporites (gypsum, anhydrite, rock salt). Karst lakes are formed by dissolution of those soluble rocks. In karst regions, infiltrating rainwater and snowmelt, which picks up weak carbonic acid from its short-term contact with the carbon dioxide in the air, has dissolved some of these rocks, at preferred sites. The dissolved rocks can form karst valley or may collapse to form sinkholes. When the valley or sinkholes are filled with water, the karst lake forms.

The Caohai lake (草海) is the largest fresh water karst lake in China.

1.7 Aeolian Lake

Wind lakes, or aeolian lake are formed in the deflation basins. Deflation and scouring by wind blown particles can erode the surface down and create a depression in arid areas. The lake is actually a part of the ground water. The surface of the lake is the water table. Aeolian lakes are all stagnant lakes that do not move, with smaller area, shallower water, changeable shape and no exit. Due to the constant movement of sand dune, the aeolian lakes are often buried and become the underground lakes.

The Crescent Moon Lake (月牙泉) in Gansu Province is formed due to wind deflation. It is a typical blowout.

1.8 Impact Crater Lake

Impact craters are geologic structures formed when a large meteoroid, asteroid or comet smashes into a planet. If the crater is filled with water, it is called impact crater lake, or impact lake. One of prominent features of impact crater lake is its central island caused by rebounding of Earth materials after impact (Figure 17.4).

1.9 Artificial Lake (Reservoir)

Artificial lakes are those lakes created behind manmade barriers.

2 Hydrology of Lake

Hydrology of the lake is the study of the

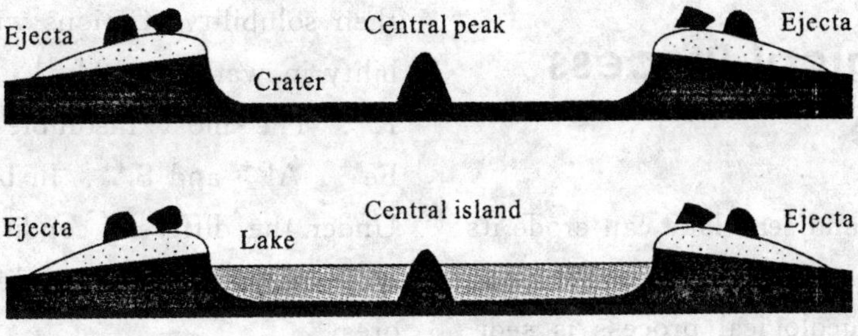

Figure 17.4 Impact crater and impact crater lake

circulation and movement of water.

2.1 Recharge

Water enters the lake as precipitation, surface-water inflow, and ground-water inflow. Sometimes glacier-melted water and residue seawater can also enter the lake. Generally speaking, lakes on highlands, such as crater lake on the top of mountains, is often recharged by precipitation. Lakes on lowlands are controlled by surface water, groundwater, as well as precipitation. Lakes in humid areas are dominated by precipitation and river. Lands in arid areas receive their water mainly from precipitation, groundwater and glacier-melted water.

2.2 Discharge

Water leaves the lake as evaporation, surface-water outflow, and ground-water outflow.

In arid and semi-arid areas, lakes often have no outlets. In this case, the lakes are discharged mainly by evaporation and seepage. Lakes in humid areas are generally discharged through numerous outlets.

Lakes without outlets are called endorheic (endo, "inside" and rhein, "to flow") lakes. The Caspian Sea is the largest endorheic lake on the Earth. Conversely, lakes drained by outflowing rivers are termed as exorheic lakes. Endorheic lakes are very dependent on the balance of inflows and evaporation. Qinghai Lake(青海湖), for example, has shrunk following a fall in precipitation. The closed nature of this water flow often leads to the concentration of salts and other minerals in the lake. With a continuing mineral input, some endorheic lakes typically become saline. On the other hand, because evaporation plus seepage are the major water outflow pathway, endorheic water-bodies also tend to be more sensitive to pollutant inputs than water-bodies that drain to the oceans.

2.3 Movement of Lake Water

Except for some small lakes with standing water, the water in most of the lakes is in constant movement. Due to its limited size and shallow water depth, the movement of lake is very weak compared with that of sea and ocean.

Like the sea, the lake can move in the form of wave, tide, current and turbidity. But the magnitude and energy of lake movement are far less than that of sea. For example, the maximum wave height and length in Michigan Lake are only 4.5m and 30m respectively. In Poyanghu(鄱阳湖),

the maximum wave height is only 1. 5m.

3 Geological Process of Lake

Like the river and sea, lake can erode its banks and transport the sediments. But the most important geological process is sedimentation. Sediments from surface water, groundwater, wind, glacier and volcano can be transported and deposited in the lake.

3.1 Mechanical Sedimentation

Clastic sediments transported by the river, glacier and wind will deposit mechanically in the lake. Especially the sediments transported by rivers will deposit in the sequence based on the grain size and density. So, the mechanically deposited sediments in lake are generally well-sorted and well-rounded. They often take a concentric pattern in the orders of grain size. The coarse sediments deposit near the banks and fine sediments will be transported to the center of the lake (Figure 17.5).

3.2 Chemical Sedimentation

Due to the chemical weathering, there are lots of different elements dissolved in the

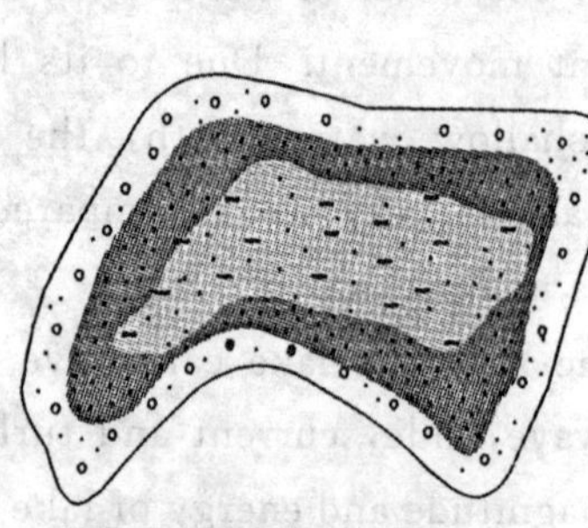

Figure 17.5 Mechanical sorting of lake-deposited sediments

lakes. Those elements can be grouped by their solubility. Cations with highest solubility in water are Mg^{2+}, Ca^{2+}, Na^{+} and K^{+}. The most insoluble cations include Fe^{3+}, Al^{3+} and Si^{4+}. In between is Fe^{2+}. Under the different conditions, these ions will precipitate and form different rocks and ores.

The main factor controlling the chemical sedimentation is climate. Therefore, the chemical sedimentations in lakes can be classified into two types:

- **Chemical Sedimentation in Humid Climate**

In humid climate, the soluble cations (Mg^{2+}, Ca^{2+}, Na^{+} and K^{+}) are extremely difficult to precipitate due to the ample recharge from rainfall and surface water. On the other hand, the insoluble elements (Al, Fe) are the main sources of deposition.

Let's take the iron as example. The dissolved Fe^{3+} in the water may react with dissolved OH^{-} form ferric hydroxide [$Fe(OH)_3$]. In the environment with abundant organic materials, the iron bicarbonate [$Fe(HCO_3)_2$] transported into lake will be oxidized into $Fe(OH)_3$:

$$4Fe(HCO_3)_2 + O_2 + 2H_2O \rightarrow 4Fe(OH)_3 + 8CO_2$$

This kind of iron ore is called bog iron ore (湖铁矿、沼泽铁). Due to dehydration, sometimes, deposited iron hydroxide alters into iron oxide in a long time.

O_2 is rapidly depleted in sediments rich in organic carbon due to exhaustive aerobic decay. The aerobic decay within the upper few millimeters of organic-rich sediment will maintain anoxic reducing conditions. Given anoxic conditions and reduced iron, the primary factor that determines whether pyrite will form is the presence of sulfide (H_2S and HS^-).

$Fe(HCO_3)_2 + 2H_2S \rightarrow FeS_2 + 3H_2O + CO_2 + CO$

(Pyrite)

Or: $FeSO_4 + 2H_2S \rightarrow FeS_2 + 3H_2O + CO_2 + CO$

(Pyrite)

Under the condition of weak oxidation and cold and wet climate, siderite(菱铁矿) will form through the combined effects of iron reduction and bacteria process by the following reaction:

$Fe(HCO_3)_2 \rightarrow Fe(CO_3) + H_2O + CO_2$

- **Chemical Sedimentation in Arid Climate**

In the arid area, the lake is recharged mainly by snow water and discharged by evaporation. This kind of lake often has no outlet to the ocean or river. Excessive evaporation of water in arid area leads to high concentration of dissolved solids. And the fresh water lake will eventually turn into saline lake. The main precipitated solids in saline lake is those salts composed of high soluble cations. Many salt lakes in extremely arid regions are ephemeral, containing water only during part of the year or only during years of exceptionally high precipitation. A good example of this is Lake Luobubo(罗布泊).

The salts dissolved in the saline lake precipitate in the orders of their solubility. An evaporite is a sediment that forms through the evaporation of saline water. Evaporites are often sorted in inverse order of solubility, meaning that the least soluable salts precipitate first (lowest) and the most soluable salts last (highest). The first precipitated solid is $CaCO_3$. That is because $CaCO_3$ is the least soluble component compared with other salts.

(1)Precipitation of Carbonate Salt

The first precipitated solid is carbonate. That is because carbonate is the least soluble component compared with other salts. At this stage, calcium carbonate $CaCO_3$ precipitate first, followed by magnesium carbonate, sodium carbonate (Na_2CO_3)(碳酸钠、苏打、纯碱) and Sodium bicarbonate ($NaHCO_3$), or sodium hydrogen carbonate (碳酸氢钠、重碳酸钠、小苏打). Potassium carbonate is the last one to precipitate. The saline lake with above depositions is also called alkaline lake.

(2)Precipitation of Sulphate Salt

The next evaporate deposit is sulphate. With the increasing of salinity, the sulphate with relatively high solubility will precipitate. The result products include gypsum ($CaSO_4 \cdot 2H_2O$)(石膏), Glauber's salt ($Na_2SO_4 \cdot 10H_2O$,芒硝)and sodium sulfate (Na_2SO_4无水芒硝). This kind of lake is often called bitter lake.

(3)Precipitation of Chloride Salt

After the precipitation of sulphate salts, the bitter lake is finally changed into brine lake (盐湖). Brine is the water saturated or nearly saturated with salt. It has a salinity of 24‰ or 25‰. At this stage, the most soluble salts, that is, the chloride salts, will precipitate. Common chloride minerals include halite (sodium chloride, NaCl, 岩盐), sylvite (potassium chloride KCl, 钾盐,钾石盐), and carnallite (potassium magnesium chloride hexahydrate, $KMgCl \cdot 6H_2O$, 光卤石). As chloride salts are highly soluble minerals, they are commonly redissolved by the influx of new salt-water. To be preserved, they must be covered over quickly by an impervious layer.

In our country, the famous brine lakes include Chaer Lake(茶卡盐湖), Keke Lake (可可盐湖) and Cherhan Lake (察尔汗盐湖) in Qinghai Province.

Evaporites are characterized in a vertical sequence by orders of rock types, such as dolomite-anhydrite-halite. Sometime salts are arranged in a"bull's eye"pattern, such that the most soluable salts occur in the center of the evaporating basin. For example, some of the small lakes on the verge of Qaidam basin (柴达木盆地) show a concentric distribution of different salts from fringe to center. Along the outer fringe are carbonate rocks. Within this outer fringe are gypsum and/or anhydrite, followed by halite, and finally by the more soluable salts in the center of the dessicated area — salts such as carnallite and sylvite.

4 Wetland

According to the Convention on Wetlands (1871, 国际湿地公约), wetland encompasses"areas of marshes, fen, peatland(泥炭地、泥炭沼泽) or water, whether natural or artificial, permanent or temporary, with water that is flowing or static, fresh, brackish or salty, including areas of marine water the depth of which at low tide does not exceed six metres". This definition is really in a broad and general way. But it states clearly that wetlands are those areas that are inundated or saturated by surface or groundwater at a frequency and duration sufficient to support, and that under normal circumstances do support, a prevalence of vegetation typically adapted for life in saturated soil conditions. Wetlands generally include swamps, marshes, bogs, rice field and similar areas. Wetlands are not only essential for ecological functions such as maintaining biodiversity(生物多样性), controlling floods, and removal of pollutants, but serve necessary economic functions in rice and fish production, transport, hydropower energy. There are various classifications about wetland. And the type of classification system chosen depends on the particular scientific, management, or regulatory application of interest.

It is important to note that many Chinese scientists include lakes and rivers among natural wetland areas, a practice not followed in the U. S and other countries. For the sake of simplicity and convenience, we classify the wetland into two general types: natural wetland and artificial wetland (rice field,人工湿地).

Mire is a main kind of natural wetland in which the current vegetation and surface waters are isolated from underlying mineral sediments by a sufficiently thick layer of organic soils and whose current natural vegetation and surface conditions promote the accumulation of a kind of organic sediment called peat. Mire usually is a soft wet area of low-lying land. In regions with abundant precipitation, little evaporation, and slight runoff, various kinds of swamps, marshes, fens, and bogs may form. Each of the all the above-mentioned terms has definite meaning in English, but unfortunately there are no broad-accepted corresponding Chinese terms. In Chinese, all the swamps, marshes, fens and bogs are usually collectively termed as"沼泽". Nowadays it seems that all these English words are united under the general term mire. So, in this textbook, the term"mire"is largely corresponding to"沼泽".

Mire has a number of different types. But up to now, there is no universally accepted classification including all the varieties scientifically. So, here just some of the most common mires with well-defined features are described.

• **Swamp**

Swamps (tentative translation: 木本沼泽) are stagnant or slow-flowing pools dominated by trees. Abundant water results in a high nutrient content. Swamps can support trees, tall shrubs, herbs and mosses. Peat accumulation is limited in swamps due to the high decomposition rates.

• **Marsh**

Marshes(tentative translation: 草本沼泽) are open, non-peat forming wetlands usually found near a river, lake or tidal waters. They are dominated by grasses and other herbaceous plants or shrubs, and are periodically or permanently flooded. Relatively high amounts of water result in a high nutrient content. But a marsh can completely dry up in draught. High decomposition rates limit the peat accumulation. Both trees and mosses are generally lacking or not abundant. The Three Rivers plain (三江平原) is mainly covered by marsh.

• **Fen**

Fen (tentative translation: 潜水泥炭沼泽) is a poorly drained area where the groundwater intercepts the soil surface. It is usually covered by mats of moss, which slowly decompose in successive layers and eventually form a material called peat.

• **Bog**

Bog (tentative translation: 季节泥炭沼泽) is a poorly drained area where precipitation saturates the soil for a season or longer. It is dominated by mats of moss and characterized by a high content of acidic materials. Peat is an organic material that forms in the waterlogged, sterile, acidic conditions of mires. These conditions favour the growth of mosses, especially sphagnum. As plants die, they do not decompose. Instead, the organic matter is laid down, and slowly accumulates as peat because of the lack of oxygen in the mire.

1. Answer the Following Questions

(1) Define the endorheic and exorheic lakes.

(2) Define the term"wetland".

(3) How is the lake recharged?

(4) What is tectonic lake?

(5) List and describe briefly the main types of mire.

(6) How does the salt form the lake?

(7) Describe the mechanical sorting of lake-deposited sediments.

(8) List and describe the main lake types.

(9) What is the mire?

(10) Describe briefly the evolution of mire.

(11) What is swamp? What is the difference between swamp and marsh?

(12) What is peat? How is it formed?

2. Multiple Choice

(1) The largest lake in the world is ______.

A: Caspian Sea

B: Baikal

C: Qinghai Lake(青海湖)

D: The Great Lake

(2) The Crescent Moon Lake in Gansu Province is ______.

A: Crater lake　B: Glacial lake

C: Karst lake　D: Aeolian lake

(3) The Caohai lake (草海) is ______.

A: Crater lake　B: Glacial lake

C: Karst lake　D: Aeolian lake

(4) The most soluble salt is ______.

A: Potassium chloride

B: Iron hydroxide

C: Sodium sulfate

D: Sodium carbonate

(5) Qinghai Lake(青海湖) is an endorheic lake.

A: True　B: False

(6) Which statement about wetland is not right?

A: Three elements are common to all wetlands: wetland hydrology, wetland soils and wetland vegetation.

B: River and lake are not to be considered as a wetland in other west countries.

C: Wetland is always saturated with water.

D: Wetlands are essential for ecological functions such as maintaining biodiversity.

(7) Which one has a high content of peat?

A: Marsh　B: Swamp

C: Bog　D: None of them

(8) The least soluble salts precipitate last.

A: True　B: False

(9) Endorheic water-bodies also tend to be more sensitive to pollutant inputs.

A: True　B: False

(10) Rice paddy is not a wetland.

A: True　B: False

(11) Which one is a saline lake?

A: 太湖　B: 青海湖

C: 鄱阳湖　D: 洞庭湖

(12) Which statement about wetland is not right ?

A: Swamps are forested, wooded or shrubby non-peaty wetlands.

B: Bogs are peatlands that receive their surface water only from precipitation.

C: Marshes and wet meadows are dominated by grasses and peat.

D: Wetlands provide habitat for animals and plants.

(13) Which mineral will precipitate first in a lake?

A: Potassium chloride

B: Sulphate

C: Halite

D: Carbonate salt

(14) Which material will be depostied at the center of a lake?

A: Sand　B: Pebble

C: Mud　D: Cobble

(15) Which one is crater lake?

A: 新疆天池　B: 长白山天池

C: 滇池　D: 杭州西湖

(16) Carbonate rocks are frequently deposited ______.

A: Near the bank

B: At the center

C: Uniformly on the lake bottom

D: All of them

(17) Brine is the water with a salinity of ______.

A: 35‰ B: 24‰

C: 35% D: 24%

(18) Fluvial lakes are commonly found on the floodplain.

A: True B: False

(19) Which type of lake is featured by a central island?

A: Crater lake B: Glacial lake

C: Karst lake D: Impact crater lake

(20) Which statement about lake deposition is not right?

A: Excessive evaporation of water in arid area will eventually turn a freshwater lake into saline lake.

B: Evaporates are characterized in a vertical sequence by orders of rock types.

C: From the fringe to center, the deposited rocks in the saline lake are carbonate, gypsum and/or anhydrite and followed by halite.

D: Halite is always the last precipitated mineral due to its high solubility.

Chapter 18 The Solar System

Key words

Solar system	Planetary geology	Planet	Terrestrial planets	Jovian planets	Asteroid belt
太阳系	行星地质学	行星	类地行星	类木行星	小行星带
Comet	Meteorite	Mercury	Venus	Mars	Maria
彗星	流星	水星	金星	火星	月海
Lunar highland	Jupiter	Saturn	Uranus	Neptune	
月陆,月球高地	木星	土星	天王星	海王星	

Planetary geology is the study of surface and interior processes on solid objects in the solar system. Our solar system consists of an average star(恒星) we call the Sun, nine planets, their satellites (moon) and numerous small bodies, including comets, asteroids, and meteoroids. The exploration of other objects in the solar system can help us to understand the evolution history of Earth. The dawn of the space age saw the birth of a new way to study the various bodies of the solar system. Over the last 40 years, various spacecrafts have investigated all the planets except Pluto, various moons, asteroids, and comets.

1 Introduction

Why is it important to study the geology and surface processes of planets? The answer is that the exploration of the solar system is an extension of humanity's exploration of the Earth.

1.1 Sun

The Sun is a star located at the center of our solar system. An estimated 99.85 percent of the mass of our solar system is contained within the sun, while the planets collectively make up most of the remaining 0.15 percent.

The Sun is a huge ball of gases, at present, about 70% hydrogen and 28% helium by mass. This leaves about 2% that is stuff like carbon, nitrogen, oxygen, neon, magnesium, silicon, sulfur, iron and over 50 other things. This changes slowly over time as the Sun converts hydrogen to helium in its core. The element helium was named after the Sun (called "Helios" in Greek) because it was first discovered on the Sun. Helium is plentiful on the Sun but rare on the Earth.

The Sun is a star. Stars are giant fireballs of gas. Stars are alive. They are born, they live and they die. It is thought that the Sun is about 5 billion years old. In the 5 billion years since the Sun formed, it has used up about half of its initial hydrogen supply. The Sun will start to die in about 4～5 billion years.

The Sun is the powerhouse for the entire

Solar System. The basic energy source for the Sun is nuclear fusion reaction, which uses the high temperatures and densities within the core to fuse hydrogen, creating energy and producing helium as a by-product. It is this energy that gives us heat and light. Without the Sun's light and heat, there would be no life on the Earth.

The Sun's magnetic field is very strong. Magnetic field of the Sun reverses once for each 11-year sunspot cycle. Sometimes, we can see little dark spots on the Sun called sunspots(太阳黑子). They occur when strong magnetic fields emerge through the solar surface and allow the area to cool slightly, from a background value of 6 000℃ down to about 4 200℃.

1.2 Planet

Planet is an object orbiting a star that is bigger than an asteroid. All the planets orbit the Sun in elliptical paths. In our solar system it is really a historical definition which simply means"a planet is a member of the set (Mercury, Venus, Earth, Mars, Jupiter, Saturn, Uranus, Neptune)".

All of the planets in the solar system are named after Roman gods, except for the Earth which was not seen as a planet by the ancients, and Uranus, which is named for a Greek god.

According to compositon, planets with solar system can be divided into two groups:

- **Terrestrial or Rocky Planets**

These planets are very silimar to the Earth, with bodies largely composed of rock. They include Mercury, Venus, Earth and Mars.

- **Jovian Planets**

These planets are very silimar to Jupiter, with bodies largely composed of gas or ice. They include Jupiter, Saturen, Uranus Nepturne and Uranian.

The planets, in order from the Sun, are Mercury, Venus, Earth, Mars, Jupiter, Saturn, Uranus and Neptune (Figure 18.1).

1.3 Asteroid

Asteroid, meaning like stars, is a rocky space object which can be from a few hundred feet to several hundred km wide (Figure 18.2). Most asteroids in our solar system orbit the Sun in a belt between Mars and Jupiter, called the asteroid belt (Figure 18.3).

Table 18.1 Solar system physical data modified after Xia Bangdong(夏邦栋) et al, 1995

	Mercury	Venus	Earth	Mars	Jupiter	Saturn	Uranus	Neptune	Sun	Moon
Mass	0.055	0.815	1	0.108	317.8	95.2	14.4	17.2	332 830	0.012
Density ($g \cdot cm^{-3}$)	5.44	5.2	5.52	3.93	1.3	0.69	1.28	1.64	1.434	3.36
Rotational period	59d	243d	23.9h	24.6h	9.9h	10.4h	10.8h	16h	27d	27.3d
Diameter (km)	4 880	12 104	12 756	6 787	142 800	120 000	51 800	49 500	1 390 000	3 460
Distance($\times 10^6$ km)	58	108	150	228	778	1 427	2 870	4 497		
Orbital period (a)	0.24	0.62	1	1.88	11.86	29.5	84	164.9		
Moon	0	0	1	2	16	18	15	8		0

Mass: Earth equals one.

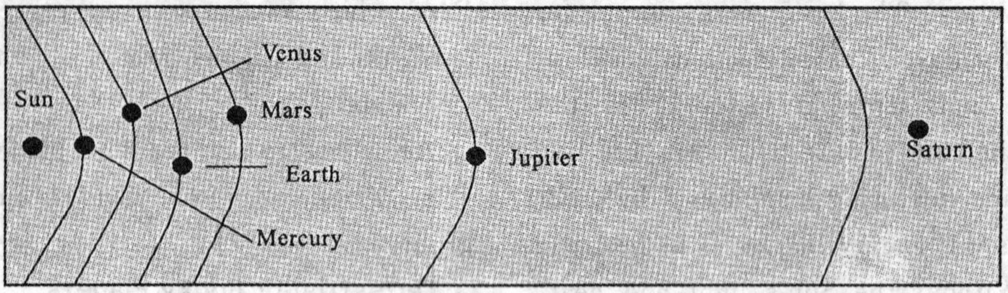

Figure 18.1 The distances of planets from the Sun

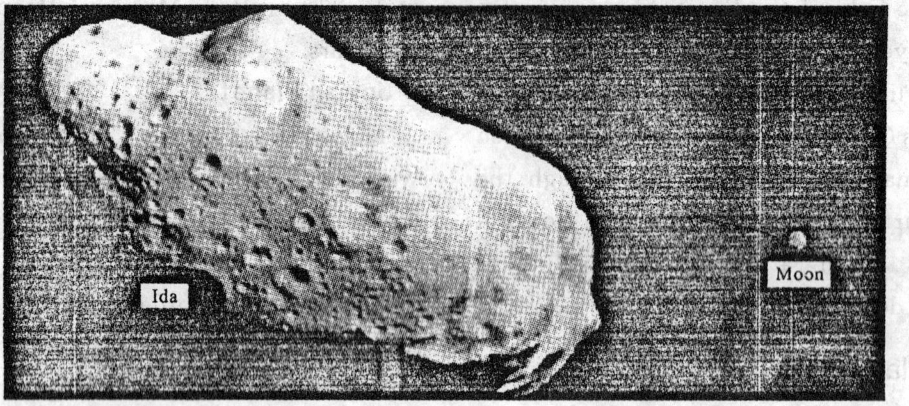

Figure 18.2 The asteroid Ida and its moon, Dactyl.
Image credit: NASA and the NSSDC. Public domain

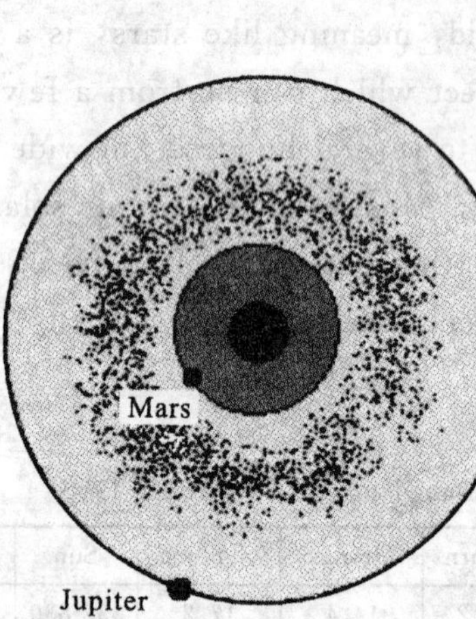

Figure 18.3 Asteroid belt

1.4 Meteoroid

A meteoroid(陨星群,宇宙尘)is matter revolving around the Sun or any object in interplanetary space that is too small to be called an asteroid or a comet. When entering a plant's atmosphere, the meteoroid heats up and partially or completely vaporizes. The trail of glowing vapor is called a meteor, or shooting star(流星). The "shooting stars" that are commonly seen in the night sky are mostly dust-sized objects striking the Earth's atmosphere. If any portion of the meteor survives to reach the ground, it is then referred to as a meteorite (陨石)(Figure 18.4). Meteorites have proven difficult to classify, but the three broadest groupings are stony, stony iron, and iron. The most common meteorites are stony meteorites(石陨石). The majority of meteorites are believed to be fragments of asteroids.

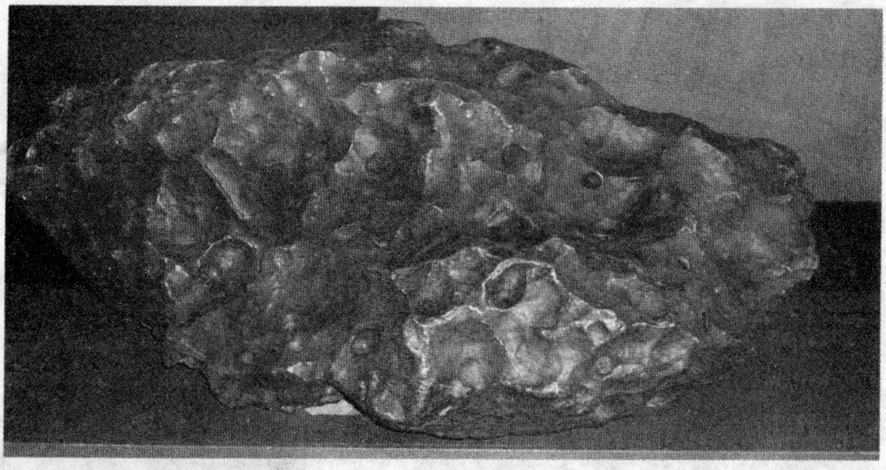

Figure 18.4 Meteorite. Photo by Dr. Yu Shan (俞姗)

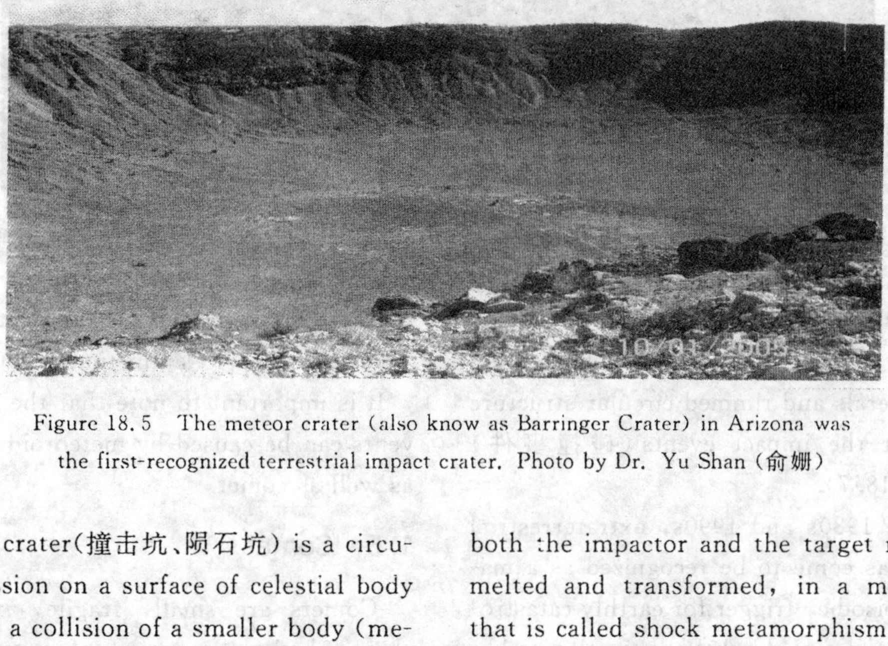

Figure 18.5 The meteor crater (also know as Barringer Crater) in Arizona was the first-recognized terrestrial impact crater. Photo by Dr. Yu Shan (俞姗)

Impact crater(撞击坑、陨石坑) is a circular depression on a surface of celestial body caused by a collision of a smaller body (meteorite) with the surface (Figure 18.5). It is often featured by a central peak and rimmed by fragmental materials ejected from crater. The central peak is formed by rebounding bedrock after the impact (Figure 18.6). Sometimes a crater lake fill the crater. The depth-to-diameter ratio of impact crater may range from 1 : 5 to 1 : 20.

Due to the very high temperature and pressure reached during the impact event, both the impactor and the target rocks are melted and transformed, in a mechanism that is called shock metamorphism (冲击变质). The extremely high pressures have been shown to be uniquely and unambiguously associated with meteorite impact craters. Some high-pressure mineral such as stishovite (斯石英) can be produced during impacting and stishovite in turn can be used as an indicator of impact event.

Acturally, due to the erosion, most of impact craters have been destroyed or buried beneath the ground. Scientists use the

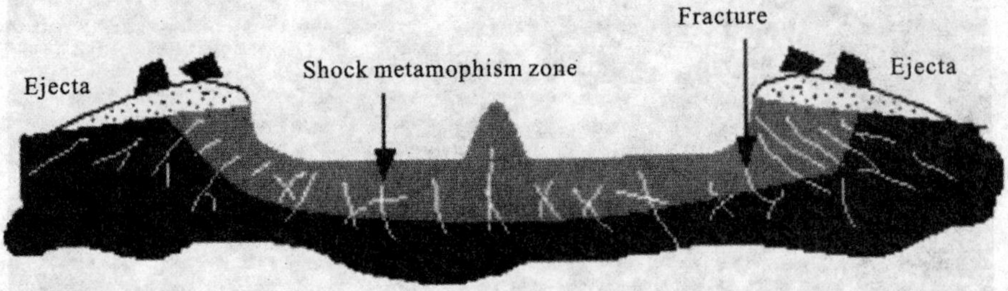

Figure 18. 6 Structure of an impact crater

Figure 18. 7 The meteor crater in Arizona is rimmed by ejectas. Photo by Dr. Yu Shan (俞姗)

various shock metamorphic effects, such as magnetic abnormal (磁异常), high pressure minerals and rimmed circular structure to detect the impact events (碰撞事件) (Figure 18. 7).

In the 1980s and 1990s, extraterrestrial impact has come to be recognized as a major, if episodic, trigger for earthly catastrophe and a force behind the extinction and evolution of organisms and ecosystems. It is believed that the five mass extinctions (生物大灭绝) in the past 600 million years were the result of impact events. The largest mass extinctin at the end of the Permian periiod 250 million years ago killed off 90% of all species. This is the Permian-Triassic (PT) extinction event. Probably the most famous is the extinction that finally saw the end of the dinosaurs reign on the Earth, 65 million years ago. This is the Cretaceous — Tertiary (K — T) extinction event.

It is important to note that the impact events can be caused by meteoroid, asteroid as well as comet.

1.5 Comet

Comets are small, fragile, irregularly shaped bodies composed of a mixture of non-volatile grains and frozen gases.

2 Terrestrial Planets

The inner planets, Mercury, Venus, Earth and Mars are called terrestrial planets. This means that they have a hard surface (rock) to stand on. All of the terrestrial planets have rocky surfaces. Most of the rocks on the surfaces of terrestrial planets came from volcanic eruptions pouring out molten lava. All of the terrestrial planets have been struck by large asteroids, leaving behind craters. The asteroids that caused these craters probably came mostly from the asteroid belt.

2.1 Mercury

Mercury is the closest planet to the Sun and the eighth largest. Mercury was named after the Roman god of travel and commerce, who had wings on his heels to carry him swiftly from the Earth to the heavens. It was named this because it appears to move very quickly through the night sky. Its orbit is said to be "inferior", as it lies between the Sun and the Earth. Mercury always appears in the sky near to the Sun, being visible a little before sunrise or a little after sunset as the morning or evening star.

The diameter of Mercury is about 4 880 km. It is the smallest planet in our Solar System. It is slightly larger than the Earth's Moon. If the Earth was the size of a baseball, Mercury would be the size of a ping pang ball. The rotational period is 59 days. It takes 88 days to complete its orbit around the Sun. It is the fastest moving planet in our solar system with an average orbital velocity of 47.87 km/s. Mercury is the densest planet in our Solar system apart from the Earth, with a mean density of 5.4 g/cm^3. There are no seasons on Mercury. Seasons are caused by the tilt of the axis relative to the planet's orbit. Since Mercury's axis is directly perpendicular to its motion (not tilted), it has no seasons. There is no air on the Mercury because Mercury's weak gravity and high surface temperature (700 K) cannot retain a permanent atmosphere.

The photos taken by spacecraft "Mariner 10" in 1974 and 1975 showed that Mercury is a heavily cratered planet (Figure 18.8). The landscape of Mercury consists of numerous large craters and rolling plains between highlands. Cratering on Mercury triggered volcanic eruptions that filled much of the surrounding area. From the number density of these craters, it seems that for about 4 billion years there was no volcanism on Mercury. Mercury is unique in that it is

Figure 18.8 Impacted craters on the surface of Mercury.
Image credit: NASA and the NSSDC. Public domain

crisscrossed by enormous thrust faults that formed as the planet contracted while it cooled.

Mercury does have a weak magnetic field that is about 1% as strong as Earth's. The weak magnetic field suggests that part of Mercury's core may be molten.

2.2 Venus

Venus, the second planet from the Sun, is the hottest world in the solar system. Venus (Greek: Aphrodite; Babylonian: Ishtar) is the goddess of love and beauty. The planet is so named probably because it is the brightest of the planets known to the ancients. Venus is also known as the "morning star" or the "evening star" since it is visible and quite bright at either dawn or dusk. It is only visible at dawn or dusk since it is closer to the Sun than we are.

Venus' mass and volume are quite similar to that of the Earth. Its mass is about 81.5% of the Earth's and volume 86% of our planet's. Venus is about 12 104km in diameter. This is about 95% of the diameter of the Earth's. Traditionally, Venus is so similar to the Earth that it is sometimes called Earth's sister planet. Venus, however, has no satellites, nor does it have a magnetic field. Venus sluggishly rotates on its axis once every 243 Earth days, while it orbits the Sun every 225 days.

The surface of Venus is covered by a dense and rapidly spinning clouds. The atmospheric pressure at the surface of Venus is 90 times that of Earth's. The atmosphere consists mainly of carbon dioxide (about 98%), droplets of sulfuric acid solid and sulphur particles. Below Venus' clouds, the atmosphere contains about 0.1% to 0.4% water vapor and small quantity of free oxygen. The presence of water and free oxygen indicates that Venus may have had abundant water early in its history, which was later lost.

Venus' atmosphere is so thick and massive that it produces a very pronounced greenhouse effect. Atmosphere allows the Sun's heat in but does not allow it to escape, resulting in extremely high surface temperatures over 450℃, hotter than the surface of the planet Mercury, which is closest to the Sun.

Because Venus is shrouded in clouds, our best views of the surface have been obtained by radar. Much of our knowledge about Venus' formation and past comes from radar mapping by many spacecrafts, such as Magellan orbiter and Venera 7～9 probes. Radar map show that the surface of Venus can be summarized as three basic types: lowland plains (20%), rolling uplands (70%), and highlands (10%). But Venus has no ocean. The highland regions have mountain ranges and rift systems. Mountain ranges, long faults, and deep troughs indicate that horizontal surface movement has occurred. All three areas show evidence of extensive volcanism. Venus also displays unique forms of volcanism. Much of the volcanic activity on Venus occurs as basaltic eruptions that flood large areas. In addition, Venus' volcanoes are like what we call "shield" volcanoes on the Earth, some of which are thought to be currently active. Recently announced findings indicate that Venus is still volcanically active, but only in a few hot spots. Generally, it has been geologically quiet for the past few hundred million years. Venus' mantle and crust are thought to be composed of silicates and its core is thought to be nickel-iron.

Impact craters are distributed randomly but uniformly over the surface. All places on the surface of Venus seem to have approximately the same age — about 600～800 million years.

2.3 Mars

Mars, the fourth planet from the Sun in the solar system, is the closest neighbor to the Earth. It is named after Mars, the god of war in Roman mythology. Mars (Figure 18.9) is also referred to as"the Red Planet" because of its red color as viewed in the night sky. Its red appearance is caused by iron oxide (rust) on its surface. The soil of Mars was sampled by *Viking* and found to consist of clays and iron oxides. From the viewpoint of scientific investigation, Mars is by far the most interesting object in the solar system. Many important questions about Mars evolution are still unanswered. For example, what is the history of water on Mars? How has the martian climate changed over time? Were past conditions on Mars conducive to the development of life? Mars is also interesting because of its potential to be future home for humanity.

Mars year is 687 days long and it rotates on its axis once every 1.026 Earth days. Because its axis is tilted at about 25 degrees, Mars experiences seasonal changes similar to those on the Earth. The Martian atmosphere consists of carbon dioxide (95.3 %), nitrogen (2.7 %), and argon (1.6 %). No measurable magnetic field has been detected, which indicates that the core is solid. Like the Earth, Mar has a large iron core, rocky silicate mantle and a largely basaltic crust. Its surface is marked by ancient volcanoes and a"grand canyon"that's as long as the China wide. Mars has polar ice caps that contain frozen water and carbon dioxide.

Based on the present data, at least six different geological processes either currently operate on Mars or have operated during Martian history. These include the aeolian, cratering, hydro, landslides, tectonic, and volcanic processes. Here we only introduce briefly four of them:

- **Cratering Process**

Cratering was an important process shaping the Martian surface. Martian crater morphology is extremely similar to that of the Moon and Mercury . Mars is unique in

Figure 18.9 Martian surface. A feature called"Lemon Rind", a thin surface layer covering portions of outcrop rocks. Image credit: NASA/JPL/Cornell. Public domain

that it has craters with ejecta deposits that have a unique"flowing"appearance. It is believed that the presence of groundwater or ice is responsible for these features.

Craters on Mars may be weathered by water and the atmosphere. They also appear to experience more erosion or other degradation than those on the Moon and Mercury. This is believed to be the combined result of a dynamic atmosphere, more active volcanism throughout the history of the planet, and more weathering due to processes associated with the possible existence of water.

• **Volcanic Process**

One of the most prominent structures on Mars is shield volcano. The largest volcano, Olympus Mons, may also be the largest volcano in the solar system. It towers 27 km above the surrounding desert. It rises about 25 km from the surrounding plain. Such a vertical relief is greater than the distance from the top of the Himalayas to the bottom of the Mariana Trench, which is only about 22 km. In addition to giant volcanoes, smaller volcanic shields are scattered all over the planet.

• **Hydro Process**

The surface of Mars shows features that resemble dry riverbeds and gullies. These might have been created by rainfall and runoff, but they might also have been made by subsurface water that seeped to the surface. In either case, the channels are evidence that temperatures were warmer in the past — warm enough for water to exist in liquid form.

It is possible that Mars once had an atmosphere somewhat like the Earth's, and almost certain that water once covered part of the Martian surface, in rivers, lakes, and perhaps small seas. At present, however, the thermodynamic (i. e., the temperature and pressure) conditions on Mars do not allow liquid water to exist. Although no water flows over the Martian surface today, frost covers the north and south poles. These polar ice caps are made mostly of frozen carbon dioxide (dry ice), although the north polar cap contains much more water ice than the southern cap.

2.4 Moon

The Moon (Figure 18.10) is the only place off-Earth that humans have visited, and it is the most likely first destination for an off-world colony. On July 20, 1969, Neil Armstrong became the first man to step onto the surface of the Moon. So far, 12 astronauts have walked upon its surface and brought back 382 kg of lunar rock and soil to the Earth.

The diameter of Moon is about 3 476km. The Moon is 384 403km distant from the Earth. Both the rotation of the Moon and its revolution around the Earth takes 27d,

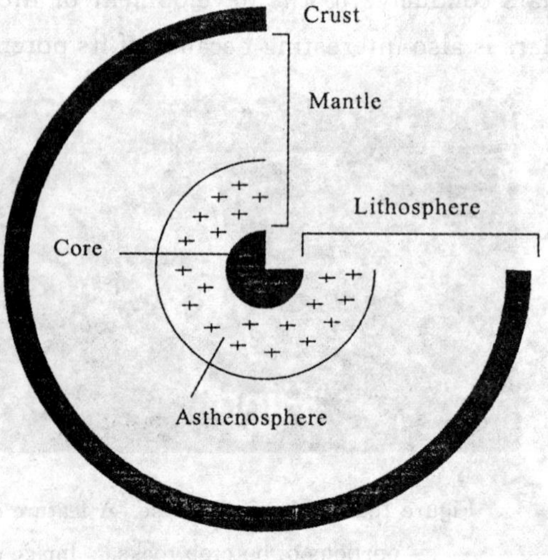

Figure 18.10 The structure of Moon

7h, and 43min. In fact, it rotates at just the perfect speed so that it keeps the same side pointed towards the Earth at all times! The Moon's gravity is one-sixth that of the Earth's. Tides (high tide and low tide) on the Earth are partially caused by the Moon's gravity pulling on the Earth. The Moon is less dense than Earth (3. 34 g/cm^3 versus 5. 52 g/cm^3). The Moon has no magnetic field (It's not a magnet).

• Surface Feature of Moon

By simply viewing with the naked eye, one can discern two major types of terrain: relatively bright highlands and darker plains. Dark regions on the Moon are called maria, which is Latin for "seas" (Figure 18. 11). These regions are smooth, low-lying plains with relatively few craters. Maria gets their color from a type of rock (called basalts) formed by lava from volcanoes. Maria occupies a smaller area, dominantly on the Earth facing side of the moon. About 3. 8 to 3. 1 billion years ago several huge, asteroid-like objects struck the Moon. This was shortly followed by lava flows which filled in the basins and formed the dark maria. Light-colored regions are called lunar highlands, consisting of a type of ultramafic igneous rock called anorthosite (斜长岩). They are hilly regions with many craters. The old, heavily cratered highlands formed during a heavy bombardment period early in the Moon's history, when huge meteoroids and asteroids were striking the surface.

• The Lunar Interior

The current model of the interior of the Moon derived from seismometers left behind during the manned Apollo program missions. These instruments were used to measure the propagation of seismic waves through the interior due to quakes and impact. From the data collected by those instruments, we can find that the moonquakes are very mild. It may indicate that there are no plate tectonics on the Moon.

The results have shown that the lunar crust thickness varies from tens of kilometers in depth (under mare basins) to more than 100km in some highland regions, with an average thickness of about 70km. The cr-

Figure 18. 11 Apollo 17 astronaut Harrison Schmitt standing next to boulder

Image credit: NASA and the NSSDC. Public domain

ust on the lunar far side is thicker than that on the side facing the Earth.

The Moon may have a small iron core perhaps 300 km in radius. Near the center, the temperature may be as low as 1 500K, too cool to melt rock. The absence of a magnetic field on the Moon also indicates that it does not have a metallic core.

The core is surrounded by a roughly 400-km-thick inner mantle of semisolid rock having properties similar to the Earth's asthenosphere. Above these regions lies an outer mantle of solid rock, some 900～950 km thick, topped by a 60 ～ 150km crust (considerably thicker than that of Earth). Together these layers constitute the Moon's lithosphere.

• The Lunar Rocks, Breccias and Regolith

The main rock type found on the Moon is igneous rock. They can be classified into 3 types. The first one is anorthosite (Feldspar — only rock). These rocks are very old (4. 0～4. 4 Ga). The existence of anorthosites on the Moon implies chemical differentiation(化学分异) of the crust, in the course of which heavier elements such as iron were separated from lighter ingredients. The second type is the highland basalt enriched with potassium. Both of them constitute the highlands. The third one is the basalt enriched with iron and titanium. This kind of basalt is found only in lunar maria. It was generated by shock-induced heating associated with the series of gigantic meteor impact events that formed the maria basins. The ages of maria basalt varies from 3. 0 Ga to 3. 8 Ga.

Perhaps the most common rock type on the Moon today, however, is impact breccias(冲击角砾岩). They are made up of angular grains of various minerals. These rocks were formed when angular fragment of preexisting crystalline rocks were welded together by the collision, sintered together by the heat of collision, or are cemented together by infiltrating impact melt after they first solidified. Breccias are mainly found on the highlands.

The regolith, or lunar soil(月壤), is unconsolidated mineral grains, rock fragments, and combinations of these which have been welded by impact-generated glass. Nearly the entire Moon's surface is covered by regolith, with the exception of steep crater and valley walls. It is 2 to 8m thick on the maria. The lunar brown glass, spherical glass droplets formed during impact events, is a significant component of lunar regolith.

• The Origin of the Moon

How did the Moon come to be? The Moon's origin has been the subject of scientific speculation since Galileo in 1609 shows that the Moon is a rocky body like the Earth. There are many different theories of its origin.

The present leading theory is that a Mars-sized body collided with the Earth soon after it's formation. The collision was so great that the Earth was simply tilted off its axis. After the impact, a fraction of the matter drifted off into space, and finally the drifted matter began orbiting the Earth. This matter then aggregated(加积) over time into a spherical planetoid: the Moon. Scientists believe that the Moon was formed approximately 4. 5 billion years ago (the age of the oldest collected lunar rocks).

3 Jovian Planets

Jupiter, Saturn, Uranus and Neptune are collectively called the Jovian planets, since they are all a lot like Jupiter (Jove was another name for the Roman god, Jupiter) (Figure 18. 12). They are also called gas giants. This is because they are made of gas and liquid (but not water). The gas planets are composed primarily of hydrogen and helium and generally have low densities, rapid rotation, deep atmospheres, rings and lots of satellites.

3. 1 Jupiter

Jupiter is the fifth planet from the Sun. Jupiter (Greek Zeus) was the king of the Gods, the ruler of Olympus and the patron of the Roman state. It has been known since prehistoric times as a bright "wandering star". This gas giant is the largest planet in the solar system. In fact, over 1 300 Earths could fit inside Jupiter! Its diameter is about 142 984km. For all its size, Jupiter is barely denser than water, at 1. 33 g/cm^3. Jupiter has really faint rings. One of the most prominent features of Jupiter is its "The Great Red Spot" (Figure 18. 13). The Great Red Spot acturally is a vast thunderstorm that has raged on the planet's surface for over 300 years. The spot is over twice the size of the Earth.

Jupiter is mostly made of helium (10%) and hydrogen (90%) with traces of methane, water, ammonia and "rock". This is very close to the composition of a small star. In fact, if Jupiter had been between

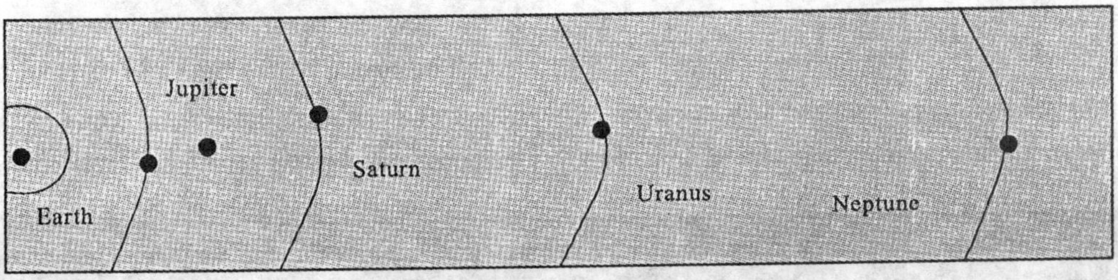

Figure 18. 12 Jovian planets

Figure 18. 13 Jupiter's Great Red Spot
Image credit: NASA and the NSSDC. Public domain

fifty and one hundred times more massive, it would have become a star rather than a planet. Jupiter probably has a rocky core. The core is surrounded by the main bulk of the planet in the form of liquid metallic hydrogen. Above the liquid hydrogen layer lies the outermost layer which is composed primarily of ordinary molecular hydrogen and helium.

Jupiter is orbited by 61 moons, only four of which have significant size: Ganymede, Callisto, Io, and Europa. The big four satellites were discovered by Galileo in 1610. Collectively, these four moons are known today as the Galilean satellites.

Io was a maiden who was loved by Zeus (Jupiter). Io is the most volcanically active body in the solar system. Volcanoes spew out vast amounts of sulfurous material which make the Io's landscape yellowish or orange-red. In contrast to most of the moons in the outer solar system, Io and Europa may be somewhat similar in bulk composition to the terrestrial planets, primarily composed of molten silicate rock. Callisto has the oldest, most cratered surface of any body yet observed in the solar system. Ganymede is the largest satellite in the solar system.

3.2 Saturn, Uranus, and Neptune

In Roman mythology, Saturn is the god of agriculture. The associated Greek god, Cronus, was the son of Uranus and Gaia and the father of Zeus (Jupiter). Like Jupiter, Saturn is composed of hydrogen and helium with traces of water, methane, ammonia and "rock". Saturn's interior structure is also similar to that of Jupiter.

Saturn's ring system makes the planet one of the most beautiful objects in the solar system. The rings are split into a number of different parts (Figure 18.14). The ring may be composed of ice and/or snowballs from a few centimeters to a few meters in size. The origin of the rings of Saturn (and the other Jovian planets) is unknown.

Uranus is named after the ancient Greek

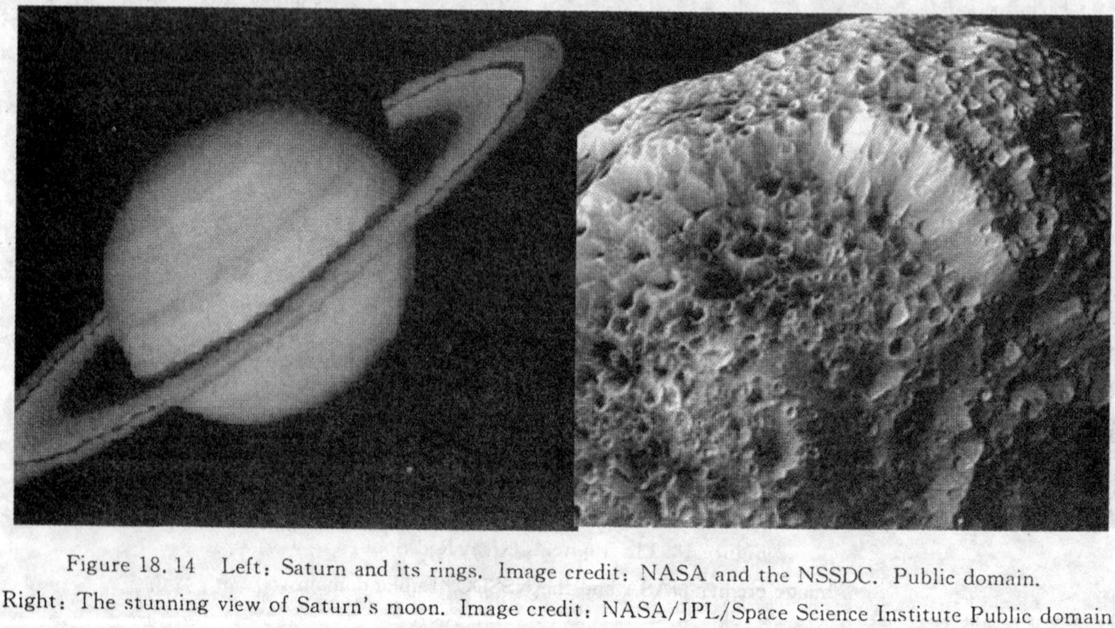

Figure 18.14 Left: Saturn and its rings. Image credit: NASA and the NSSDC. Public domain. Right: The stunning view of Saturn's moon. Image credit: NASA/JPL/Space Science Institute Public domain

deity of the Heavens, Uranus. Most of the planets spin on an axis nearly perpendicular to the plane of the ecliptic(黄道面), but Uranus' axis is almost parallel to the ecliptic. Uranus is composed primarily of rock and various ices, with only about 15% hydrogen and a little helium. The atmosphere of Uranus is composed of 83% hydrogen, 15% helium, 2% methane and other hydrocarbons.

Neptune is the outermost planet of the gas giants. In Roman mythology Neptune was the god of the Sea. Neptune's composition is probably similar to Uranus'.

4 Origin and Evolution of the Solar System

The earliest accounts of how the Sun, the Earth and the rest of the solar system were formed are to be found in early myths, legends and religious texts. Any successful theory for the origin of the solar system will need to take into account the following unusual features of the solar system:

(1) All planets revolve around Sun counterclockwise.

(2) The Sun spins slowly and only has 1% of the angular momentum(角动量) of the solar system, but 99.9% of its mass. The planets carry the rest of the angular momentum.

(3) All planets except Venus rotate on axis counterclockwise.

(4) All planets orbits lie in plane of ecliptic.

(5) Axes of planets except Uranus are perpendicular to ecliptic.

(6) The formation of the terrestrial planets with solid cores.

(7) The formation of the gaseous giant planets.

There have been many attempts to develop theories for the origin of the solar system. None of them can be described as totally satisfactory and it is possible that there will further development which may better explain the known facts. Here we only introduce the nebular hypothesis proposed by Kant(康德) and Laplace(拉普拉斯) in the 18th century and current theory.

4.1 The Nebular Hypothesis

German philosopher Immanuel Kant in 1755 and French mathematician Laplace in 1796 hypothesized the origin of the solar system as beginning with a rotating gaseous nebula out of which condensed globular bodies that became the Sun and planets — all revolving in the same direction. The main flaw of this hypothesis is that according to the law of conservation of angular momentum, the Sun would have to rotate more rapidly than it does (25 days) after collapse of the primeval dust cloud.

4.2 Contemporary Theories

Current theory is actually based on Kant and Laplaces "nebula theory". It can be summarized as follows:

• The primordial cloud of gas and dust (called a nebula, 原始星云) is disturbed and begins to collapse under its own gravity (Figure 18.15). The disturbance could be, for example, the shock wave from a nearby supernova or density waves in spiral galaxies.

• Because of the competing forces associated with gravity, gas pressure, and rotation, the contracting nebula begins to flat-

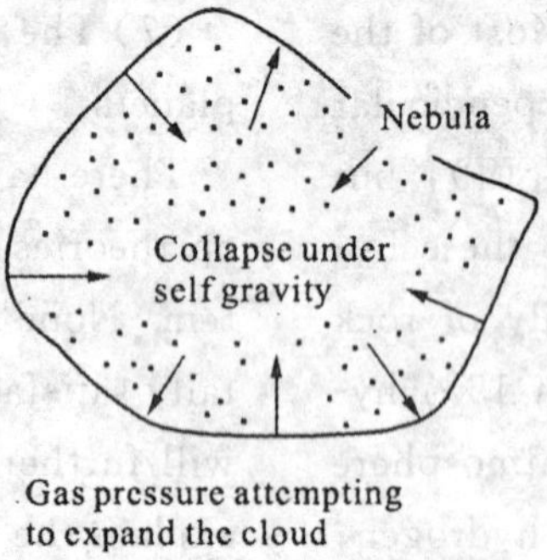

Figure 18.15 Primordial cloud

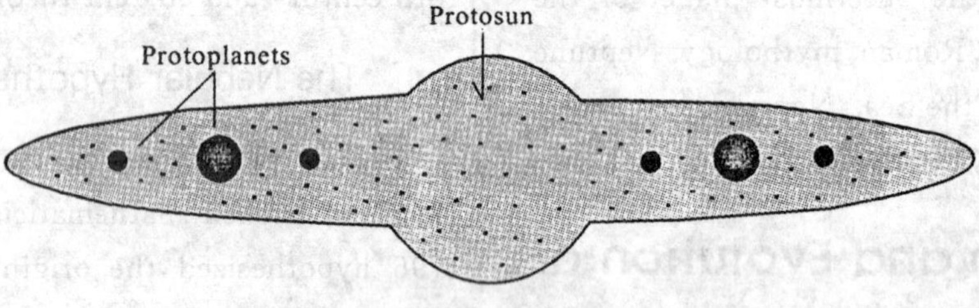

Figure 18.16 Side view of solar system

ten into a spinning pancake shape with a bulge at the center.

• As the collapse proceeds, the temperature and pressure within the cloud increase. The center compresses enough to become a protostar and the rest of the gas orbits/flows around it.

• The centrifugal force prevents some of rotating gas from reaching the forming star. The gas cools off enough for the metal, rock and (far enough from the forming star) ice to condense out into tiny particles.

• The dust particles collide with each other and form into larger particles. Finally, the particles will grow and turn into protoplanet.

• In the hot inner portion of nebula, rock and metallic elements condensed out first, resulting inner planets are "terrestrial" in composition (Figrue 18.16). The outer portion of nebula is cooler, so gaseous phases such as H, He, methane, ammonia condensed out, resulting in the formation of Jovian in composition.

1. Answer the Following Questions

(1) Differentiate between jovian planets and terrestrials planets.
(2) What is the maria of the Moon?
(3) Briefly outline the history of the Moon.
(4) Describe in a sketch of impact crater.
(5) What is shock metamorphism ?
(6) What is lunar soil?
(7) What is comet?
(8) List the planets in order of distance from the Sun.

2. Multiple Choice

(1) The ring of Saturn may be composed of ______.
A: Granite B: Basalt
C: Ice D: All of them
(2) Which one of the followings is a jovian planet?
A: Mars B: Earth
C: Venus D: Jupiter
(3) Which one of the followings is a terrestrials planet?
A: Saturn B: Uranus
C: Neptune D: Mars
(4) The asteroid zone lie between the orbits of ______.
A: Earth and Sun B: Earth and Mars
C: Mars and Jupiter D: Venus and Earth
(5) The age of Sun is ______ billion years old.
A: 2 B: 3
C: 4 D: 5
(6) The closest planet to the Sun is ______.
A: Mars B: Earth
C: Venus D: Mercury
(7) Moon may have a small molten iron core.
A: True B: False
(8) Which planet is very similar to the Earth?
A: Saturn B: Uranus
C: Neptune D: Mars
(9) The densest planet in our solar system apart from Earth is ______.
A: Mars B: Earth
C: Venus D: Mercury
(10) The Moon is less dense than Earth.
A: True B: False
(11) The extinction of dinosaurs occurred during ______.
A: PT B: KT
C: DP D: KQ
(12) Which planet is named after the ancient Greek deity of the Heavens?
A: Saturn B: Uranus
C: Neptune D: Mars
(13) Which planet is composed of gas or ice?
A: Mars B: Earth
C: Venus D: Jupiter
(14) Which planet rotate on axis counterclockwise?
A: Mars B: Earth
C: Venus D: Mercury
(15) Dark regions on the Moon are called ______.
A: Lunar highland B: Moon upland
C: Maria D: Jupiter
(16) The primordial cloud of gas and dust is ______.
A: Protostar B: Protosun
C: Protoplanet D: Nebula
(17) The Moon has no magnetic field .
A: True B: False
(18) The farthest planet from the Sun (usually) and by far the smallest.
A: Saturn B: Uranus
C: Neptune D: Mars
(19) Which planet was named after the god of the Sea?
A: Saturn B: Uranus
C: Neptune D: Mars
(20) Which planet has the most beautiful ring system?
A: Saturn B: Uranus
C: Neptune D: Mars
(21) Which planet is referred to as "the Red Planet"?
A: Mars B: Earth
C: Venus D: Mercury
(22) Jupiter is mostly made of ______.
A: Oxygen B: Helium
C: Ice D: Hydrogen
(23) The Moon's gravity is ______ that of the Earth's.
A: 1/3 B: 1/4

C: 1/5　　D: 1/6

(24) Second only to the Moon in brilliance in the night sky is ______.

A: Mars　　B: Neptune

C: Venus　　D: Mercury

(25) Geological processes either currently operating on Mars or having operated during Martian history include ______.

A: Landslides　　B: Tectonic

C: Volcanic processes　　D: All of them

(26) The gas planets are composed primarily of ______.

A: Carbon dioxide

B: Water gas

C: Melted rocks

D: Hydrogen and helium

(27) The atmosphere of Venus consists mainly of ______.

A: Carbon dioxide

B: Water gas

C: Melted rocks

D: Hydrogen and helium

(28) Most planets revolve around Sun counterclockwise.

A: True　　B: False

(29) Moon only rotates ______.

A: Once a year

B: Once in a month

C: Once in a day

D: In a period of about 30 days

(30) The main rock type found on the Moon is ______.

A: Sedimentary rock

B: Metamorphic rock

C: Igneous rock

D: All of them

Chapter 19 Resource and Sustainable Development

Key words

Renewable resources	Nonrenewable resources		Resource	Reserve	Petroleum
可再生资源	不可再生资源		资源	储量	石油
Reservoir rock	Trap	Coal	Bituminous	Lignite	Anthracite
储油层	圈闭	煤	烟煤	褐煤	无烟煤
Placer deposit	Hydrothermal ore deposit		Black smoker	Hydrothermal veins	
砂矿	热液矿床		黑烟囱	热液脉	

Natural resource can be defined as concentration of naturally occurring solid, liquid, or gaseous material in or on the Earth's crust in such forms and amount that economic extraction of a commodity from the concentration is currently or potentially feasible.

Natural resources are commonly divided into two broad categories: renewable and nonrenewable resources. Renewable resources are those that can be restored and replenished by nature in a reasonable amount of time. Renewable resources are plants, animals, surface water, soil, and sunshine. Certain basic minerals (such as salt or clay or the silica used to make glass) are also sometimes considered renewable because they are extremely abundant. Nonrenewable resources are resources that cannot be readily replaced once they are used. Nonrenewable resources cannot be replenished by nature within our lifetimes, and are therefore in limited supply. Nonrenewable resources include fossil fuels, metals, other minerals, and groundwater. Once these resources have been used-up there are no more.

Because the Earth is finite, physical stocks of nonrenewable resources must shrink over time. With roaring global demand, energy prices reaches to record levels. Some of the important resources, such as crude oil, iron ore, and bauxite will be run out of in tens of or hundreds of years.

The overall introduction of various resources is beyond the scope of this course. Here, we will introduce mainly on the geological resources and the associated social problems. The term"mineral resources"(矿产资源) are extensively used to refer to "geological resources", but it is not accurate. Actually, many geological resources, such as coal, gas and oil, are not minerals at all. Geological resources can be grouped into three categories: energy resources (oil and coal), metals and nonmetallic resources. All of them are nonrenewable.

Resource (资源) is a concentration or occurrence of geological material in or on the Earth's crust in such form and quantity and of such a grade or quality that it has reasonable prospects for economic extraction. It includes both the geological material that can be extracted economically currently and those that may be extracted economically in the future. Reserve is that part of geological resource which can be economically extracted from the Earth with current technology at current prices. In other words, reserve (储量)is discovered resource that can be extracted economically at present conditions. Because the definition of "mineral reserve" (or "ore") involves technological, legal and economic considerations, a particular mineral deposit may change its status from reserve to non-reserve, or vice versa, depending on circumstances. If a known mineral deposit cannot be mined, processed and marketed economically, it does not constitute a resource in any practical sense.

1 Oil and Gas

Oil is often referred to as "black gold" due to its great value to mankind. The products derived from oil are the "lifeblood" of modern society. It touches all of our lives. Today, over 6 000 products are produced wholly, or in part, from petroleum. The petroleum industry is so fundamentally important to modern human society that it even can lead to war. In *The Dynamic Earth*, written by Brian Skinner and Stephen Porter, a wise and insightful paragraph can be found: "Iraq invaded Kuwait in 1990 in order to control the gigantic Kuwaiti oil pools; the United Nations went to war in 1991 to expel the Iraqi invaders. How sad it is that we humans find it necessary to kill each other because of tectonic events that happened millions of years ago."

In oil industry, petroleum (from Greek petra, rock; and oleum, oil, 石油) is a broad term that includes not only crude oil — the liquid phase, but also natural gas — the gaseous phase. Crude oil (原油) is a thick, dark brown liquid consisting mainly of complex mixture of various hydrocarbons (烃,碳水化合物). Natural gas(天然气)is a mixture of hydrocarbon gases consisting principally of simplest hydrocarbon molecule methane (CH_4).

Oil and natural gas are produced by the same geological process. Therefore, they are commonly found together in the field. But that's not always the case. Actually natural gas is also generated in swamps, marshes and even landfill sites (垃圾填埋场).

Current theory holds that oil was formed from the bodies of marine plants and animals, especially plankton. Over the years, the remains of animals and plants were covered by layers of mud before decomposition, and finally the heat and pressure "cooked" them into crude oil and gas.

In order for a substantial natural-gas or oil deposit to form, three geologic conditions must be met:

1.1 Source Rock

The rock in which oil originally formed is the source rock. Oil is made primarily of phytoplankton, bacteria and plant material (algae) from the ocean. So the source rock is either black shale or limestone formed in the ocean basin.

1.2 Reservoir Rock

Once crude oil and natural gas have formed, they migrate upward because they weigh less than the water. And finally, the oil formed in the shale will migrate and concentrate to place where it is stored. A reservoir rock is a place that oil migrates to and is held underground. A porous and permeable sandstone has plenty of room inside itself to hold oil, just like a sponge has room inside of itself to soak up spills in your kitchen. Sandstones and carbonate rocks are most common reservoir rocks. It is important to remember that oil is not all by itself in some sorts of underground cave, but is, instead, contained within solid rock, which has enough room within it to actually soak up oil (Figure 19.1).

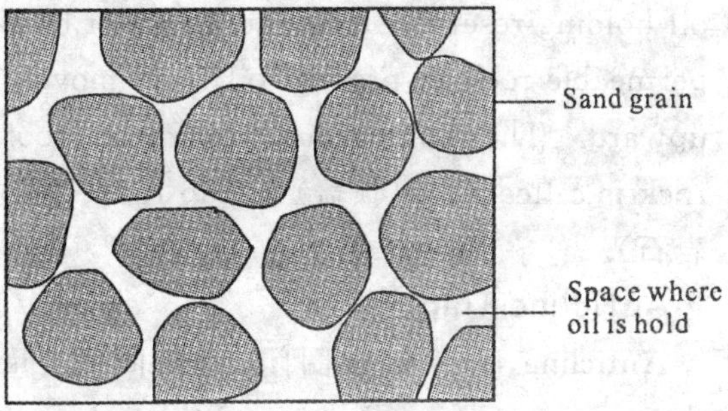

Figure 19.1 Reservoir rock

1.3 A Trap

The storing of oil in reservoir needs some conditions. Otherwise, they will rise to the Earth's surface and no oil will be preserved! The geological structure in which the oil will accumulate into large quantities is called a trap(圈闭，储油构造). There is a variety of geologic traps (Figure 19.2). All of them are composed of two basic units:

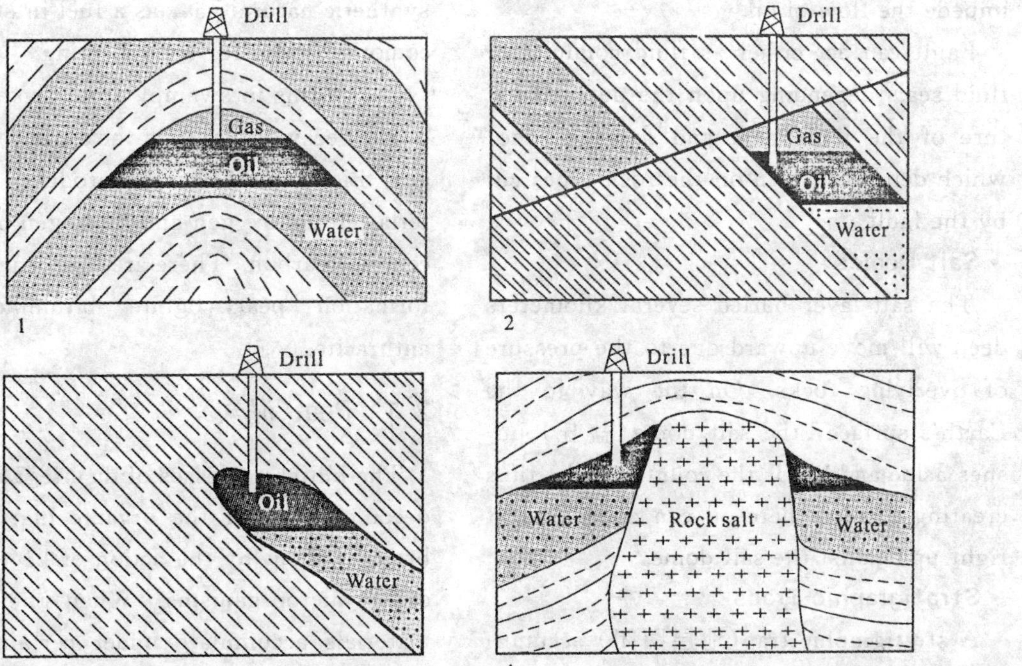

Figure 19.2 Various traps

1. Anticline trap; 2. Fault trap; 3. Stratigraphic trap ; 4. Salt dome trap

oil-holding reservoir rock and a roof of non-permeable rock to prevent oil from moving upward. The non-porous, overhead roof rock is called a cap(顶盖) or cap rock(顶板岩石).

• Anticline Trap

Anticline trap(褶皱圈闭、背斜圈闭) is the most common oil trap. When the oil moves upward in porous sandstone reservoir rock and reaches the core of the anticline, the overlying shale acts as cap to prevent the upward migration of oil. A reservoir rock that isn't completely filled with oil usually contains large amounts of water beneath and gas above.

• Fault Trap

Fault traps form when an impermeable layer above and a fault zone (generally impermeable because the friction along a fault plane creates poorly sorted, broken rock) impede the flow of oil.

Fault can act either as fluid conduits or fluid seals depending upon the detailed nature of the fault rock and damage zone, which depend in turn on the rock-types cut by the fault.

• Salt Domes

The salt layer buried several kilometers deep will move upward due to the pressure of overlying rocks. On the way to the Earth's surface, the salt dome(盐丘) pushes aside and bends the country rock, thus creating a trap where oil can come to rest right up against the salt dome.

• Stratigraphic Traps

A stratigraphic trap(地层圈闭) accumulates oil due to changes of rock character rather than faulting or folding of the rock. There are many stratigraphic traps. For example, the sandstone lens occurring in between the shales can make a perfect stratigraphic trap.

An oil field (油田) is an area with an abundance of oil wells extracting petroleum (oil) from below ground. There are more than 40 000 oil fields around the globe, on land and offshore. The largest are the Ghawar Field in Saudi Arabia and the Burgan Field in Kuwait, with more than 60 billion barrels estimated in each. The world's top six crude oil-producing countries are: Saudi Arabia, Russia, United States, Iran, China and Mexico.

2 Coal

Coal is an abundant source of energy. Coal supplies more than 70% of the electricity consumed by China. Coal is also used in steel blast furnaces, the manufacture of synthetic natural gas, as a fuel in steam locomotives, and for some heating.

Coal forms in swampy areas as the result of the decay of plants in the absence of oxygen. Biochemical changes produced by bacteria release oxygen and hydrogen and concentrate carbon. There are four stages in its formation: peat, lignite, bituminous and anthracite.

2.1 Peat(泥炭)

The initial processes of disintegration and decomposition of the organic matter were brought about by the action of bacteria and other microorganisms. When the organic materials accumulate under certain conditions, such as swamps, part of biomass will undergo decomposition due to the action of bacteria. As soon as the dissolved oxygen in the sediment is used up, decay stops and remaining organic matter is preserved as peat.

Peat is not coal but represent the initial stage of coal development. Peat is generally considered to be partly decomposed biomass (organic materials). This forerunner of coal is a cheap source of fuel but does not contain enough carbon compared to the other elements to be an efficient energy producer.

2.2 Lignite(褐煤)

Over time these areas with peat accumulation experienced cyclical periods of subsidence and reemergence. During the periods of subsidence, the relatively high temperature and pressure will turn the peat into lignite. During this process, moisture and other materials are squeezed out, leaving a high carbon concentration. Compared to peat, lignite has a higher density and higher content of carbon. At this stage, the coal still belongs to sedimentary rock.

Lignite is the youngest type of coal. And some lignite is merely 1 million years old. It is soft and brown in color and mainly used for electric power generation. It is subject to spontaneous combustion as it oxidizes in air.

2.3 Bituminous(烟煤、沥青煤)

With the increasing depth, lignite will undertake metamorphism. The higher hydrostatic pressure due to the overlying materials and the rustling higher temperature will finally metamorphic the lignite into bituminous. At this point, a series of physical and chemical processes take place and the chemical composition and inter structure of lignite are totally altered. Bituminous has a higher concentration of carbon than lignite.

Bituminous usually is called soft coal. It is an lower-grade metamorphic rock. It is used primarily for generating electricity and making coke for the steel industry. Bituminous are dusty to handle, ignite readily, and burn with a smokey flame.

2.4 Anthracite (无烟煤)

When bituminous is subjected to further increasing pressure and temperature, it is converted to anthracite. The anthracite (Figure 19.3) is the final product of coalifi-

Figure 19.3 Anthracite and bituminous. Image from "Minerals in Your World", a cooperative project between the U.S. Geological Survey and the Mineral Information Institute. Public Domain

Table 19.1 Varieties of coal

Rank	Color	Hardness	Products of burning	Heat potential
Lignite	Brown	Softest	Greatest amount of ash and smoke	Least
Bituminous	Black	Soft	Tar, ash and smoke	Less
Anthracite	Black	Hard	Little smoke or ash	Great

cation(成煤作用). It is actually a high-grade metamorphic rock and has the highest carbon content. Anthracite usually is called hard coal(硬煤). It is denser and contains less moisture and gases and has a higher heat value than lower-ranking bituminous and lignite.

Anthracite, which is nearly pure carbon, is very hard and black and is extensively used for home heating. Anthracite has a semi-metallic luster and is capable of burning with little smoke.

Coal range from lignite through bituminous to anthracite, reflecting the progressive response of individual deposits of coal to increasing heat and pressure.

Coal is found in beds(煤层)interstratified with shales, clays, sandstones, or (rarely) limestones. Individual coal layers, known as "seams"(煤线), may range in thickness form less than a few cm up to 30m. They are characteristically tabular or lenticular in form and extensive in area but may be interrupted by features such as sand-filled channels that were cut into the peat immediately after its deposition, and by later structural features such as faulting and folding (Figure 19.4). Coal deposits are characteristic of thick non-marine sedimentary basins that gave dominated certain areas of the lobe at different times. Coal was formed chiefly in the Carboniferous period of geologic time, but valuable deposits date also from the Permian, Triassic, Jurassic, Cretaceous, and Tertiary periods. The major coal-producing fields in China are in Shanxi Province (山西省).

Figure 19.4 Faulted and folded coal

There are two general ways to extract the coal. Coal found close to the surface can be uncovered and removed by large machines in a process that is called strip mine (露天开采、露天煤矿). Sometimes the terms open-pit, open-cast, or surface mining is used in the same sense. Unless the strip mine is filled and revegetated, surface water runoff from the mined area can flush sediments and sulfur-bearing compounds (acid drainage) into nearby streams and rivers, endangering plant and wildlife communities.

Underground mining methods are used where the coal seam is too deep or the land too hilly for surface mining. There are two ways of underground mining: longwall mining (长壁开采法)and room and pillar mining(房柱式开采法) (Figure 19.5). In longwall mining, a cutting head moves back and forth across a panel of coal several hundred

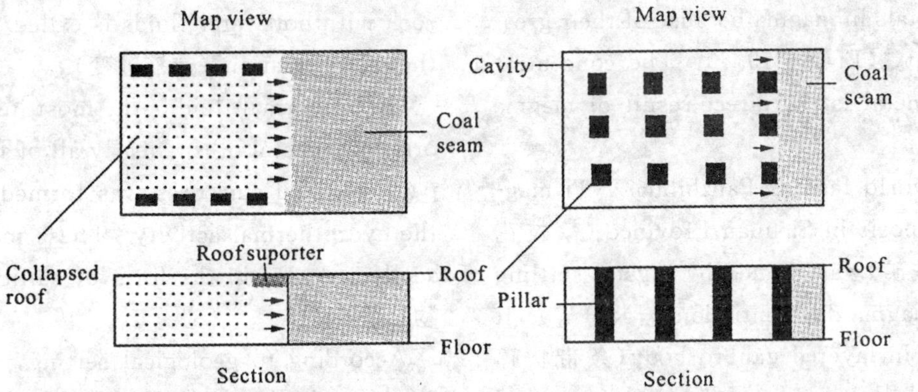

Figure 19. 5 Underground mining. Left: Longwall mining. Right: Room-and-pillar mining

feet wide. The cut coal falls onto a flexible conveyer for removal. Hydraulic roof supporters are advanced as the seam is cut. The roof in the mined-out areas falls as the supports advances. Longwall today accounts for about one-third of total underground coal tonnage. Room-and-pillar mining means that during mining the pillars must be left to support the hanging wall.

Extracting and using coal produce impacts environment seriously. Underground mining can led to ground subsidence. Coal is one of the most impure fuels. Its impurities range from trace quantities of many metals, including uranium and thorium. Products of coal combustion include the oxides of carbon, nitrogen, and sulfur. The burning of coal with high S content can result in the acid rain. New coal technologies should be developed to meet the world's energy needs in a sustainable way.

3 Metals and Ores

An ore is a naturally occurring mineral deposit containing a metal or other valuable resource that can be extracted economically. The term ore is commonly used for reserves of metallic minerals that can be mined for profit. Whether or not a mineral (or rock) is considered an ore depends on the percentage of extractable metal and the potential profits from mineral. Nearly all rock deposits contain some metallic minerals, but in many cases the concentration of metal is too low to justify mining the ore. The percentage of metal is not the only feature to define an ore. Some rocks have a high percentage of certain metals, but the metal is tightly locked in the crystalline structure of minerals and thus difficult to be extracted economically. In this case, the metal-bearing rock is not considered as an ore. Profit or the current price of metal is an indispensable part of the ore definition.

The metal ore can be formed in many ways. But in most cases, there are closely associated with igneous activity. Some of the valuable metal ores can also be formed by surface process, such as weathering and leaching. Here are only some of the most important ore types.

3.1 Origin of Metallic Ore Deposits

3.1.1 Crystal Settling Crystal settling(结晶分离) is the sinking of early-forming mineral crystals in magma because of their greater density (Figure 19.6). The concentration of metal is the direct result of magma cooling.

The world famous Panzhihua V-Ti magnetite deposit in Sichuan Province (攀枝花钒钛磁铁矿) was formed by crystal settling during magma differentiation(岩浆分异). It occurred in layered gabbro body(条带状辉长岩) which consist of layers of mineral crystals deposited in magma. Most of the chromite ore (铬铁矿) in the world was formed during magma crystallization. They generally occurred in layers near the bottom of a cooling magma body.

3.1.2 Magmatic-hydrothermal Ore Deposits The magmatic hydrothermal fluid (岩浆热液) is the heated water containing a large amount of valuable elements. There are two possible sources. One source can be the magma itself. During the final stage of cooling, the magma exsolves the water. Another source is the groundwater circulation. Anyway, the hot water is the result of igneous activity, including both intrusive and volcanic activity. The heated water contains many elements leached from intrusive, and thus reacts with country rocks to form new rocks or ores. The process in which the mineralogy of original rocks has been changed as a result of interaction of these rock with hot water fluids is called "hydrothermal alteration"(热液蚀变).

Hydrothermal fluid is the most important origin for metal ores. Nearly all of the copper, lead and zinc ores was formed due to the hydrothermal activity. These metals usually occurred in the form of sulfide(硫化物).

According to geological setting, the hydrothermal ores can be largely classified into four different types:

(1) Contact Metamorphism. If the country rock is limestone or dolostone (白云岩), the hydrothermal fluid from the magma will react with these rocks, thus altering completely or partially the country rock and forming new rock or ore. The mineralogical change of limestone is actually the result of hydrothermal alteration. The newly-forming rock is called skarn(矽卡岩). Skarn is an very important indicator of skarn-type iron ore (矽卡岩型铁矿). The famous Daye iron ore (大冶铁矿, 湖北) was formed by this way.

(2) Hydrothermal Veins. Hydrothermal veins(热液脉) are tabular ore bodied ores formed by the injection of hydrothermal fluid along the joints and faults in country rock (Figure 19.7). Most hydrothermal veins are metallic sulfides, often mixed with milky quartz. Veins may be very large and

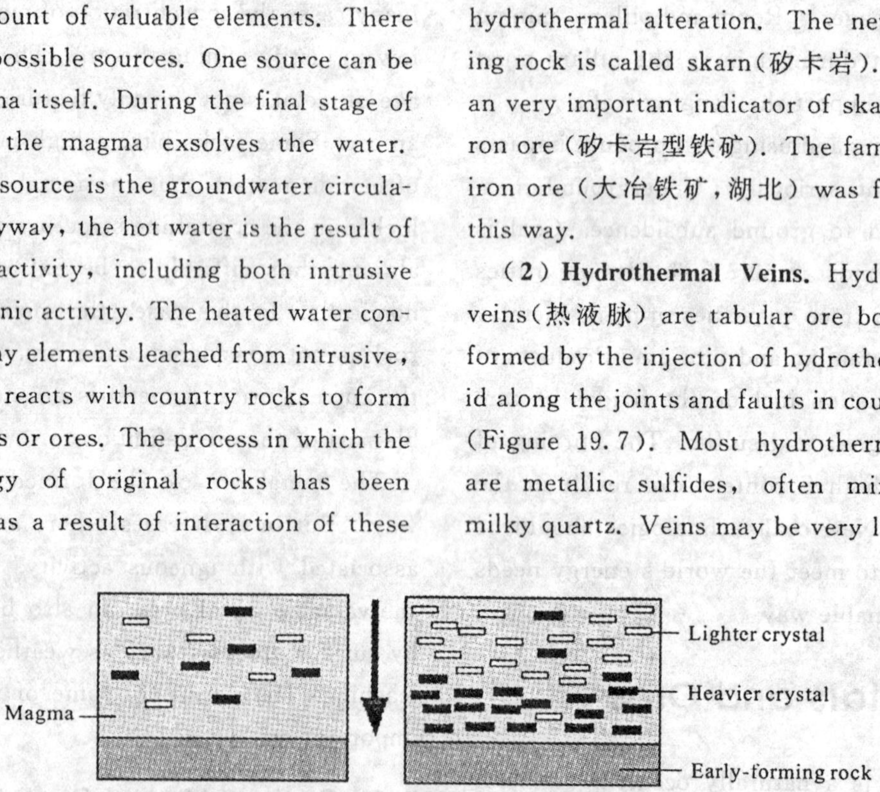

Figure 19.6 Crystal settling during magma cooling

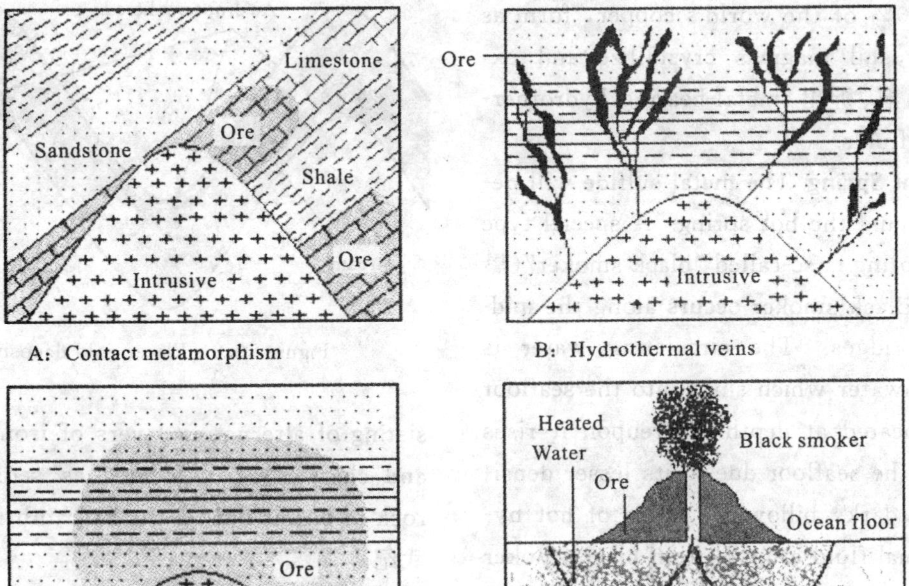

Figure 19.7 Various hydrothermal ore deposits

complex. Some veins measure several hundred meters in outcrop length and tens of meters in thickness. Most of the gold ore deposits mined in Xiaoqinling(小秦岭) belongs to gold-bearing quartz veins(含金石英脉). These deposits generally fall in the category of low tonnage(吨位), high grade types of deposits. Minerals with no commercial value, called gangue minerals(脉石、矸石、废石), are usually found mixed with the ore(矿石) in the vein.

(3) **Disseminated Deposits.** Disseminated deposits (浸染型矿床) are featured by uniform distribution of metallic sulfides within or above the igneous bodies. The ore is generally in low concentration but occur through large volumes of rocks.

Sedimentary rock-hosted disseminated Au deposits are often called Carlin-type gold ore deposits (卡林型金矿). Disseminated

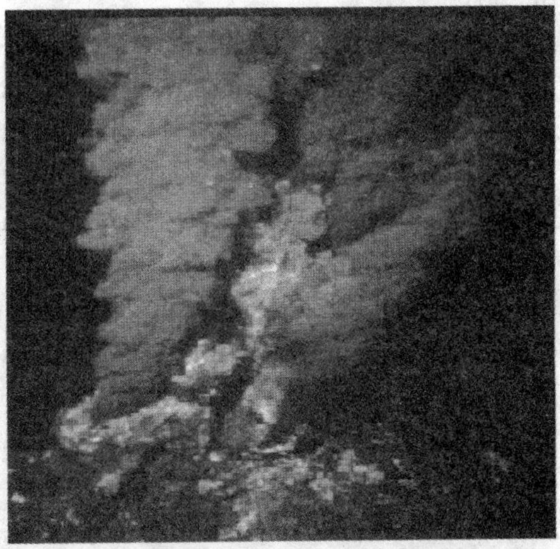

Figure 19.8 Black smoker. Courtesy of NOAA(Public Domain)

Cu deposits are also called porphyry copper deposits(斑岩铜矿) because they are usually associated with porphyritic rocks(斑岩). Porphyry copper deposits, which supply nearly 60% of the world's copper, form as upper-crustal magmas crystallize and exsolve(出溶、析出) metal-bearing hydrothermal ore fluids.

(4) **Hot Spring.** The metal sulfide will deposit around the hot spring. A special type of hot spring is so-called "black smoker"(黑烟囱). Black smoker occurs along the mid-oceanic ridges. The circulating water is from seawater which sinks into the seafloor and is heated at depth whereupon it rises back to the seafloor due to its lesser density. The dark, billowing clouds of hot hydrothermal fluid erupted from black smoker are rich in dissolved metals. As they rise into the cool ocean water, they precipitate to form sulfide ores along the sides of the vent, thus creating a "chimney" that builds over time. Hydrothermal fluids at depth in black smoker reach temperatures in excess of 350℃.

3.1.3 Placer Deposit(砂矿) As valuable minerals gradually is eroded away from its original place and washed off the slopes and into streams, it becomes sorted or stratified by flowing water, and tend to accumulate in certain common locations. These accumulations are called "placer deposits"(砂矿). Gold is commonly found to accumulate in placer deposits with large pebbles and other heavy minerals, among which magnetite is almost invariably present (Figure 19.9).

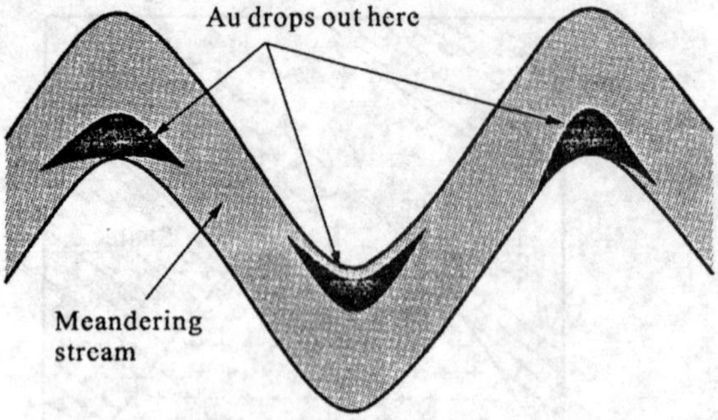

Figure 19.9 Placer gold deposit

3.1.4 Chemical Precipitation in Layers(化学沉积矿床) Few metal ores, except iron, precipitated directly from solution. Banded iron beds are the most important commercial source of iron ore. Most world-class high-grade iron ore deposits are the products of enrichment of Precambrian banded iron formations. The banded iron ores, consisting of alternating layers of iron mineral and chert (燧石), formed as sedimentary rock in basins during the Precambrian(前寒武纪).

3.1.5 Concentration by Weathering(风化矿床) The most important ore formed by weathering is aluminum ore, called bauxite (铝土矿) (Figure 19.10). The major bauxite deposits of the world are found in the tropical and subtropical regions. During the weathering, the soluble materials were taken away by water and the least soluble mineral, aluminum oxide would accumulate on the top of wreathing rocks, thus forming bauxite in the form of soil.

3.2 Some Important Metals

• **Iron** (铁)

Iron is the basis for the modern industrialized society. Iron is essential for the production of steel, which in turn is essential to maintain a strong industrial base.

Iron ore almost always consists of iron oxides, the primary forms of which are magnetite (Fe_3O_4) (磁铁矿) and hematite (Fe_2O_3). Hematite (赤铁矿) is the most important ore of iron. The amount of hematite needed in any deposit to make it profitable to mine must be in the tens of millions of tons. The typical reddish color of hema-

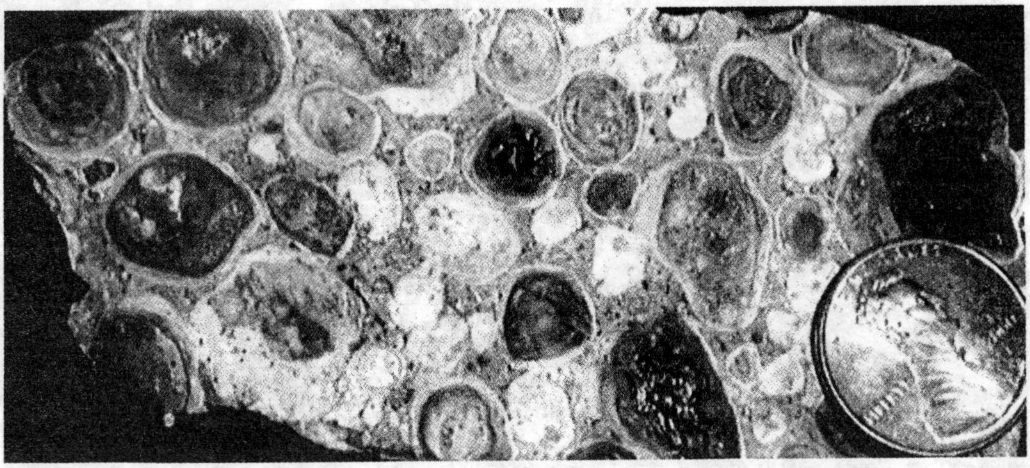

Figure 19. 10 Bauxite

Image credit: from the"Minerals in Your World"project, a cooperative effort between the United States Geological Survey and the Mineral Information Institute. Public Domain

tite is responsible for the red coloration of many sedimentary rocks. Large ore bodies of hematite are usually of sedimentary origin. Magnetite typically occurs in igneous rocks.

The concentration of a metal in an ore is called its grade(品位). Grade is usually expressed as a weight percentage of the total rock. Most of the iron ores in our country is low-grade ores (20%~40%).

China is now the biggest importer of iron ore in the world. In 2004, China imported 208 million tons of iron ore, accounting for one third of the world's total maritime trade volume.

- **Copper(铜)**

Copper can occur as native elements(自然元素) in the fields. But the most important copper ores are sulfide, among which the chalcopyrite($CuFeS_2$)(黄铜矿) is the primary copper-bearing mineral. The disseminated porphyry copper ore accounts for 60% of total ore copper ore.

Nearly all of the copper ores are of hydrothermal fluid origin. Sometimes, the original hydrothermal will be reformed by weathering and groundwater. The Cu is leached out of upper part of the ore and deposited at the lower part of ore body. This process can lead to the further concentration of Cu ore. The iron left on the upper part of ore forms a rusty cap called a gossan(铁帽) which is a very important indicator of copper ore below (Figure 19.11).

Jade Dragon ore in Tibet (西藏玉龙铜矿) is the largest copper ore deposit in China and has not been exploited due to its remoteness . It is a typical porphyry copper ore deposit formed along the margin of a stock. World-class Dexing copper ore in Jiangxi (江西德兴铜矿) is also a porphyry deposits.

- **Lead and Zinc (铅和锌)**

Lead is one of the most widely used metal and over 60% of all lead produced is used in lead-acid batteries for the storage of energy. Zinc is the third most used non-ferrous metal(有色金属) after aluminum and copper. Zinc are intensively used in making alloys, such as brass (copper and zinc ,黄 铜) and

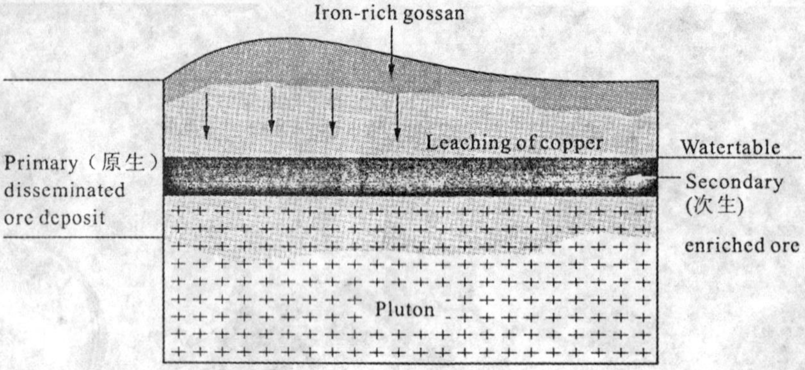

Figure 19.11 Secondary enrichment (次生富集) of copper ore

bronze(copper ,tin and zinc,青铜).

Both the lead and zinc ores are of hydrothermal origin. Galena(方铅矿) is the most important mineral for extracting lead and sphalerite(闪锌矿) is the major zinc ore. Galena is found closely associated with sphalerite in the field, and thus they are often collectively referred to as lead-zinc ore deposit(铅锌矿).

China is the fourth zinc-lead producer in the world.

• **Aluminium**(铝)

Aluminium is the third most abundant element in the Earth's crust after oxygen and silicon. Mankind has been using copper, lead and tin for thousands of years and yet today more aluminium is produced than all other non-ferrous metals combined. It is widely used in food packaging, construction and the manufacture of car, airplanes and electrical cable. With the rapid development of economy in China, the use of aluminium is ever increasing.

Bauxite is the main source of aluminium. Bauxite (typically $Al_2O_3 \cdot 2H_2O$) contains Fe_2O_3, SiO_2, and other impurities. The purified dry ore, called alumna, is aluminum oxide — Al_2O_3. Bauxite is relatively abundant in nature. At the current rate of extraction, known bauxite reserves will last for hundreds of years.

• **Gold** (金)

As a native element and a precious metal, gold has been used by mankind for thousands of years. Gold is widely used today in coins, decoration, electronics (where its superb properties as a conductor help offset its tremendous cost), medical and industrial technology. Gold can be found in a variety of geological settings ranging from quartz vein, placer deposit and disseminated gold deposit. Gold flakes or nuggets (狗头金) are often found in placer deposits.

China is the fourth-largest producer of gold in the world and Shandong (山东) is the largest gold producer in China. China mined over 212t of gold in 2004. Of that total 64.5t came from Shandong Province alone in 2004.

4 Nonmetallic Resources

Examples of nonmetallic resources are common clay, sand, shell, gravel, and crushed or broken stone, including limestone, dolomite, trap rock, sandstone, quartz, quartzite, slate and shale. Sand and gravel on the beach or any other places can be used for manufacture of concrete. Limestone is the raw material of cement.

5 Resources and Sustainable Development

Everybody knows that resources of the Earth are finite and sooner or later we will run out of oil, coal and most of the metals. And when it happens, the consequences will be unimaginably catastrophic. It will be so if we do nothing at all to forestall these consequences.

Very little in nature is wasted, and waste from one organism almost always is useful to another. We can say that nature is self-sustaining. Humans have chosen a non-sustainable approach. Our life-style depends on a constant flow of natural resources from the Earth. These precious resources are limited and not renewable. Providing affordable and environmentally sustainable energy to a future world population of up to ten billion is a major challenge. We must learn to eliminate waste, recycle resources, and develop sustainable sources of energy.

5.1 Sustainable Energy

It has been more than twenty years since energy has been a frequent topic of news headlines. Today, however, we hear frequent references to a new"energy crisis".

Sustainable energy includes energy efficiency, combined heat and power and fuel poverty. Carbon dioxide, a greenhouse gas produced when electricity is generated or gas is burned, is a major contributor to climate change. Improved energy efficiency is therefore a key element of sustainable energy.

• A large number of players have a role in delivering energy efficiency — the UN, national governments, regional and local authorities, industry etc. Only a combination of measures at the various levels will realize the full potential to be exploited. In addition, further international interactions are needed.

• To promote the use of renewable energy and advanced clean energy technologies among residential, commercial, institutional and industrial customers. Although currently alternative fuels still make up a small fraction of total energy supplies they are growing rapidly. Just as George Sterzinger, executive director of the Renewable Energy Policy Project in Washington, says,"This is a period of high fossil-fuel costs and low-interest rates, which gives renewable energy an advantage unlike it's had over the last 20 to 25 years."

• To optimize the pattern of energy consumption.

• To promote energy conservation. If everyone were truly conservation-minded, we would save thousands of megawatt-hours of electricity every day.

5.2 Pollution Prevention

All pollution is waste. Waste is simply raw materials that are not being used efficiently. Pollution is not only expensive to treat, it also takes energy and labor to make in the first place! In the long run, it costs less to avoid waste than to dispose of it.

5.3 Recycle

Many of the resources concerned are either renewable or recyclable. They include batteries, plastic, paper, glass, aluminum, steel, motor oil, tires, toxics, car, refrigerators and computer printers. The increasing use of recycled materials saves both energy and mineral resources needed for primary production.

China faces many problems during its modernization. The conflict between rapid economic growth and voluminous consumption of resources and ecological deterioration is by far the most serious problem. In 1992, the Chinese government published China's Agenda 21 *White Paper on China's Population, Environment and Development in the 21st Century*. That document marked the beginning of the sustainable development process in China. To help implement the sustainable development strategy, the *Program of Action for Sustainable Development in China in the Early 21st Century* was promulgated in 2003. All of the actions taken by Chinese government show the determination to curb the environment deterioration and achieve sustainable development. Anyway, in order to construct a harmony society, we should first learn how to shape a harmony relationship with nature.

1. Answer the Following Questions

(1) What is peat?

(2) What is the youngest type of coal?

(3) Describe in your own words the formation of coal from plant matter to anthracite.

(4) List and describe the traps.

(5) Describe in sketch the black smoker.

(6) What is the hydrothermal vein?

(7) Describe in sketch the placer deposit.

(8) Differentiate the resource and reserve.

(9) List and describe the main types of hydrothermal ore deposits.

(10) What is the main environmental problem of coal combustion?

2. Multiple Choice

(1) Which metal is frequently mined in placer deposit?

A: Copper B: Iron

C: Lead D: Gold

(2) Which type of coal has the highest carbon content?

A: Peat B: Hard coal

C: Lignite D: Soft coal

(3) In what type of rock is coal formed ______.

A: Igneous rock B: Metamorphic rock

C: Sedimentary rock D: All of them

(4) Which type of coal has the highest heating value?

A: Peat B: Hard coal

C: Lignite D: Soft coal

(5) Which sentence about coal is not right?

A: The extensive accumulation of thick coals involve slow rates of basin subsidence.

B: Anthracite, which is nearly pure carbon, is very hard, black, and lustrous.

C: Coal forms in ocean basin by the accumulation of decayed plant materials.

D: Bituminous ignite readily and burn with a smoky flame.

(6) Nugget is often found in ______.

A: Quartz vein

B: Disseminated deposit

C: Placer deposit

D: All of them

(7) Oil is formed in ______.

A: Shale B: Sandstone

C: Both of them D: None of them

(8) Which period were the most of the coal deposits in China formed in?

A: Permian B: Triassic

C: Tertiary period D: Carboniferous

(9) Galena is the most important mineral for extracting ______.

A: Zinc B: Copper

C: Gold D: Lead

(10) Which metal ore deposit can be formed by chemical precipitation?

A: Copper B: Iron

C: Lead D: Gold

(11) Most of the metal ore deposits was formed by ______.

A: Crystal settling

B: Concentration by weathering

C: Magmatic-hydrothermal fluid

D: Placer deposit

(12) Lignite is soft and subject of spontaneous combustion as it oxidizes in air.

A: True B: False

(13) The major coal-producing fields in China are in ______.

A: 河北省 B: 陕西省

C: 山西省 D: 河南省

(14) Peat is not coal but represent the initial stage of coal development.

A: True B: False

(15) Which rock can act as a reservoir rock?

A: Shale B: Granite

C: Gneiss D: Sandstone

(16) Which one is not a energy resource?

A: Coal B: Oil

C: Gas D: Metal

(17) Which one is not a structural trap?

A: Anticline trap

B: Fault trap

C: Salt dome trap

D: Sandstone lens in shale

(18) Black smoker can be found at ______.

A: Mid-oceanic ridges B: Trenches

C: Hawaii Island D: Continental shelf

(19) Which sentence about the trap is not right?

A: Oil that finds its way into a reservoir rock that has been bent into an arch will flow to the crest of the arch, and get stuck.

B: Stratigraphic traps form where changes in rock type can retain hydrocarbons.

C: Salt which is buried kilometers below the surface of the Earth can move upward.

D: Oil is stored in some sort of underground cave.

(20) The most commonly found hydrothermal vein in the field is ______.

A: Gold B: Quartz

C: Iron D: Copper

(21) Which sentence about the coal is not right?

A: We use the term "coal" to describe a variety of fossilized plant materials.

B: The carbon content of coal supplies most of its heating value.

C: Anthracite rank below bituminous.

D: Lignite ranks the lowest and is the youngest of the coals.

(22) Most of the metallic ore deposits are associated with ______.

A: Sedimentary rock B: Igneous rock

C: Metamorphic rock D: All of them

(23) Peat is generally considered to be completely decomposed biomass.

A: True B: False

(24) Which sentence about the oil is right?

A: There must be a source rock that will generate the gas and oil.

B: There must be a reservoir rock to hold the natural gas and oil.

C: There must be a trap in the underground reservoir rock to concentrate.

D: All of the above sentences are right.

(25) The most important iron ore is ______.

A: Pig iron B: Hematite

C: Magnetite D: Steel

(26) Sooner or later we will run out of oil.

A: True B: False

Chapter 20 Atmosphere and Environment

Key words

El Nino	La Nina	Ozone hole	Greenhouse effect	Global warming
厄尔尼诺	拉尼娜	臭氧洞	温室效应	全球变暖

This world has been plagued with hurricane, El Nino and global warming. The greenhouse effect has been describe as the potentially most dangerous environmental problem facing mankind, with consequences second only to the nuclear war. Preserving our environment is becoming the most serious challenges facing us today.

1 El Nino and La Nina

El Nino and La Nina are two global scale climate change events. They refer to temperature fluctuations in surface waters of the tropical Eastern Pacific Ocean. In Spanish, El Nino means"the little boy", a reference to the Christ child. La Nina means"the little girl". El Nino has occurred for centuries and even millennia. Mankind was not fully aware of this phenomenon until this century.

El Nino originally referred to warm, southward, seasonal current that appears along the Pacific coast of South America each year around Christmas. But now the term has been applied to the abnormal warming of surface ocean waters in the eastern tropical Pacific (Figure 20. 1).

In the normal year, strong westward-blowing trade wind(信风) pushes water away from the western coast of South America and piles up warm water in the Western Pacific. This wind allows cool water in the eastern Pacific to well up along the South American coast, carrying nutrients that support large fish populations.

In an El Nino year, the trade wind subside and the warm water slowly moves back

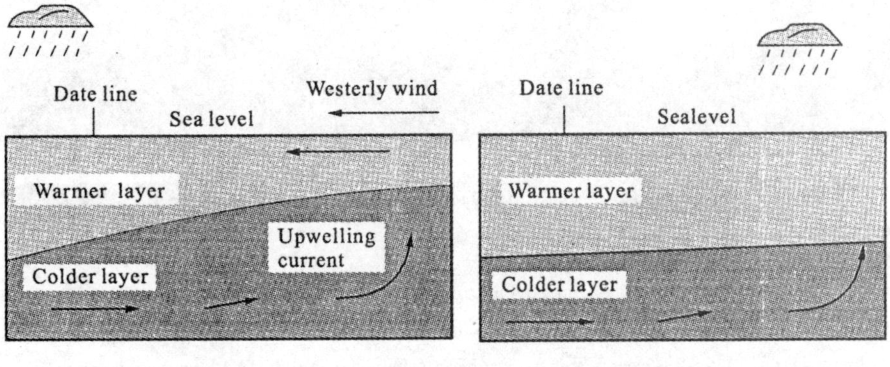

Figure 20. 1 Left: Normal year. Right: El Nino year

eastward across the Pacific. This eastward shift of warm water makes the equatorial cool water upwelling suppressed, and thus resulting the fish die and climatic changes affecting many parts of the world. At the same time, the piling-up of warm water will lead to an abnormal warming of surface ocean waters in the eastern tropical Pacific. When the east-shifted warm water reaches the South American coast, we call it El Nino (Figure 20. 2). El Nino occurs every two to seven years. The changes in air pressure between El Nino and non El Nino periods are referred to as the Southern Oscillation (南方振荡). The climate anomalies resulted from warming ocean water is one part of the Southern Oscillation. Scientists have found that the El Nino Southern Oscillation (ENSO) is the main force behind the change in global rain patterns. ENSO is a global event that arises from large-scale interactions between the ocean and the atmosphere.

El Nino events disrupt weather patterns worldwide. Its effects on global weather change may be positive and negative. The negative effects are summarized here:

• Drought and bush fires in southwestern Pacific regions, including Australia, India, Indonesia and Philippines. Even India and Sri Lanka are affected seriously. In this regions, crops fail, coral reefs die and starvation follows.

• Downpours, flooding and mud slide in South and North American coastal regions, including U. S. , Gulf states, Peru. Even the Southern Africa is affected by drought, disease and malnutrition. The fishing industry in South American coastal regions have been seriously affected.

• The worst effect of El Nino may be the global warming. Corresponding with global warming, the worldwide glaciers have been retreating, which in turn leads to sea levels rise.

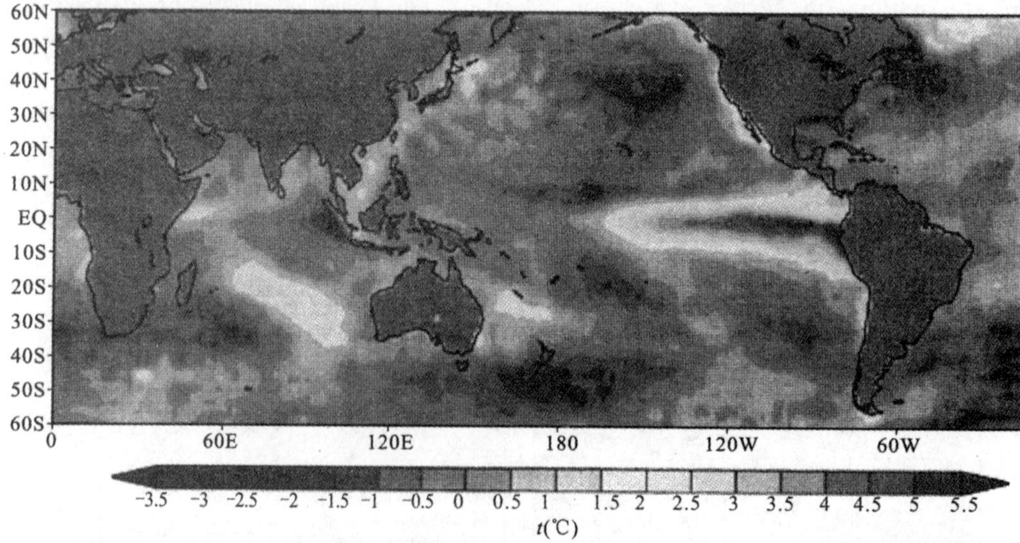

Figure 20. 2 Chart of abnormal ocean surface temperatures observed in December 1997 during the last strong El Nino (source: National Centers for Environmental Prediction, US).

Image credit: National Oceanic and Atmospheric Administration (NOAA), public domain

• The frequency of extreme climate events has increased remarkably in recent year. The destructive flooding in China in 1995 and hurricanes hitting the U. S. more frequently in 2005 may be some examples.

Is there a good side to El Nino? There can be. One of the most important positive effects may be that the deserts and arid areas in Middle East have received more precipitations in recent years. The abnormally warm ocean waters off the east coast of South America could result in the appearance of unusual marine species.

La Nina is just the opposite of El Nino. it is featured by unusually cold ocean temperatures in the eastern equatorial Pacific. During a La Nina, the trade winds are stronger and cold, nutrient-rich water occupies much of the tropical Pacific Ocean. Most of the precipication occurs in the western tropical Pacific Ocean, so rain is abundant over Indonesia.

2 Air Pollution and Ozone Zone

Air pollution is the contamination of air by the discharge of harmful substances. It can be separated into two different groups. First one is primary pollutant. It included the materials that enter the atmosphere directly. The main primary pollutants are carbon monoxide(一氧化碳), sulfur oxides(二氧化硫), nitrogen oxides(二氧化氮) and volatile organic compounds (生物挥发物质) and other natural and human-made matters (Table 20. 1). The secondary pollutants are those materials resulted from the interaction between primary pollutants or between the primary pollutants and the basic components of atmosphere. The main secondary pollutant is photochemical smog (光化学雾). Air pollution began in the 19th century, the time corresponding to the Industrial Revolution in Europe. But now it has become a global problem that is gradually affecting the world.

In the past decades, the most hotly discussed topics about air pollution include acid rain(酸雨), ozone hole (臭氧洞) and greenhouse effects.

2. 1 Acid Rain

It is a general term referring to the deposition of acidic components in the form of rain, snow, fog, dew, or dry particles. The more proper name should be acid precipitation.

Natural rain and snow is slightly acidic because carbon dioxide (CO_2) from the air is dissolved in rainwater to create a weak acid, carbonic acid (H_2CO_3):

$$H_2O + CO_2 \longrightarrow H_2CO_3 \longrightarrow H^+ + HCO_3^-$$

Rain + carbon dioxide ⟶ carbonic acid ⟶ hydrogen ion bicarbonate ion

The natural rain has pH of average 5. 6. This kind of acid in rain is actually beneficial because it helps dissolves minerals in

Table 20. 1 The main primary and secondary pollutants

Primary	Carbon monoxide, Carbon dioxide, Sulfur dioxide (SO_2), Nitrogen oxides(氮的氧化物), Volatile organic, Chlorofluorocarbons (CFCs) (氯氟烃,氟里昂), Lead
Secondary	Photochemical smog, Ozone(O_3)

the soil that both plants and animals need.

Acid rain is defined as rain with a pH of below 5. 6. It is caused by pollution. The excessive acid in acid rain comes from two kinds of air pollutants — sulphur dioxide (SO_2) and nitrogen oxides (NO_x). The SO_2 and NO_x come from the burning of fossil fuels such as, petrol, diesel, and coal, as well as manufacturing exhaust. Volcano eruption can also emit the SO_2 and NO_x. These pollutants stay in the atmosphere and eventually react with the moisture in the air to form the new compounds sulfuric acid (H_2SO_4), ammonium nitrate (NH_4NO_3) and nitric acid (HNO_3).

The damage of acid rain includes:

- Damaging buildings and statues made of limestone.
- Damaging aquatic life in lakes.
- Damaging the crops and vegetation.
- Harming human health.

2. 2 Ozone Depletion

Ozone is a molecule that contains three atoms of oxygen and thus has the formula O_3. It occurs naturally in the atmosphere.

Naturally occurred ozone is formed when intensive ultra-violet radiation from the Sun breaks down O_2 into two oxygen atoms, because oxygen molecules are very efficient absorbers of high energy ultraviolet radiation. These free oxygen atoms will collide with O_2 to form O_3. The ozone (O_3) is a relatively unstable molecule and an efficient absorber of the middle and lower energy parts of the ultraviolet radiation. When UV with lower energy strike the O_3, the O_3 will split into one molecular oxygen (O_2) and one atomic oxygen (O). The escaped free oxygen atom will react further with other O_3 to form 2 oxygen molecules. About 90 percent of all ozone in the atmosphere is formed in this way. Through these processes, ozone molecules in the stratosphere are constantly being created and destroyed, over and over, thus keeping a delicate dynamic balance (Figure 20. 3).

Most of the ozone is formed in tropical re-

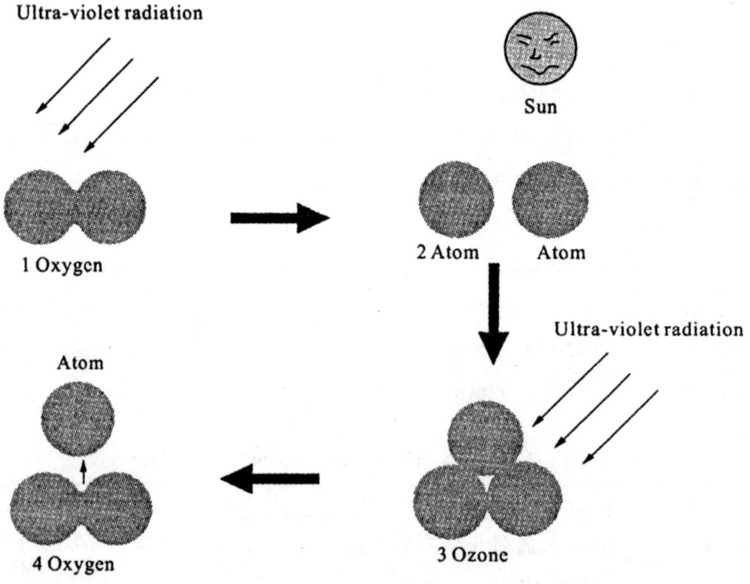

Figure 20. 3 Formation and destruction of ozone

gions because solar radiation is strongest in there. The ozone formed in tropical regions will spread to poles, where it accumulates because the weak solar radiation. This is the formation of ozone layer in the stratosphere. Ozone layer is thinnest in the tropics (because the highest rate of destruction) and denser towards the poles. About 90 percent of the Earth's ozone actually is in the stratosphere. The ozone layer is important because it shields the entire Earth from much of the harmful ultraviolet radiation that comes from the Sun. Ozone absorbs ultraviolet radiation from the Sun that can damage DNA in living systems.

Besides the ozone layer in stratosphere, another zone accumulation, called ground-level ozone or bad zone, exists in the troposphere. In addition to the ultraviolet radiation, other natural (volcanic eruptions, fewer sunspots) or man-made compounds containing nitrogen, hydrogen, and chlorine can destroy ozone molecules(Figure 20. 4). For example, the chlorine atom can break down the ozone molecule into a chlorine monoxide and an oxygen molecule. Ozone in the troposphere comes from human activities that involve burning fossil fuels and using certain chemicals, such as solvents. The zone "smog" plagues many other major urban areas in the world.

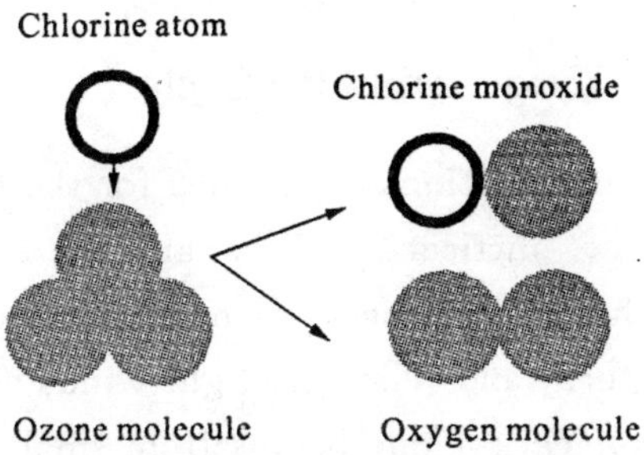

Figure 20. 4 Ozone can be destroyed by chlorine

Chlorofluorocarbons, or CFCs, once widely used in spray cans and refrigerators, can deplete the Earth's protective layer, because in the upper atmosphere UV light breaks off a chlorine atom from a CFC molecule. The chlorine monoxide, the product of reaction between chlorine atom and ozone, can react readily with free oxygen atom, forming an oxygen molecule and freeing the chlorine to begin the process all over again. That is why CFCs destroys the ozone layer so efficiently.

The ozone hole over the South Pole is actually an annual thinning of the ozone layer during the southern hemisphere spring months of September and October (Figure 20. 5). Ozone hole is caused by the release of CFCs and other man-made compounds such as halogen compounds and nitrogen oxides.

The ozone hole was first noticed in the 1970s by a research group from the British Antarctic Survey (BAS). According to the observation during 2005, the ozone hole is gradually healing in recent years due to the reduction of CFCs release.

3 Tropical Cyclone

China is frequently hit by tropical storms. In September 2005 alone, typhoons Talim, Khanun, Damrey and Longwang killed more than 200 people across the southern part of the country. These typhoons wrecked great havoc and affected millions of people lived in Southeastern China. On October 2, 2005, 86 members of paramilitary police brigade were swept away by flood when Typhoon Longwang(台风龙王)slammed into Fujian Province(福建省).

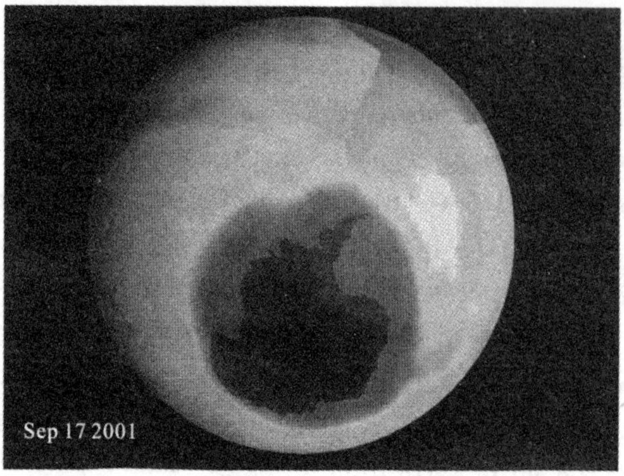

Figure 20. 5 Antarctic ozone hole on September 17, 2001
Image credit: NASA-GSFC, public domain

However, the typhoon also brought plentiful rainfalls necessary to the areas and relieved the droughts there.

In August 2005, the United States experienced the worst hurricane season in history. Hurricane Katrina and the flooding that followed devastated many homes, buildings and, in some cases, entire neighborhoods. The New Orleans was totally flooded and destroyed by the Hurricane Katrina.

What are the typhoon and hurricane? And why are they so powerful and destructive? This is the topic we will talk about in this section.

Both typhoon and hurricane are tropical cyclone (热带气旋). A tropical cyclone is a low-pressure system that derives its energy primarily from evaporation from the sea in the presence of high winds and lowered surface pressure and the associated condensation in convective clouds concentrated near its center. Hurricanes and typhoons are the same thing, but in different places. The name "hurricane" is derived from the aboriginal Caribes' (indigenous peoples of the West Indies at the time of the arrival of Christopher Columbus) expression for "evil spirit". Typhoon simply is from Chinese. It is called hurricane in Atlantic Ocean and typhoon in the Pacific. In India Ocean, it is simply named as cyclone. Hurricane and typhoon develop from tropical disturbance(热带扰动), which is the initial stage of tropical cyclone. Based on the maximum sustained surface winds, tropical cyclone can be described in terms of tropical disturbance, tropical depression (热带低压槽), tropical storm(热带风暴) and hurricane/typhoon. Sustained surface winds are defined as a 1 min average wind measured at about 10m above the surface (Table 20. 2).

Here we will take the hurricane as an example to explain the development of tropical cyclone.

3. 1 Origin and Life Cycle

Four conditions are needed for the formation of hurricane. They are pre-existing weather disturbance, warm tropical oceans, moisture, and relatively light winds aloft.

Tropical cyclone begins from small tropical disturbances. If the ocean is warm enou-

Table 20.2 Types of tropical cyclone

Name	Speed of wind* (m/s)	Region
Tropical depressions	<17	World wide
Tropical storm	17～33	World wide
Hurricane	>33	The North Atlantic Ocean, the Northeast Pacific Ocean east of the dateline, or the South Pacific Ocean east of 160E
Typhoon	>33	The Northwest Pacific. Ocean west of the date-line. The South China Sea
Tropical cyclone	>33	Indian Ocean

* Speed of wind refers to maximum sustained surface winds

gh to provide excessive heat and moisture to the air, the initial rotary motion will occur due to the unsettled weather and Coriolis effect. The warm water is necessary. That is why tropical cyclone is restricted to the North Atlantic, the Northeast Pacific and the India Ocean. The equator regions are never hit by the hurricane because of weak Coriolis effect. The above explanation is just in a very simple and general way. However, the exact mechanism of the initial disturbance is still not well understood.

The initial disturbance will transform into tropical depression and tropical storms under favorable conditions(Figure 20.6). At this moment, a typical spiral movement of the air will take place. About 100 tropical storms are formed worldwide every year, but only about half of them eventually become hurricanes or typhoons. When the tropical disturbances finally develop into full-fledged hurricane or typhoons, the violent winds, incredible waves and torrential rains will be produced.

Once the hurricane moves away from the warm ocean, it will diminish in intensity. One of main causes is that the storm's source of warm moisture is cut off. The life span of a hurricane or typhoon ranges from a few hours to two weeks.

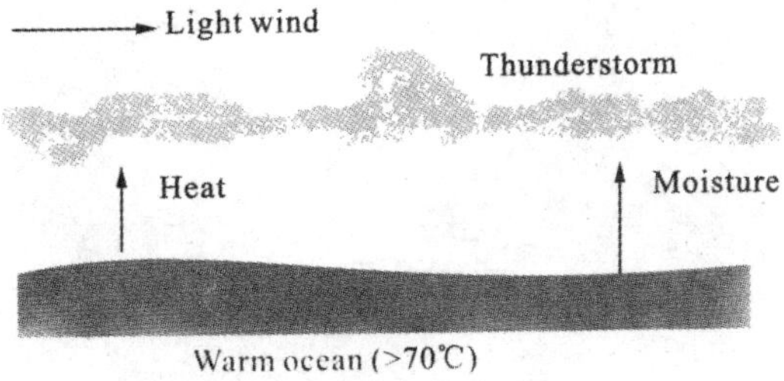

Figure 20.6 Formation of tropical cyclone

3.2 Structure

Due to the spiral movement of air, tropical cyclone generally has typical structural patterns. It is interested that air spirals in toward the center in a counter-clockwise pattern in the northern hemisphere and clockwise pattern in the southern hemisphere. Hurricane always spiral in a counter-clockwise pattern. Typical hurricanes are about 300 miles wide although they can vary considerably (Figure 20.7).

• **Eye(飓风眼)**

The very center around which the winds whirl is termed as the "eye" of the hurricane or typhoon (Figure 20.8). It is a circular region with little wind, no rain and often a

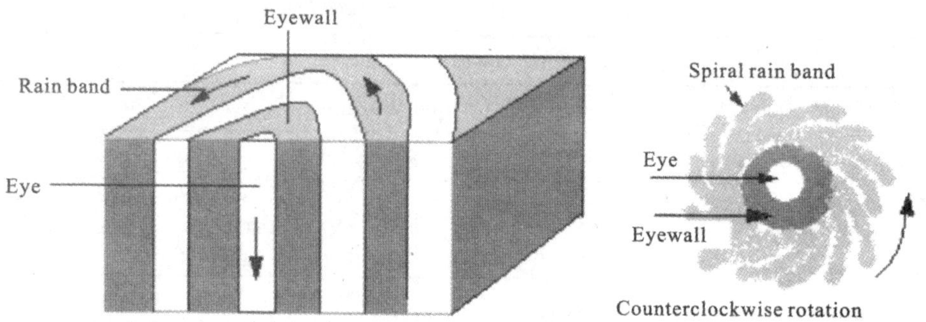

Figure 20.7 Structure of hurricane. Right: Map view. Left: Diagram

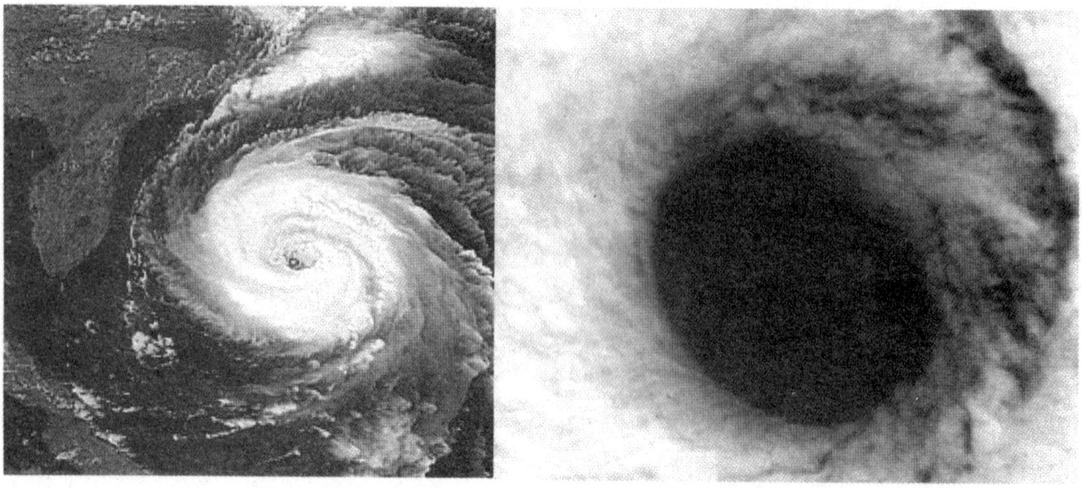

Figure 20.8 Satellite images of hurricane
Image credit: NASA - GSFC(left), NASA - JSC - ES&LA(right), public domain

clear blue sky. The air with the eye gradually sinks, forming the cloud-free eye. The eye of hurricane is typically 32～64 km across. The eye can grow or shrink in size.

• **Eyewall(飓风眼墙)**

The placid eye is surrounded by a dough-nut-shaped "eyewall" of furious, thunderstorm-type clouds and the fiercest winds. It is in the region of eyewall, the hurricane reaches its highest wind speed and heaviest rainfall.

• **Rain Bands(雨带)**

Rain bands are large curved bands of clouds and precipitation that spiral around the storm. They are generally in a spiral fashion, thus frequently being termed as spiral rain bands. These bands are capable of producing heavy bursts of rain and wind, as well as tornadoes. They usually range in width from a few miles to tens of miles and are 50 to 300 miles long. There are sometimes gaps in between spiral rain bands where no rain or wind is found.

3.3 Destruction

Damage cause by hurricane can be summarized into three groups: storm surge, wind action and flooding.

The storm surge(风暴潮)is the most dangerous part of a hurricane and most of the death and destruction during hurricanes is caused by the storm surge. It is actually the abnormal rise of water levels along a coastline caused by wind and pressure forces of an approaching hurricane or other intense storm. The piling up of water is harmless in the deep ocean because the dome of water can sinks and flows away readily. But once the storm makes landfall, the rising sea floor blocks the water's escape and it comes ashore as deadly storm surge.

For example, on November 13, 1970, the delta region of Bangladesh was inundated by a deadly storm surge, leaving a death toll up to 20 000. In August 2005, the surge slams into levees east of New Orleans and pours over them, flooding eastern New Orleans. A month later, the Hurricane Katrina sent white water rushing over makeshift dikes that had been hastily built, reflooding the low-lying urban neighborhoods.

Freshwater flood is mainly caused by torrential rains. Wind often affects much larger areas than storm surge.

4 Global Warming

The warming of the Earth has been the subject of intense debate and concern for many scientists, policy-makers, and citizens for at least the past decade. Tracking the world's average temperature from the late 19th century, people in the 1 930s realized there had been a pronounced warming trend.

4.1 Greenhouse Effect

The greenhouse effect is the rise in temperature that the Earth experiences because certain gases in the atmosphere trap energy from the Sun. The greenhouse effect and global warming are not the same thing. The Earth has a natural greenhouse effect due to trace amounts of H_2O and CO_2 that naturally occur. The Earth's "greenhouse effect" is what makes this planet suitable for life as we know it (Figure 20.9). Without greenhouse effect, the Earth will be too cold for survival, and the surface will freeze. Enhanced greenhouse effect or "bad greenhouse effect" refers to the augmentation of these natural gases by human activities. If the greenhouse effect becomes stronger, it could make the Earth warmer than usual.

In order to stable temperate on the Earth's surface, the amount of radiation reaching and warming the Earth must balance the amount leaving and cooling the Earth (Figure 20.10). This delicate balance is known as the radiation budget . It is the

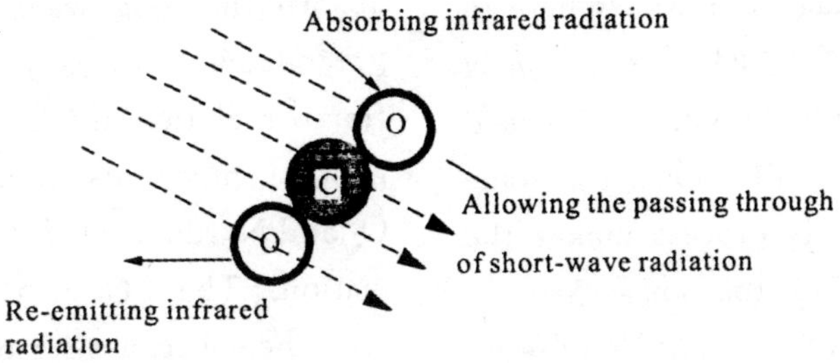

Figure 20.9 Carbon dioxide can absorb infrared radiation

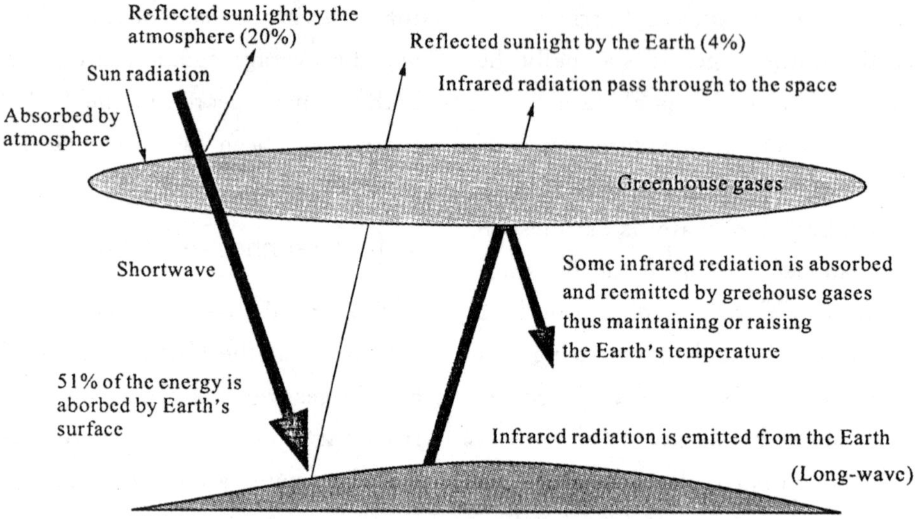

Figure 20. 10 Earth radiation budget

balance between incoming energy from the Sun and the outgoing long-wave (thermal) and reflected short-wave energy from the Earth. Most of the radiant energy from the Sun is concentrated in the visible and near-visible parts of the spectrum. They are short wave length radiation. When the solar radiation reach to the Earth, 50% of the total solar energy will be absorbed by the Earth's surface and about 20% will be reflected by the Earth and the atmosphere. The warmed surface will emit long wavelength, infrared (heat) radiation to the atmosphere. So the air is not heated directly by the Sun, but is heated indirectly from the Earth's surface. Some of the infrared radiation passes through the atmosphere to the outer space (外层空间). And nearly 90% of the infrared radiation is absorbed and reemitted in all directions by some greenhouse gases. This process makes the Earth's temperature"not too hot as Venus, not too cold as Mars, just right"for life on the Earth's surface to exist.

The greenhouse effect can be sketched as follows: Radiation from the Sun → absorbed by the Earth → some re-emitted to space→ some trapped by the atmosphere. Thanks to greenhouse gases, the atmosphere retains the Sun's radiation and warms up the planet. Without greenhouse effects, the Earth would have a uniform temperature in the atmosphere of about −18℃.

The so-called greenhouse gases mainly include CO_2, water vapor (H_2O), carbon dioxide methane (CH_4), and nitrous oxide (N_2O), sulphur hexafluoride and hydrofluorocarbons (HFCs) (Table 20. 3). These gases trap the heat because they allow incoming short wavelength radiation to pass through the atmosphere and intercept and absorb the long wavelength radiation. So greenhouse gas is any gas that absorbs infrared radiation in the atmosphere. The two major components of atmosphere, that is, O_2 and N_2, don't absorb long wavelength radiation. The CO_2 is the primary gas that traps heat because its high concentration in the atmosphere. Although methane is actually a more potent greenhouse gas than

Table 20.3 Greenhouse gases and their sources

CO_2	Burning of fossil fuels
CH_4	Anerobic bacteria in rice field, cows and sewage. Food production
CFCs	Refrigeration and spray cans
N_2O	Burning of fossil fuels, fertilizer
SO_2	Volcanic cruption, burning of fossil fuels
Water vapor	Various industrial activity

CO_2, it plays a less important role in global greenhouse effect due to its low content.

In nature, carbon is plentiful and dynamic. It is cycled through a series of natural process. For example, the plants consume the carbon dioxide and produce the oxygen. And the animals consume the oxygen and produce dioxide. So a subtle balance between the amount of CO_2 in the atmosphere and the amount of plants is maintained. This carbon cycle is thought to have been in a relatively steady state for thousands of years until the industrial revolution.

Due to the human activities, it is estimated that Earth's surface temperature has risen by about 0.555 6℃ in the past century. The original chemical composition of the atmosphere has been greatly altered through the buildup of greenhouse gases. Humans disturb the carbon cycle by clearing forests and burning fossil fuels (coal, oil, and natural gas), which increased the CO_2 content of the atmosphere. Increased agriculture, deforestation, landfills, industrial production, and mining also contribute a significant share of carbon dioxide emissions. Ice core samples show that both carbon dioxide and methane levels are higher than at any time in the past 160 000 years. Further study suggests that the combustion of fossil fuels is a major contributor to the increase in the carbon dioxide concentration, such contributions being 2 to 5 times the effect of deforestation.

4.2 Potential Effects of Global Warming

The impacts of global warming on human being may be disastrous. Global warming will affect the Earth's evolution as well as our daily life in the future. Here are just some of the unsystematically observations about the effects of global warming. Anyway, global warming is bringing other unpredictable changes.

- **Health**

Human health depends largely on local climate. With the increasing of temperature, a number of diseases, such as skin cancer, malaria and the hanta virus will be more actively attacking the human beings.

- **Agriculture**

It is not clear about the exact effect of global warming on plant agriculture. Some suggests that the high content of CO_2 and high temperature will be helpful to the growing of plants. In addiction, the warming climate may lead to a high precipitation, which in turn, support the plant growing with plenty of water. But other scientists don't think it will help plants. They think that high temperature means high evaporation. Instead of plenty of water, many regions will be hit by drought. The entire world has to face the situation of starvation.

It seems that the biggest effects on the agriculture is that we don't know the effects at all. Or the only thing we are sure is that global warming will certainly affect agriculture very much.

• **Glacier melting**

One of the direct results of glacier melting is the flooding. Many lakes in Nepal and Bhutan are formed by glacier melting, a process accelerated by global warming. They are dammed lakes due to the accumulation of moraines. The burst of these fast-growing lakes will have a catastrophic result. Actually, there is thousands of valley lakes formed by glacier melting. And the risk of burst exists for every such lake.

• **Sea Level Rising**

Sea level rising is the result of global glacier melting. Sea level is currently rising at around 2mm a year, which is at a faster rate than over the last 5 000 years (less than 1mm a year), but still slower than the average rate (5 mm a year) predicted for the next 80 years. It is likely that much of the rise in sea level has been related to the concurrent rise in global temperature over the last 100 years. The studies in recent years also show the huge ice sheets of Greenland and Antarctica contribute much to the sea level rising. Many of the coastal regions will be drowned in the next century. Rising sea level increases the salinity of rivers, bays, and groundwater tables. Additionally, such a rise would substantially increase the risks from erosion and storms in most coastal areas and upset important wetland ecosystems. Scientific evidence suggests there is a link between global warming and the power — not frequency — of hurricanes.

4.3 How to Curb the Global Warming?

1997 Kyoto Protocol(京都协定书)(*United Nations Framework Convention on Climate Change*) marked the first step for the international society to taken actions to curb the global warming. In recent years, many new and innovative programs to increase energy security, improve environmental quality, and reduce greenhouse gas emissions throughout the world have been proposed. But to curb the global warming, there is a still long way to go.

• **Energy Efficiency**

The most important action we can take to curb global warming is to consume energy more efficiently. For example, methane (CH_4, 甲烷,沼气) is a large contributor to global warming, second only to carbon dioxide. Thus, changes in methane emissions have a potential to contribute strongly to global climate change in the future.

Landfills are the biggest source of methane. Landfills produce methane as organic wastes decompose over time. Another big source is livestock and manure.

$$CH_4 + 2O_2 \longrightarrow CO_2 + 2H_2O$$

Although , during the combustion , CO_2 is emitted. But because each molecule of methane is twenty times more effective than each molecule of carbon dioxide at trapping infrared radiation, methane is a relatively effectively way to reduce the overall greenhouse effects. Methane can also be use to generate electricity. Harvesting methane emissions as fuel is one of most economical ways to use energy efficiently.

- **Clean, Renewable Energy**

Renewable energy source include wind, clean biomass, sun and geothermal resources. At present, only 2% of our electricity comes from clean, renewable sources. These energies do not pollute our air or our water and will never run out, unlike coal, natural gas and other fossil fuels. Using current and cutting-edge technology to develop clean renewable energy is the most efficient ways to reduce our use of fossil fuels and curb global warming.

By switching to energy efficient cars and trucks and buildings and homes that conserve energy we can reduce dramatically the emissions of greenhouse gases. To protect the wellbeing of current and future generations, we must take actions to curb the global warming by using any available ways.

Anyway, the global warming is fact. The time for debate is over. We must act now.

1. Answer the following questions

(1) What is hurricane?

(2) What's the difference between a hurricane and a typhoon?

(3) Discuss with your classmates the affects El Nino has had on coastal communities, including environment, wildlife, and humans.

(4) What is greenhouse effect?

(5) List the main greenhouse gases and describe their sources.

(6) What is ground level ozone?

(7) How does ozone layer protect us?

(8) Describe the possible ways to destroy the ozone layer.

(9) What are the potential effects of global warming?

(10) What is your opinion about the future of the Earth?

2. Multiple Choice

(1) Hurricane always spiral in clockwise pattern.

A: True B: False

(2) Which of the following statements about El Nino is not true?

A: El Nino is thought to occur due to changes in the normal patterns of trade wind circulation.

B: It is named for the Christ child, because it first appears around Christmas.

C: Torrential rains hit Australia.

D: El Nino occurs every two to seven years.

(3) In South Pole, the solar radiation is ______.

A: Strong B: Weak

C: Zero D: None of them

(4) The heaviest rainfall occurs in ______.

A: Eye B: Eyewall

C: Rain band C: All of them

(5) Which of the following statements about greenhouse effect is not true?

A: There is a greenhouse effect, but, if there were not, we would all be dead!

B: The greenhouse effect is the name applied to the process which causes the surface of the Earth to be warmer than it would have been in the absence of an atmosphere.

C: The greenhouse effect is an increase in the temperature of a planet as heat energy from sunlight is trapped by the gaseous atmosphere.

D: The carbon dioxide can trap all of the solar energy.

(6) The natural rain has average pH of ______.

A: 5.6 B: 7

C: 7.3 D: 4.2

(7) Which of the following statements about ozone is not true?

A: Ozone layer is thinnest in the tropics and denser towards the poles.

B: Ozone is a primary pollutant.

C: CFCs can deplete the Earth's protective layer, because it can free chlorine atoms.

D: Ozone can be produced by solar radiation.

(8) The greenhouse gases trap short wavelength radiation.

A: True B: False

(9) At the center of the a cyclone, air will ______.

A: Rise B: Fall

C: Remain steady D: All of them

(10) Basically, hurricanes and typhoons form over water and are huge, while tornados form over land and are much smaller in size.

A: True B: False

(11) The terms "hurricane" and "typhoon" are regionally specific names for a strong "tropical cyclone".

A: True B: False

(12) A tropical cyclone is a high-pressure system.

A: True B: False

(13) Greenhouse gas is ______.

A: CH_4 B: CFCs

C: N_2O D: All of them

(14) CFCs is produced mainly by ______.

A: UV B: Vocanic

C: Livestock D: Refrigeration

(15) Typhoon spiral in a counterclockwise pattern.

A: True B: False

(16) The atmosphere allows a large percentage of

the rays of visible light from the Sun to reach the Earth's surface and heat it.

A: True B: False

(17) El Nino has occurred for about 20 years.

A: True B: False

(18) In India Ocean, the tropical cycle is called ______.

A: Tropical cycle B: Hurricane

C: Typhoon D: Tornado

(19) The warmed Earth's surface will emitted short wavelength radiation.

A: True B: False

(20) The ingredient for the formation of hurricane is ______.

A: Initial climate disturbance

B: Warm ocean water

C: Light wind

D: All of them

References

1 夏邦栋. 普通地质学. 北京:地质出版社,1998

2 吴泰然,何国琦. 普通地质学. 北京:北京大学出版社,2003

3 Skinner B J, Porter S C. The Dynamic Earth: An Introduction to Physical Geology. 5th edition. Hoboken, NJ: John Wiley & Sons Inc, 2003

4 Plummer C C, McGeary D. Physical Geology. Dubuque, Iowa: Wm. C. Brown Publisher, 1991

5 Lutgens F K, Tarbuck E J. Foundations of Earth Science. New Jersey: Prentice-Hall Inc, 1996

6 Longwell G R, Flint R F, Sanders O E. Physical Geology. Hoboken, NJ: John Wiley & Sons Inc, 1969

7 Press F, Siever R. Earth. New York: W. H. Freeman and Company, 1978

8 Sanders J E, Anderson Jr A H., Carola Rr. Physical Geology. New York: Harper & Row Publishers Inc, 1976

9 Tarbuck E J, Lutgens F K. Earth Science. Columbus, Ohio: Charles E. Merrill Publishing Company, 1988

10 Robinson E S. Basic Physical Geology. Hoboken, NJ: John Wiley & Sons Inc, 1982

图版 I

Figure 1　Special karst topography in Huanglong（四川黄龙），
Courtesy of Chen Fuxing（谌福兴）

Figure 2　Incised meander in the upper reaches of Changjiang（长江），
Courtesy of Huangsheng（黄盛）

Figure 3　Arête and horn in Jade Dragon Mountains（玉龙雪山，丽江），
Courtesy of Huangsheng（黄盛）

图版Ⅱ

Figure 1　Cross bedding in Walnut Canyon, Arizona, U. S. Courtesy of Dr. Yu Shan(俞姗)

Figure 2　The Meteor Crater (also know as Barringer Crater) in Arizona, U. S.

Figure 3　The horizontal layers and incised meander in the Great Canyon in U. S. Courtesy of Dr. Yu Shan(俞姗), 2005

图版Ⅲ

Figure 1　Yardangs (Devils'Town) in Gansu Province (甘肃省)

Figure 2　A mesa in Arizona, U.S. Courtesy of Dr. Yu Shan(俞姗)

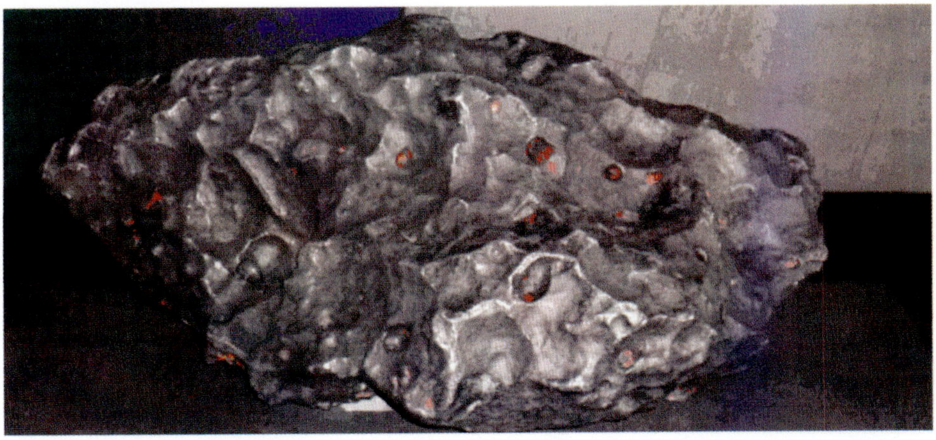

Figure 3　Meteorite. Photo by Dr. Yu Shan

图版Ⅳ

Figure 1　A variety of minerals and rocks. Image credit: U. S. Geological Survey, the Department of the Interior. Public domain